Sustainable Development and Biodiversity

Volume 24

Series Editor

Kishan Gopal Ramawat
Botany Department, Mohanlal Sukhadia University, Udaipur, India

This book series provides complete, comprehensive and broad subject based reviews about existing biodiversity of different habitats and conservation strategies in the framework of different technologies, ecosystem diversity, and genetic diversity. The ways by which these resources are used with sustainable management and replenishment are also dealt with. The topics of interest include but are not restricted only to sustainable development of various ecosystems and conservation of hotspots, traditional methods and role of local people, threatened and endangered species, global climate change and effect on biodiversity, invasive species, impact of various activities on biodiversity, biodiversity conservation in sustaining livelihoods and reducing poverty, and technologies available and required. The books in this series will be useful to botanists, environmentalists, marine biologists, policy makers, conservationists, and NGOs working for environment protection.

More information about this series at http://www.springer.com/series/11920

Kishan Gopal Ramawat
Editor

Biodiversity and Chemotaxonomy

Springer

Editor
Kishan Gopal Ramawat
Botany Department
Mohanlal Sukhadia University
Udaipur, India

ISSN 2352-474X ISSN 2352-4758 (electronic)
Sustainable Development and Biodiversity
ISBN 978-3-030-30745-5 ISBN 978-3-030-30746-2 (eBook)
https://doi.org/10.1007/978-3-030-30746-2

This Springer imprint is published by the registered company Springer Nature Switzerland AG
The registered company address is: Gewerbestrasse 11, 6330 Cham, Switzerland

Preface

In developed countries, almost all plants are known and explored, but in developing countries and tropical hot spots, all plants are not well known. We are finding ways and means to know these plants and identify and conserve them for future generations. Plant classifications are based on morphological characters, and it is difficult particularly in small plants and grasses to identify these below generic level on the basis of these characters using a dissecting microscope. Modern tools of molecular biology are being used to identify and correlate the species. Plant species have intra- and inter-specific variation in secondary metabolites which can be utilized as marker compounds for identification and classification of plants. Secondary metabolites are produced as a result of primary metabolism, and the production of these compounds not only involves several genes but also is an energy-dependent process. Hence, these products cannot be considered as insignificant for the plant and the environment. Modern tools of molecular biology and secondary metabolites present in them can definitively decide about phylogeny and classification of plants. Incorrect identification of plant leads to many problems in resource utilization. Due to wide availability of these tools, interest has revived in systematics and correct classification of plants based on these parameters for their sustainable utilization and resource management.

The purpose of this book is to assess the potential of phytochemical and molecular tools in the systematic and classification of plants. The topics covered include basics of biodiversity and chemotaxonomy markers (alkaloids, flavonoids, isoquinoline alkaloids, polyketides, carotenes, cuticular wax, volatile oils, biodiversity of corals, metazoans, Ruta, sedum, and Echinocereus) for several species and groups.

The book will provide comprehensive and broad subject-based reviews, useful for students, teachers, researchers, and all others interested in the field. The field has been kept wide and general to accommodate the wide-ranging topics. Well-recognized international specialists in their respective fields of research contributed these chapters. This book will be useful to agriculturists, chemists, botanists, industrialists, and those involved in planning of cultivation and utilization of plants.

Udaipur, India — Professor Kishan Gopal Ramawat

Contents

Contributors

Salvador Arias Jardín Botánico, Instituto de Biología, Universidad Nacional Autónoma de México, Coyoacán, Ciudad de México, Mexico

Debora Cristina Baldoqui Departamento de Química, Universidade Estadual de Maringá, Maringá, Paraná, Brazil

Hachemi Benhassaini Laboratory of Plant Biodiversity, Conservation and Enhancement, Faculty of Natural and Life Science, University of Djillali Liabes Sidi Bel-Abbes, Sidi Bel-Abbes, Algeria

Zineb Bennaoum Laboratory of Plant Biodiversity, Conservation and Enhancement, Faculty of Natural and Life Science, University of Djillali Liabes Sidi Bel-Abbes, Sidi Bel-Abbes, Algeria

Punyasloke Bhadury Integrative Taxonomy and Microbial Ecology Research Group, Department of Biological Sciences & Centre for Climate and Environmental Studies, Indian Institute of Science Education and Research Kolkata, Nadia, West Bengal, India

Marta Regina Barrotto do Carmo Departamento de Biologia Geral, Universidade Estadual de Ponta Grossa, Ponta Grossa, Paraná, Brazil

Erick E. Dokalahy Pharmacognosy Department, College of Pharmacy, Cairo University, Cairo, Egypt

Anderson R. dos Santos Department of Chemistry, Laboratory of Natural Products and Chemical Ecology, Federal University of Paraná, Curitiba, Paraná, Brazil

Mohamed Ali Farag Pharmacognosy Department, College of Pharmacy, Cairo University, Cairo, Egypt

H. R. El-Seedi Pharmacognosy Group, Department of Medicinal Chemistry, Uppsala University, Biomedical Centre, Uppsala, Sweden;
International Research Center for Food Nutrition and Safety, Jiangsu University, Zhenjiang, China

Geetanjali Department of Chemistry, Kirori Mal College, University of Delhi, Delhi, India

Eoin Gillespie Department of Environmental Science, School of Science, CERIS, Centre for Environmental Research, Innovation and Sustainability, Institute of Technology Sligo, Sligo, Ireland

Ivan R. Green Department of Chemistry and Polymer Science, University of Stellenbosch, Stellenbosch, South Africa

Hidayat Hussain Department of Bioorganic Chemistry, Leibniz Institute of Plant Biochemistry, Halle (Salle), Germany

Snežana Č. Jovanović Department of Chemistry, Faculty of Science and Mathematics, University of Niš, Niš, Serbia

Manish Mathur ICAR-Central Arid Zone Research Institute, Jodhpur, India;
Department of Botany, Center of Advanced Studies, JNV University, Jodhpur, India

Dónal Mc Gee Department of Environmental Science, School of Science, CERIS, Centre for Environmental Research, Innovation and Sustainability, Institute of Technology Sligo, Sligo, Ireland

Adriano Borges Meniqueti Departamento de Química, Universidade Estadual de Maringá, Maringá, Paraná, Brazil

Poonam Department of Applied Chemistry, Delhi Technological University, Delhi, India

Kishan Gopal Ramawat Botany Department, Mohanlal Sukhadia University, Badgaon, Udaipur, India

Anderson Valdiney Gomes Ramos Departamento de Química, Universidade Estadual de Maringá, Maringá, Paraná, Brazil

Daniel Sánchez CONACYT—Laboratorio Nacional de Identificación y Caracterización Vegetal, Centro Universitario de Ciencias Biológicas y Agropecuarias, Universidad de Guadalajara, Zapopan, Jalisco, Mexico

Maria Helena Sarragiotto Departamento de Química, Universidade Estadual de Maringá, Maringá, Paraná, Brazil

Barbara Schulz Institut für Mikrobiologie, Technische Universität Braunschweig, Brunswick, Germany

K. R. Shivanna Ashoka Trust for Research in Ecology and the Environment, Bengaluru, India

Ram Singh Department of Applied Chemistry, Delhi Technological University, Delhi, India

Gordana S. Stojanović Department of Chemistry, Faculty of Science and Mathematics, University of Niš, Niš, Serbia

S. Sundaramoorthy Department of Botany, Center of Advanced Studies, JNV University, Jodhpur, India

Teresa Terrazas Colegio de Postgraduados, Texcoco, Estado de México, Mexico

Nelissa P. Vaz Department of Chemistry, Laboratory of Natural Products and Chemical Ecology, Federal University of Paraná, Curitiba, Paraná, Brazil

Monserrat Vázquez-Sánchez Departamento de Botánica, Instituto de Biología, Universidad Nacional Autónoma de México, Coyoacán, Ciudad de México, Mexico

Bojan K. Zlatković Department of Biology and Ecology, Faculty of Science and Mathematics, University of Niš, Niš, Serbia

Chapter 1
An Introduction to Biodiversity and Chemotaxonomy

Kishan Gopal Ramawat

Abstract About 10–25% species of total over 400,000 species present on the planet Earth are used for human welfare. Plants and their habitats are destroyed for human benefits, and thus, valuable diversity is lost without understanding the utility. Tropical and subtropical agroclimatic conditions support the rich biodiversity of plants which decline towards poles. Plant diversity, genetic make-up and chemical present in them need to be evaluated and identified for proper use. Classical methods are inadequate to clearly classify the plant species, and new and modern tools are always applied for their classification from cytogenetics, molecular fingerprinting to DNA barcoding. Secondary metabolites as chemical markers (alkaloids, flavonoids, terpenes and others) are added tools in this process of identification and help in preventing adulteration and collection and planning of cultivation of desired plant material for industrial use. In this brief overview, problems associated with biodiversity and use of modern tools to resolve these issues are discussed.

Keywords Biodiversity · Chemotaxonomy · Chemical markers · Barcoding · Systematics · Secondary metabolites

1.1 Introduction

Number of plants on the planet Earth is estimated to be over 400,000 which include 352,000 flowering plants, 1050 gymnosperms, 15,000 ferns and fern allies, and 33,750 mosses. In angiosperms, three major families Asteraceae, Leguminosae and Orchidaceae, contain 20% of flowering plants spread in about 62,000 species. About 25% of these are considered endangered because of various reasons. About 40,000–100,000 species of plants are used for food, medicine, shelter, fibres and wood (Ceccarelli 2009; Chapman 2009; Ramadan 2017; Osbourn 2018).

K. G. Ramawat (✉)
Botany Department, Mohanlal Sukhadia University, Udaipur, India
e-mail: kg.ramawat@gmail.com

K. G. Ramawat (ed.), *Biodiversity and Chemotaxonomy*, Sustainable Development and Biodiversity 24, https://doi.org/10.1007/978-3-030-30746-2_1

Plants thrive well in rich soil, high temperature and heavy rainfall conditions. Vegetation thrives well at equator and decline towards high latitudes (away from equator towards pole) and also high altitudes (reduced to bushes and finally inhospitable snowline). Therefore, tropics are rich in biodiversity, and multi-layers of plant canopies grow in these conditions. About 34 biodiversity-rich hot spots are recognized throughout the world and are protected by laws of the land. Hot spot of biodiversity is a biogeographical region with highly preserved biodiversity. Mosaic of biodiversity is influenced by spatial and environmental variations at the frequency by which new characters are developed. Biodiversity depends upon species–area relationship, genetic diversity, latitudinal gradient in species and species richness in different habitats (Schluter and Pennell 2017). This biodiversity is threatened because of various human activities, ever-increasing population, expanding urbanization, more land use for agriculture and economic activities (Chitale et al. 2015; Myers et al. 2000).

The term taxonomy is originated from two Greek words, Taxis—arrangement and nomos—laws. Taxonomy is also called as systematics or systematic botany and deals with identification, classification, description and naming of plants. Most of the classifications are based on morphological characters of flowers (reproductive organs) and vegetative parts. On the basis of these characters, plants have been classified by Alphonse de Candolle, Jean Batiste de Lamarck, George Benthum and Joseph Dalton Hooker. Phylogenetic classification is based on genetic relationship of plants and according to their evolutionary sequences. Such relationship is expressed in the form of a tree, and such system was advocated by HG Adolf Engler. Historically, man had tried to classify plants for their colour, taste, aroma, usefulness and relevant to various ailments (Wink et al. 2010). Animals and insects also have preferences for sweet plants over the bitter or toxic plants. Thus, this selection pressure resulted in cultivation of desirable plants and loss of unwanted germplasm.

However, chemical examination of plants with chemotaxonomic perspective started in the early twentieth century. They studied volatile oils in eucalyptus and volatiles and alkaloids in angiosperms, but rapid development took place with advent of paper chromatography in mid-twentieth century, which made the comparison of samples quick and simple (Wink et al. 2010). Rapid advancement during 1970–1980 in techniques of phytochemistry such as capillary column (or high resolution) gas–liquid chromatography (GLC), high-performance liquid chromatography (HPLC), mass spectrometry (MS, as GLC–MS, LC–MS) and nuclear magnetic resonance (^{1}H, ^{13}C–NMR) paved the way for development of this line of approach.

Distribution of structure-specific secondary metabolites is restricted to certain taxa, and this leads to the development of chemotaxonomy. It is well known that these secondary metabolites are biosynthesized by several steps, involving several enzymes and consequently genes (Zhu et al. 2014; Sharma and Ramawat 2013). Because of synthesis of similar compounds by different species, and presence of such genes in several widely related plants, it is presumed that this is because of horizontal gene transfer from microorganisms as a result of endosymbiosis (mitochondria or plastid evolution). In this brief overview, we have summarized the use of biodiversity for human welfare through identification of plants using modern

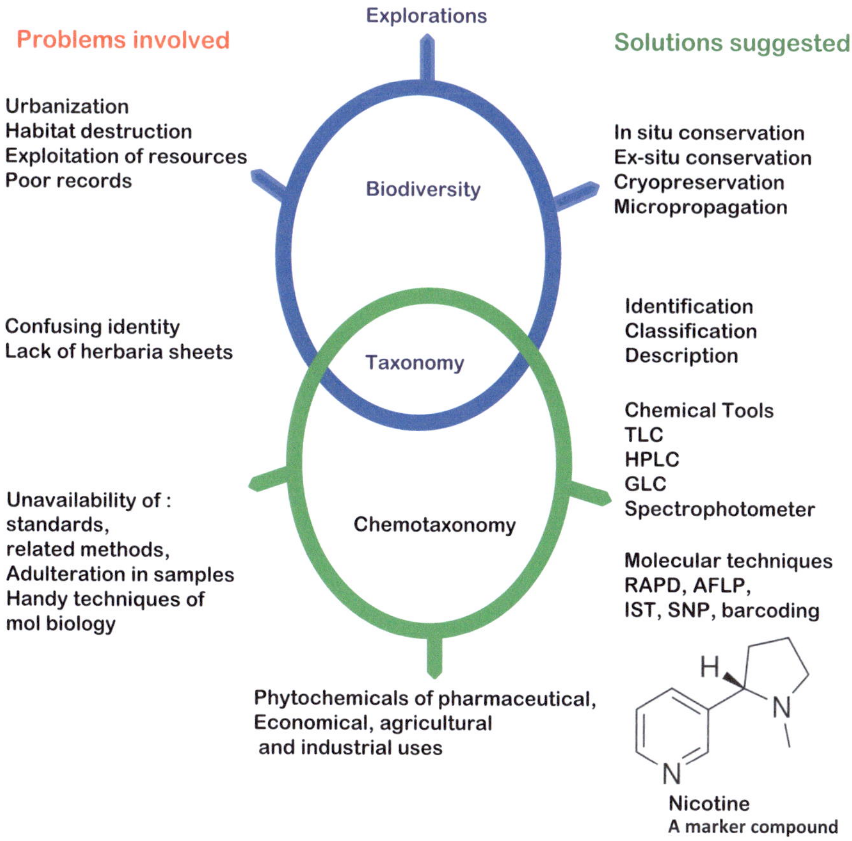

Fig. 1.1 Overview of different tools involved in conserving biodiversity, identification of plants and sustainable use of plant resources by knowing useful chemical constituents. One or more techniques are employed to solve problem of identification of plant and its compounds

tools and chemotaxonomic markers. A global scene of this interrelationship is presented in Fig. 1.1 depicting problems associated with biodiversity identification and obtaining useful metabolites by correct identification and planning the cultivation of required species.

1.2 Evolution of Diversity

Evolution of plant diversity in global context is a complex phenomenon. Plants adapt themselves to changing environmental conditions by incorporation of new characters by sexual reproduction or hybridization. Interspecific hybridization between native and invading species or between two invading species may result in a sexually reproducible new plant which is more suitable to the new conditions.

Transposable elements are large movable genetic elements that shift within genome by transposition or retrotransposition and constitute a large portion of genetic material of vertebrates (4–60%) and higher plants. The modern tools like sequencing can determine their role in evolution of organisms as these not only show transposition but also horizontal transfer (Sotero-Caio et al. 2017).

Plants have diverse chemicals for defence against herbivores. About 2500 plant species contain cyanogenic glycosides or cyanolipid as constitutive defence chemical. With onset of chewing by insects and damage to leaves, cyanogenic glucoside is cleaved by b-glucosidases and a-hydroxynitrile lyases to produce toxic hydrogen cyanide (a potent inhibitor of mitochondrial respiratory chain), thus providing plants a generalized defence against insects. Now insects have evolved themselves by obtaining a gene from bacteria by horizontal gene transfer (Wybouw et al. 2014). This is how continuous change in plant and their pest genome paved the way for adaptation to new environment and evolution of organisms. Habitat destruction by water and fire also provides new opportunity for plant diversity to flourish (Pausas et al. 2017). Therefore, constant pressure from biotic and abiotic stress factors and sexual reproduction contribute to the diversity and subsequently evolution of organisms.

1.3 Problems in Systematics

The exact number of living organisms or species either of flowering plants or of microorganisms is not known. Thus, taxonomists do not know the volume of work. Throughout the world, efforts are made to digitalize the specimens using digital camera and molecular tools to identify the DNA sequence, a part of it in the form of ATGC bases. It is equally important that new species discovered and described are simultaneously known to others by fast-communicating systems like digitalization and sequence data (Page 2016).

Identification of large flowering plants is quite easy, but identification of small flowers particularly those of grasses and microorganisms is quite a challenging task. With modernization of science, lack of interest in plant morphology resulted in shortage of trained taxonomists, improper knowledge of laws of classification, terminology and unavailability of authentic herbaria for comparison are major hurdles in the identification of plants (Komarek and Beutel 2006). To exploit the plants for economical or industrial use, correct identification and continuous supply of uniform plant material must be assured. This requires efforts by biologists, molecular tools and agriculturist to assist economic activity. The basic of all these is correct identification of the plant and removal of alterations. The chemotaxonomy is one of the important tools to resolve this problem.

1.4 Impact of Cultivation and Hybridization

Domestication is a long process involving farmers, crop adaptation and its environment. Over several centuries, plants were selected for different agroclimatic zones, and this diversity within a crop is useful diversity for further adaptation and evolution (Vigouroux et al. 2011). Natural selection and migration contribute significantly in adaptation and evolution of crop plants. Out of 452,000 plants, less than 500 are cultivated, although a number of plants or plant products used are many. Thus, domestication has resulted in fewer germplasms and loss of diversity. The principal regions of plant domestication are Mesoamerica, the southern Andes, the Near East, Africa (probably the Sahel and the Ethiopian highlands), Southeast Asia and China. Later on, cultivated plants reached from these centres to Europe and America. Examples of domestication of crop plants in some of the geographical regions are: rice and millet (7500 BC), mung bean and soybean (4000 BC) in China; potato (4500 BC) and cotton (7500 BC) in north of Latin America; sunflower (3000 BC) in North America. Cross-pollination, development of useful traits, discard of toxic plants and hybridization in recent past contributed significantly in domestication and present-day crops and gene pool (Gepts et al. 2012; Ceccarelli 2009).

To feed the ever-increasing population, crop improvement and higher production are mandatory. In 1970, green revolution changed the cropping system with high inputs of water, fertilizers and pesticides with responsive, dwarf and highly productive varieties cultivated in large acreage. Homogeneous cultivation on large area has its own drawback like vulnerability to epidemics and loss of diversity. This was followed by crop improvement by genetic engineering methods, and new genes were introduced to produce disease resistant and better in quality plants and fruits (Lange et al. 2018; Juma 2018). Therefore, selection of plants, hybridization, and genetic engineering contribute to incorporation of new characters and improvement of plants (diversity and evolution).

1.5 Application of Chemical Markers in Taxonomy

Plants produce primary metabolites (carbohydrates, proteins and lipids) as a result of photosynthesis and synthesize secondary metabolites (alkaloids, terpenes, phenolics and others) from these primary metabolites. Besides plants, secondary metabolites are known to occur in fungi, insects, prokaryotes and some vertebrates (Ramawat 2007; Goyal et al. 2017). Plant analysis for biologically active ingredients has resulted in identification of more than 21,000 alkaloids, 20,000 terpenes, 10,000 phenolics (e.g. flavonoids present from mosses, fern, gymnosperms to angiosperms), 750 polyketides, 700 non-protein amino acids (mainly in Fabaceae), 1500 polyacetylenes (e.g. in Araliaceae, Apiaceae) and fatty acids, 200 carbohydrates and 200 cyanogenic glycosides and glucosinolates (Bell and Charlwood 1980; Conn 1981;

DNP 1996; Harborne, 1993; Roberts and Wink 1998; Dewick 2002; Wink 2008, 2010a, b; Ramawat and Merillon 2013). These chemically different metabolites not only serve as defence (from pathogenic microbes) and communication molecules for the plant but also attract pollinators (for seed dispersal), defend from insects and herbivores, compete from other plants for nutrients, and are useful for adaptation of plants to various environments (Goyal and Ramawat 2019). Different plants synthesize an array of secondary metabolites. Some of these metabolites are very useful for human welfare because plants not only synthesize them but also accumulate them in great quantities. Some of the examples of such useful metabolites are shown in Table 1.1. It is because of the presence of these biologically active ingredients that plants are known as medicinal plants. Accumulated secondary metabolites also serve as chemical marker for classification of the plant and their products in quality control to check adulteration. Arrangement or classification of plants based on their chemical constituents is called as chemotaxonomy or chemical-based taxonomy. It is useful for developing industrial application of plants and resource management. Glycosides, cyanogenic glycosides, glucosinolates, alkaloids and phenolics have been used as chemotaxonomic markers (Singh 2016).

Most of these markers are unique to the group or plant and can be used as chemical marker for identification or comparison of species. Such chemical markers are used for characterization of genera like Acacia, Cannabis and Withania, and for authentication of herbs like chamomile. There are about 300 non-protein amino acids found in angiosperms. Canavanine, a close analogue of arginine, is found only in Fabaceae. In *Lathyrus*, seven intrageneric groups and in *Vicia* fourteen intrageneric groups are recognized on the basis of distribution of amino acids. The presence of cyclopropyl amino acids in two families, Sapindaceae and Aceraceae, shows their close relationship (Bell and Charlwood 1980; Roberts and Wink 1998; Wink 2010a, b).

There are many general observations about plant groups such as plants are aromatic in Juglandales, non-aromatic in Fagales; plants produce betalains in Caryophyllales, Cactaceae, Chenopodiaceae (but not anthocyanins), anthocyanins in polygonales (but not betalains); plants are tanniferous in Sapindaceae; isoquinoline alkaloids in the order Ranales, plants having alkaloids in Solanaceae and glucosinolates in Brassicaceae, etc. Primary metabolites are universally present in all plants and hence rarely used as chemical marker (Polturak and Aharoni 2018; Singh 2016; Ramawat and Merillon 2013).

1.6 Application of Modern Tools in Taxonomy

Plant taxonomy and study of secondary metabolites are two different fields of plant science. Systematics requires strong botanical knowledge, whereas the study of secondary metabolites not only requires a strong knowledge of organic chemistry but also elaborates equipments to isolate and identify the compound and determine

Table 1.1 Selected examples of plant species and their chemical markers

Plant species Family	Appearance	Bioactive molecules	References
Shiitake Mushroom, Lentinus edodes	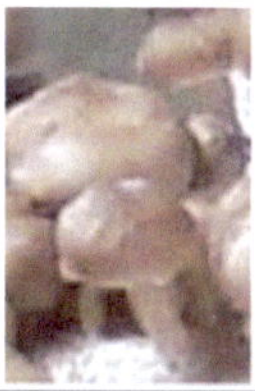	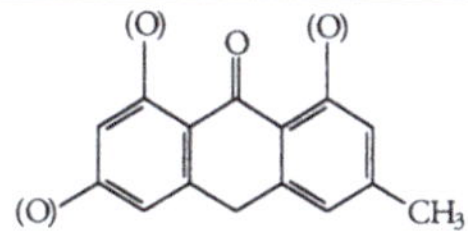 Polyketides: Tricyclic anthrone Common in fungi, many antibiotics	Goyal et al. (2017), Merillon and Ramawat (2017b)
Lathyrus odoratus Fabaceae		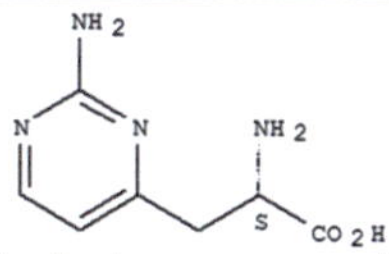 Lathyrine a non-protein amino acid	Rozan et al. (2001)
Nicotiana tabaccum, Solanaceae		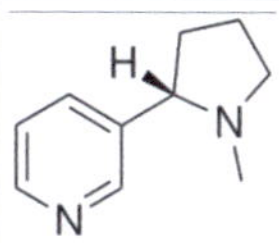 Nicotine from Nicotiana tabaccum	Nobukazu et al. (2016), Kajikawa et al. (2017)
Catharanthus roseus, Apocynaceae		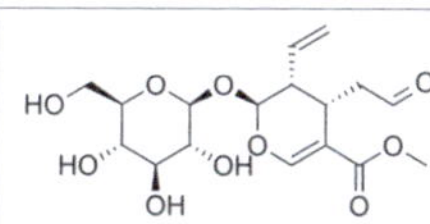 Monoterpenes: Iridoids; secologanin involved in biosynthesis of indole alkaloids	Zhu et al. (2014)
Cinchona species, Rubiaceae		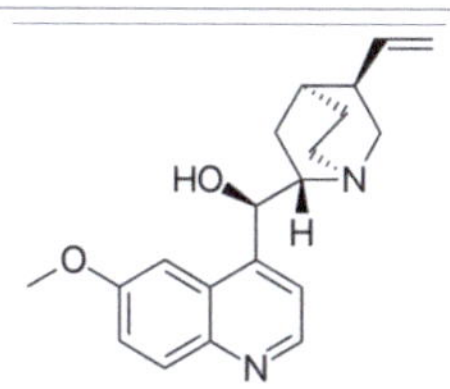 Indole alkaloids: Quinine *Cinchona* species	Talapatra and Talapatra (2015)
Glycine max, Fabaceae		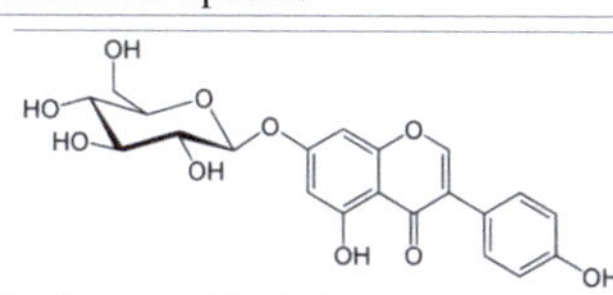 Isoflavone: Genistin Fabaceae; Pueraria, soybean and other pulses	Sharma and Ramawat (2013)

(continued)

Table 1.1 (continued)

Plant species Family	Appearance	Bioactive molecules	References
Cannabis sativa, Cannabaceae		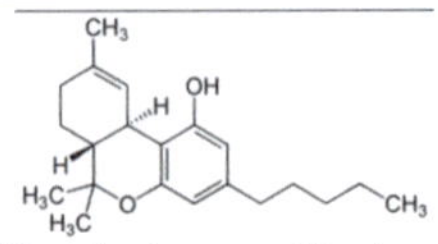 Tetrahydrocannabinol; 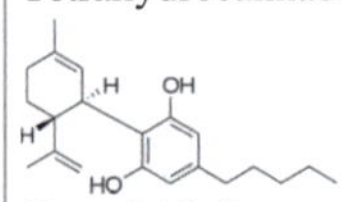 Cannabidiol	Elzinga et al. (2015)
Lupin species Fabaceae		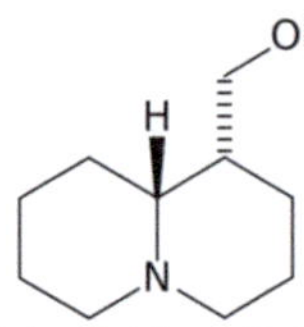 Quinolizidine alkaloids: Lupinine from *Lupin* species	Osorio et al. (2018)
Guggul, Commiphora wightii, Burseraceae		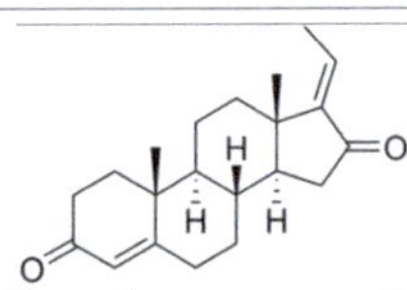 Guggulsterone can exist as either of two **stereoisomers**, E-guggulsterone and Z-guggulsterone	Cunnigham et al. (2018), Ahmed et al. (2016)
Coleus forskohlii, Lamiaceae		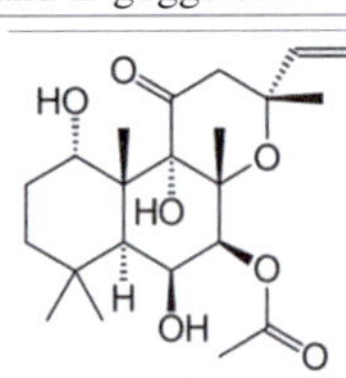 Forskolin from Coleus forskohlii	Shukla et al. (2017)
Brassica species, Brassicaceae		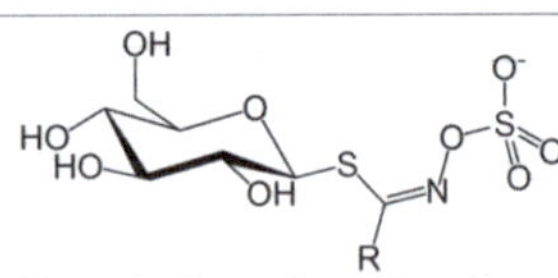 Glucosinolates from Brassicaceae	Merillon and Ramawat (2017b)

(continued)

Table 1.1 (continued)

Plant species Family	Appearance	Bioactive molecules	References
Chamomile, Matricaria chamomilla, Asteraceae		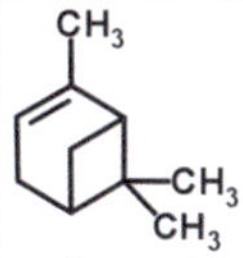 a-**pinene** (components of essential oils); camphene, b-**pinene**, sabinene, myrcene, **1,8-cineole**, y-terpinene, caryophyllene, and propyl angelate and butyl angelate	Lopez and Blazquez (2016)
Ashwagandha, *Withania somnifera,* Solanaceae		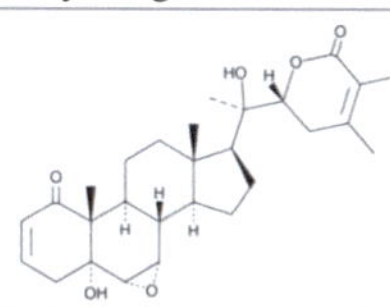 Triterpene sterol Withanolide A from *Withania somnifera*	Ganzera et al. (2003)

Methods of quantification will depend upon the chemical nature of molecule and its concentration. From: National Center for Biotechnology Information. PubChem Compound Database; CID = 166766, https://pubchem.ncbi.nlm.nih.gov/compound/166766 (accessed April 3, 2018) of plants and sustainable use of plant resources by knowing useful chemical constituents

their complete structure. Usually, methods of estimation are needed to be developed before botanists put hands of the material that requires availability of standard reference compounds. Therefore, it is collaboration between plant taxonomists and organic chemists to carry out such a study. Study and quantification of secondary metabolites require sensitive and rapid methods of analysis like thin-layer chromatography (TLC), high-performance TLC (HPTLC), high-performance/pressure liquid chromatography (HPLC), gas–liquid chromatography (GLC) and use of nuclear magnetic resonance (NMR) and mass spectroscopy (MS) for identification of compounds. This can be well exemplified by work on mountain flora of Central Asia for alkaloids producing plants reported since 1930 (Tayjanov et al. 2017). In all 37% plants contained alkaloids (4267 out of 11,551) with a good percentage showing cytotoxic and neurotoxic property required for defence in hard mountainous climate.

TLC, HPLC and GC are sensitive and rapid analytical tools used in laboratories depending upon types of compounds (chemical nature, fluorescent or not, volatile nature) and number of compounds to be separated or present in the sample (mixture of compounds). In GLC, gas is used as carrier medium (volatile nature of compounds), whereas in TLC and HPLC, liquid solvents are used as medium (soluble nature of compounds). Two representative examples are presented in Figs. 1.2 and 1.3. In TLC, overlapping bands (closely related compounds) cannot be separated and quantified, whereas in HPLC, clear separation can be achieved and is a highly sensitive system to detect microgram quantity. These compounds serve as markers for identification of compounds in different samples, detecting adulteration and comparison of species. Details of these techniques are beyond the scope of this paper and can be found elsewhere. These techniques have their own advantages and disadvantages, and limitations.

DNA barcoding is a perfect technology developed to accurately and rapidly recognize the gene fragments for the identification of plant with a possibility to upgrade the system to automation (Hubert and Hanner 2015; Hebert and Gregory 2005). This technique of molecular biology can resolve issues related to taxonomic identification in closely resembling species (Hebert and Gregory 2005; Miller 2007). Thus provide a much needed tool for solving many problems of biodiversity and has found wide applications in the field of functional ecology (Smith et al. 2007), biogeography (Kerr et al. 2009), conservation (Forest et al. 2007), wildlife forensic (Jian et al. 2014; Wong and Hanner 2008), identification of constituents in herbal formulations (Mahadani and Ghosh 2013) and socio-economic-related issues of biodiversity (Vernooy et al. 2010). However, still single locus like animal world using DNA sequence of Cox1 gene is lacking in the plant systematics.

DNA bar coding is not only helpful in plant identification but can effectively be used to resolve issues related to commercial aspect of plant material where plant

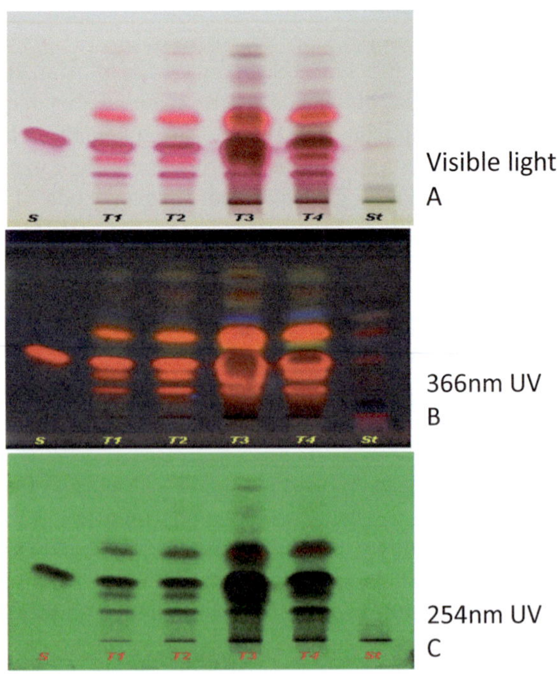

Fig. 1.2 TLC analysis of marker compounds of *Coleus forskohlii.* [Observed after derivatization with anisaldehyde sulphuric acid reagent in visible (**a**), and 366 (**b**), 254 nm UV light (**c**)]. Standard forskolin is in left (s). Photos by author

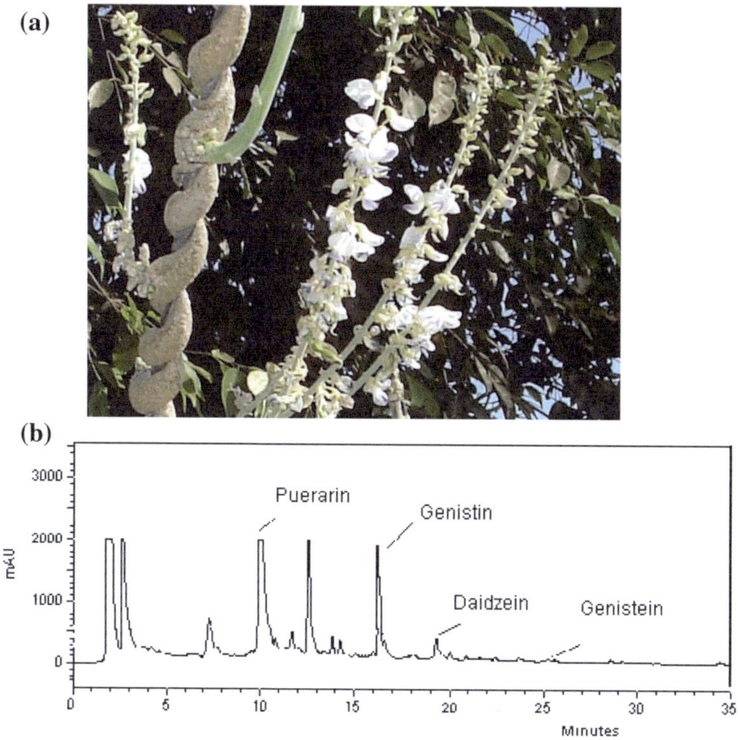

Fig. 1.3 A flowering branch (**a**) and HPLC profile (**b**) of *Pueraria tuberosa* showing isoflavonoids with major peak of puerarin. Note the concentration difference between different isoflavonoids as evident by peak area. Photo by author

parts are powdered for specific use. This is particularly useful where a single plant is used, e.g. barley tea case (Jian et al. 2014). The details of this technique and its applications are presented in this book.

1.7 Conclusion

Biodiversity and habitat conservation involve society and governmental efforts throughout the world. Traditionally, several tribes and communities protect wildlife, both flora and fauna. Nowadays, several non-governmental organizations (NGOs) are involved in protection, identification and conservation of flora. Chemical fingerprinting, isolation and identification of metabolites and DNA barcoding are additionally helpful techniques for identification and proper use of plant material as well as their planned cultivation or collection. Any economic or industrial use of plants requires proper identification of plant. Plants are renewable resources but must be exploited with caution. Though the techniques of DNA sequencing have evolved rapidly and become automated, still high cost and elaborate laboratory facilities are required for sequencing. Looking the number of species or controversial cases in taxonomy in identification, application of this technique to resolve issues is still a far cry particularly for the developing tropical countries where biodiversity is unexplored.

References

Ahmed R, Wang YH, Ali Z, Smillie TJ, Khan IA (2016) HPLC method for chemical fingerprinting of Guggul (*Commiphora wightii*)–quantification of E-and Z-guggulsterones and detection of possible adulterants. Planta Medica 82(4):356–361. https://doi.org/10.1055/s-0035-1558211

Bell EA, Charlwood BV (1980) Secondary plant products. Springer, Heidelberg

Ceccarelli S (2009) Evolution, plant breeding and biodiversity. J Agric Environ Inter Dev 103 (1/2):131–145

Chapman AD (2009) Numbers of living species in australia and the world, 2nd edn. Australian Government, Department of the Environment, Water, Heritage and the Arts. Canberra, Australia

Chitale VS, Behera MD, Roy PS (2015) Global biodiversity hotspots in India: significant yet under studied. Curr Sci 108:149

Conn EE (1981) Secondary plant products. The biochemistry of plants, vol. 7. Academic Press, New York

Cunningham AB, Brinckmann JA, Kulloli RN, Schippmann U (2018) Rising trade, declining stocks: the global gugul (*Commiphora wightii*) trade. J Ethnopharmacol 223:22–32

Dewick PM (2002) Medicinal natural products, a biosynthetic approach, 2nd edn. Wiley, New York

DNP (1996) Dictionary of natural products. CD-ROM Version 5:1. Chapman and Hall London

Elzinga S, Fischedick J, Podkolinski R, Raber JC (2015) Cannabinoids and terpenes as chemotaxonomic markers in cannabis. Nat Prod Chem Res 3:181. https://doi.org/10.4172/2329-6836.1000181

Forest F, Grenyer R, Rouget M, Davies J, Cowling RM, Faith DP et al (2007) Preserving the evolutionary potential of floras in biodiversity hotspots. Nature 445:757–760

Ganzera M, Choudhary MI, Khan A (2003) Quantitative HPLC analysis of withanolides in *Withania somnifera*. Fitoterapia 74(1–2):68–76

Gepts P, Famula TR, Bettinger RL, Brush SB, Damania AB et al (2012) Biodiversity in agriculture: domestication, evolution, and sustainability. Cambridge University Press, Cambridge, p 606

Goyal S, Ramawat KG (2019) Co-evolution of secondary metabolites during biological competition for survival and superiority: An overview. In: Mérillon JM, Ramawat K (eds) Co-evolution of secondary metabolites. Reference Series in Phytochemistry Springer, Cham

Goyal S, Ramawat KG, Mérillon JM (2017) Different shades of fungal metabolites: an overview. In: Mérillon JM, Ramawat KG (eds) Fungal Metabolites. Reference Series in Phytochemistry Springer, Cham

Harborne JB (1993) Introduction to ecological biochemistry, 4th edn. Academic Press, London

Hebert PDN, Gregory TR (2005) The promise of DNA barcoding for taxonomy. Syst Biol 54: 852–859

Hubert N, Hanner R (2015) DNA barcoding, species delineation and taxonomy: a historical perspective. DNA Barcodes 3:44–58

Jian C, Deyi Q, Qiaoyun Y et al (2014) A successful case of DNA barcoding used in an international trade dispute. DNA Barcodes 2(1):21–28

Juma C (2018) Moving beyond the green revolution in Africa's new era of hunger. African J Food Agric Nutr. https://theconversation.com/moving-beyond-the-green-revolution-in-africas-new-era-of-hunger-87310

Kajikawa M, Sierro N, Hashimoto T, Shoji T (2017) A model for evolution and regulation of nicotine biosynthesis regulon in tobacco. Plant Signaling Behav 12:6. https://doi.org/10.1080/15592324.2017.1338225

Kerr KCR, Lijtmaer DA, Barreira AS, Hebert PDN, Tubaro PL (2009) Probing evolutionary patterns in neotropical birds through DNA barcodes. PloS One 4:e4379

Komarek A, Beutel RG (2006) Problems in taxonomy and suggestions for a standardized description of new insect taxa. Entomol Probl 36(2):55–70

de Lange O, Klavins E, Nemhauser J (2018) Synthetic genetic circuits in crop plants. Curr Opin Biotechnol 49:16–22

Lopez M, Blazquez MA (2016) Characterization of the essential oils from commercial chamomile flowers and chamomile teabags by GC-MS analysis. Int J Pharmacogn Phytochem Res 8 (9):1487–1491

Mahadani P, Ghosh SK (2013) DNA barcoding: a tool for species identification from herbal juices. DNA Barcodes 35–38. https://doi.org/10.2478/dna-2013-0002

Merillon JM, Ramawat KG (eds) (2017a) Fungal metabolites. Springer International Publishing, Switzerland, pp 1001

Merillon JM, Ramawat KG (eds) (2017b) Glucosinolates. Springer International Publishing, Switzerland, pp 473

Miller SE (2007) DNA barcoding and the renaissance of taxonomy. Proc Nat Acad Sci USA 104:4775–4776

Myers N, Mittermeier RA, Mittermeier CG, Da Fonseca GA, Kent J (2000) Biodiversityhotspots for conservation priorities. Nature 403:853–858

Nobukazu S, Hayashida M, Yazaki K (2016) Secondary metabolites in plants: transport and self-tolerance mechanisms. Biosci Biotech Biochem 80(7):1283–1293

Osbourn L (2018) https://www.currentresults.com/Environment-Facts/Plants-Animals/estimate-of-worlds-total-number-of-species.php

Osorio CE, Amiard VSE, Aravena-Calvo J et al (2018) Chromatographic fingerprinting of Lupinusluteus L. (Leguminosae) main secondary metabolites: a case of domestication affecting crop variability. Genet Resour Crop Evol 65:1281. https://doi.org/10.1007/s10722-018-0613-x

Page RDM (2016) DNA barcoding and taxonomy: dark taxa and dark texts. Phil Trans R Soc B 371:20150334. https://doi.org/10.1098/rstb.2015.0334

Pausas JG, Ribeiro E, Thomas G (2017) Fire and plant diversity at the global scale. Global Ecol Biogeogr 26(8):889–897. https://doi.org/10.1111/geb.12596

Polturak G, Aharoni A (2018) "La Vie en Rose": biosynthesis, sources, and applications of betalain pigments. Mol Plant Cell 11(1):7–22

Ramadan MF (2017) Underutilized plant species and agricultural sustainability in Egypt. In: The handbook of environmental chemistry. Springer, Berlin, Heidelberg, pp 1–24. https://doi.org/10.1007/698_2017_195

Ramawat KG (2007) Production of alkaloids. In: Ramawat KG, Mérillon JM (eds) Biotechnology secondary metabolites: plants and microbes, 2nd edn. Sci Pub Enfield, USA, pp 179–207

Ramawat KG, Mérillon JM (2013) (eds) Natural products. Springer, Berlin, Heidelberg

Roberts MF, Wink M (1998) Alkaloids: biochemistry, ecological functions and medical applications. Plenum, New York

Rozan P, Kuo YH, Lambein F (2001) Amino acids in seeds and seedlings of the genus Lens. Phytochem 58:281–289

Schluter D, Pennell MW (2017) Speciation gradients and the distribution of biodiversity. Nature 546:48–55

Sharma V, Ramawat KG (2013) Isoflavonoids. In: Ramawat KG, Mérillon JM (eds) Natural products. Springer, Berlin, Heidelberg, pp 1849–1865

Shukla PK, Misra A, Kumar M et al (2017) Simultaneous quantification of forskolin and iso-forskolin in *Coleus forskohlii* (Wild.) Briq. and identification of elite chemotype, collected from Eastern Ghats (India). Pharmacogn Mag 13(Suppl 4):S881–S885. https://doi.org/10.4103/pm.pm_202_17

Singh R (2016) Chemotaxonomy: a tool for plant classification. JMPS 4(2):90–93

Smith MA, Wood DM, Janzen DH, Hallwachs W, Hebert PDN (2007) DNA barcodes affirm that 16 species of apparently generalist tropical parasitoid flies (diptera, tachinidae) are not all generalists. Proc Nat Acad Sci USA 104:4967–4972

Sotero-Caio CG, Platt RN, Suh A, Ray DA (2017) Evolution and diversity of transposable elements in vertebrate genomes. Genome Biol Evol 9(1):161–177. https://doi.org/10.1093/gbe/evw264

Talapatra SK, Talapatra B (2015) Quinine. *Cinchona* Alkaloids (*Tryptophan Derived Quinoline Alkaloids*). In: Chemistry of plant natural products. Springer, Berlin, Heidelberg

Tayjanov K, Mamadalieva NZ, Wink M (2017) Diversity of the mountain flora of Central Asia with emphasis on alkaloid-producing plants. Diversity 9:11. https://doi.org/10.3390/d9010011

Vernooy R, Haribabu E, Muller MR, Vogel JH, Hebert PDN, Schindel DE et al (2010) Barcoding life to conserve biological diversity: beyond the taxonomic imperative. PLoS Biol 8:e1000417

Vigouroux Y, Barnaud A, Scarclli N, Thuillet AC (2011) Biodiversity, evolution and adaptation of cultivated crops. Comp Rend Biol 334:450–457. https://doi.org/10.1016/j.crvi.2011.03.003

Wink M, Botschen F, Gosmann C, Schäfer H, Waterman PG (2010) Chemotaxonomy seen from a phylogenetic perspective and evolution of secondary metabolism. In: Wink M (ed) Biochemistry of plant secondary metabolism. Wiley-Blackwell, Oxford. Ann Plant Rev 40

Wink M (2008) Plant secondary metabolism: diversity, function and its evolution. Nat Prod Commun 3:1205–1216

Wink M (2010a) Functions and biotechnology of plant secondary metabolites. Wiley-Blackwell, Oxford. Ann Plant Rev 39

Wink M (2010b) Biochemistry of plant secondary metabolism. Wiley-Blackwell, Oxford. Ann Plant Rev 40

Wong EHK, Hanner RH (2008) DNA barcoding detects market substitution in north American seafood. Food Resour Int 41:828–837

Wybouw N, Dermauw W, Tirry L, Stevens C, GrbićM, Feyereisen R, Leeuwen TV (2014) A gene horizontally transferred from bacteria protects arthropods from host plant cyanide poisoning. eLife 3:e02365. https://doi.org/10.7554/elife.02365

Zhu X, Zeng X, Sun C et al (2014) Biosynthetic pathway of terpenoid indole alkaloids in *Catharanthus roseus*. Front Med 8:285. https://doi.org/10.1007/s11684-014-0350-2

Chapter 2
The 'Sixth Mass Extinction Crisis' and Its Impact on Flowering Plants

K. R. Shivanna

Abstract Human-induced environmental changes caused by habitat loss and its degradation, overexploitation of resources and climate change, have already pushed considerable number of plant and animal species to extinction, and a large number of them are at the verge of extinction. These catastrophic environmental changes have precipitated the 'sixth mass extinction crisis' in which a large proportion of the species would be lost in geologically a short time. As biological diversity and ecosystem functioning form the basis of human existence, human-induced environmental changes may eventually lead to serious repercussions on the biosphere and threaten the survival of the human race itself. Conservation of biodiversity and ecosystem functioning has, therefore, become a major challenge the humanity has to face in the coming decades. Flowering plants form the major component of plant diversity. Sustainability of the prevailing diversity depends on the ability of species/populations to reproduce and recruit new individuals to sustain populations. Recruitment is the final step in a long series of sequential events starting with the flower. Among these sequential events, pollination, a prerequisite for fruit and seed set, and seed dispersal, needed for effective recruitment of new individuals, are two of the most critical events, and both involve largely plant–animal mutualism. Human-induced environmental changes have imposed serious constraints on these mutualisms, thus seriously hampering recruitment. 'Global pollinator crisis' has been recognized as a major hazard not only for the sustainability of plant diversity but also for crop productivity and thus the food and nutritional security of human beings. Climate change, apart from inducing migration of species to higher altitudes and latitudes, brings about phenological changes particularly in the time of flowering and fruiting resulting in mismatches between the plant and animal partners involved in mutualistic interactions. There is an urgent need for concerted global action to reduce and reverse this trend of environmental degradation and thus conserve our biological diversity and ecosystem services to protect ourselves and our Planet.

K. R. Shivanna (✉)
Ashoka Trust for Research in Ecology and the Environment,
Srirampura, Jakkur Post, Bengaluru 560 064, India
e-mail: shivanna@atree.org

K. G. Ramawat (ed.), *Biodiversity and Chemotaxonomy*, Sustainable Development and Biodiversity 24, https://doi.org/10.1007/978-3-030-30746-2_2

Keywords Biodiversity loss · Climate change · Flowering plants · Global warming · Overexploitation · Pollination crisis · Human-induced environmental changes · Recruitment constraints · Seed dispersal constraints

2.1 Introduction

Biodiversity includes all heritable variations at all levels of organization (Wilson 1997). The vast biodiversity prevailing on the Planet is the result of the evolutionary processes operating since the time life originated on the Earth about 3.5 billion years ago. Biodiversity is generally considered at three levels: genetic diversity (all the genes and alleles present in the populations of a species), species diversity (all the species present on earth) and community diversity (interactions associated with biotic and abiotic components of the environment) (Gaston 2010; Bawa et al. 2011). As new species keep on evolving on the Earth as a result of natural selection acting on heritable variations, a few species that become misfits to the prevailing environment become extinct. However, the number of species that become extinct, referred to as background extinction, has been very small when compared to the number of new species that have been evolving. The rate of background extinction varies among different group of organisms. It is estimated that background extinction rate of mammal species is one extinction in 700 years (Kolbert 2014).

Human beings, right from the time of their origin, initiated their impact on biodiversity. After the industrial revolution, his ability to control the environment intensified over the years. Concomitant with this, rapid increase in the population and evolving human greed for attaining luxurious lifestyle have been putting a heavy demand on bioresources and ecosystems. According to the 2012 UN estimations, the world population was 6.2 billion in 2000, reached 7 billion in 2012, and is expected to reach 9.5 billion (±0.4) in 2050 and 11 billion (±1.5) in 2100 (Gerland et al. 2014). Human dependence on and overexploitation of bioresources, and human-induced environmental degradation steadily increased, particularly in recent decades beyond the sustainability of a large number of plant and animal species as well as ecosystems; this led to the extinction of considerable number of species and a large number of them are on the edge of extinction. Thus, anthropogenic activities in recent decades have increased the rate of extinction of species by 100 s or even 1000 s of times over background extinctions. According to many biologists and geologists, human activities have precipitated the 'sixth mass extinction crisis' (Sodhi and Ehrlich 2010; Lenzen et al. 2012; Dirzo et al. 2014; Kolbert 2014; Ceballos et al. 2017; IPBES 2018). The impact of *Homo sapiens* on the Global climate is so massive on the Earth that many geologists have started calling the present geologic epoch as Anthropocene (Crutzen 2002) and possible sixth mass extinction as 'Anthropocene extinction'.

According to many scientists we are already in the midst of the sixth mass extinction (Barnosky et al. 2011; Dirzo et al. 2014; Ceballos et al. 2017; Ripple et al. 2017). As biological diversity and ecosystem functioning form the basis of human welfare and survival, this may eventually have serious repercussions on the biosphere and threatens the survival of human race itself. In the light of these ongoing trends, conservation of biodiversity has become one of the main challenges the humanity has to face in the coming years. The present review elaborates the consequences and the drivers of the 'sixth mass extinction crisis' and discusses the impacts of human-induced environmental changes on the sustenance of flowering plant diversity, the major component of plant diversity. It highlights the impacts of the prevailing ecological changes on two important plant–animal mutualisms, pollination, and seed dispersal, which are critical for the sustenance of flowering plant species.

2.2 Consequences of the 'Sixth Mass Extinction Crisis'

As pointed out earlier, anthropogenic activities in recent decades have already pushed a large number of species toward extinction. Current predictions on the rate of extinction risks from human-induced environmental changes vary greatly. Thomas et al. (2004) assessed extinction risks of sample regions covering about 20% of the Earth's terrestrial surface and predicted, on the basis of mid-range climate-warming scenario, 15–37% of the species will be committed to extinction by 2050. According to IUCN (2017) considerable proportion of species assessed so far are under threat (Table 2.1) (see also Anonymous 2014). Based on the meta-analysis of 131 published predictions, Urban (2015) estimated that the global proportion of extinction risks increase in line with the increase in global temperature. Extinction risks increase from 2.8% at the present level of global temperature to 5.2% at a rise of 2 °C post-industrial stage. The extinction risk rises to 8.5% if

Table 2.1 Percentage of evaluated species threatened in different groups of plants and animals (IUCN version 2017-3)

Groups	No. of species evaluated	% of evaluated species threatened
Cycads	307	63
Amphibians	6576	41
Selected dicots	1781	36
Selected reptiles	342	35
Conifers	607	34
Reef-forming corals	845	33
Sharks and rays	1091	31
Selected crustaceans	2872	27
Mammals	5591	25
Birds	10,966	13
Selected gastropods	633	08
Selected bony fishes	2390	07

the earth temperature rises by 3 °C. If the present trend of global warming continues, the extinction rate rises to 16% (one in six species). According to this model, South America, Australia, and New Zealand have the highest risks whereas North America and Europe are likely to have the lowest risk. According to Birdlife International (Anonymous 2017a) also 13% of the total bird species are threatened with extinction. Amphibians are now considered to be the most endangered class of animals. Their extinction rate is estimated to be as much as 45,000 times higher than the background extinction (Kolbert 2014). A fungal disease, caused by *Batrachochytrium dendrobatidis*, is considered to be a proximate driver for global amphibian decline; it is thought to have wiped out a third of all frog species (Skerratt et al. 2007; O'Hanlon et al. 2018).

As early as 1992, 1700 world's leading scientists including 99 of the 196 living Nobel laureates realized the seriousness of the situation and warned the humanity to curtail environmental destruction that threatens life support system of this Planet (see Ripple et al. 2017). They felt that humanity was pushing Earth's ecosystems beyond their capacities to support the web of life. The scientists identified ozone depletion, freshwater availability, marine life depletion, ocean dead zones, forest loss, biodiversity destruction, climate change, and continued human population growth as causes for current, impending or potential damage on the planet Earth. After 25 years of the first warning, as many as 15,364 scientists from 184 countries gave a second warning to humanity (Ripple et al. 2017). They warned that since the first warning, "with the exception of stabilizing the stratospheric ozone layer, humanity has failed to make sufficient progress in generally solving these foreseen environmental challenges, and alarmingly, most of them are getting far worse". They highlighted the 25-year negative global trends—a 26% reduction in the availability of freshwater per capita and a loss of nearly 300 million acres of forestland combined with a collective 29% reduction in the numbers of mammals, reptiles, amphibians, birds and fish, and a 75% increase in the number of ocean dead zones. Although world scientists recognized the progress made in stabilizing the ozone layer (Ripple et al. 2017), a recent report (Montzka et al. 2018) highlights a persistent increase in global emission of CFC-11 (trichlorofluoromethane), one of the important ozone-depleting chlorine, despite its production being banned following the Montreal protocol.

Based on analyses of data of a sample of 27,600 vertebrate species, Ceballos et al. (2017) concluded that the sixth mass extinction crisis is more severe than perceived. They feel that the window of corrective action to tackle the crisis is very short, at the most 20–30 years. Based on the analyses of 31 carbon isotopic events during the past 542 million years that includes the previous five mass extinctions, Rothman (2017) identified significant changes in carbon cycle associated with the past five mass extinctions. If the rise in carbon emission continues at the present rate, he predicts that the threshold level would reach by 2100 leading to the beginning of mass extinction. According to Commonwealth Academies of Science consensus statement on climate change (Anonymous 2018) "Changing agricultural conditions, ocean warming and acidification, rising sea levels, and increased frequency and intensity of many extreme weather events are impacting infrastructure environmental assets and human health.”

2.3 Drivers of the ‘Sixth Mass Extinction Crisis’

The major drivers of human-induced environmental changes are habitat loss and its degradation, overexploitation of bioresources and climate change. All three of them act together and difficult to study their effects independently. Earth has witnessed five mass extinctions in the past; in each of these events, a significant proportion of species became extinct. After each mass extinction, the earth was able to restore biodiversity over millions of years and continued to increase the number of species reaching the maximum in recent geologic epoch. Unlike earlier mass extinctions which were due to natural catastrophes, the ‘sixth mass extinction’ is exclusively the result of human activities.

2.3.1 Habitat Loss and Degradation

Extensive areas of natural habitats in almost all countries are lost or degraded due to a number of causes. Major drivers of habitat loss and degradation include conversion of natural habitats including forest lands to agricultural and plantation lands, and for urbanization, industrialization and mining. Other drivers include increasing levels of pollution, rampant use of pesticides and spread of alien invasive species, forest fires and grazing by domestic animals (Tilman et al. 2001; Defies et al. 2004, see Murray et al. 2009; Ghazoul and Sheil 2010; IPBES 2018). Pesticides contaminate the soil, water, and vegetation, and eventually enter non-target organisms such as beneficial insects, birds, fishes and plants leading to loss of biodiversity. Beketov et al. (2013) analyzed the effects of pesticides on invertebrate richness in the streams of Europe and Australia and found that pesticides caused up to 42% losses in species and family richness in both the locations. Invasive alien species alter the composition and functioning of native ecosystems. A large number of invasive species have become a major driver for displacement of native species both in terrestrial and aquatic habitats (Bawa et al. 2011). This is particularly severe in oceanic islands (Sax and Gaines 2008). A number of such invasive species which have become difficult to control are causing major problems in all the countries.

Recently, Intergovernmental Science-Policy Platform on Biodiversity and Ecosystem Services (IPBES 2018) with a membership of 129 states released the summary of a 3-year assessment report on land degradation prepared by over 100 experts from 45 countries and scrutinized by over 200 external reviewers. The report projects a very gloomy picture for the future unless drastic corrective measures are taken promptly by the Governments and the public. The following are some of the highlights of the report:

- Land degradation is worsening over the years and has now reached a 'critical' level in many parts of the world. It is the main driver of the loss of species diversity and ecosystem services. High-consumption lifestyles (in the developed economies), and rising consumption (in emerging economies) are the underlying drivers for land degradation. By 2050, less than 10% of the Earth's land surface would have escaped substantial impacts of human activity. The Earth has lost 87% of wetland areas of which 54% are lost since 1900.
- Land degradation undermines the well-being of 3.2 billion people. It is a major contributor to mass human migration and increased level of conflict.
- Land degradation is also a major contributor to climate change; deforestation alone is contributing about 10% of all human-induced greenhouse gas emissions. It has also resulted in the release of carbon stored in the soil; up to 4.4 billion tons of CO_2 is estimated to have been released globally between 2000 and 2009.
- Global crop yield is likely to be reduced by an average of 10% by 2050 by the combination of land degradation and climate change.

2.3.2 *Overexploitation of Resources*

Hunting has been one of the major and ancient activities of humans responsible for driving a large number of species to extinction. Most of the larger mammals were hunted down even when man was still a hunter/gatherer (Diamond 2005). This has continued in a much larger scale and extended to a range of animal species (Corlett 2007a). Animals are hunted/poached for a number of uses—for their meat, hide, medicinal/ornamental value. They are also hunted as game species or they happen to be agricultural pests or predators of domestic animals. Some of them fetch huge amount of money in international markets. Many live animals particularly birds are collected for pet trade. Hunting and poaching activities of humans combined with the loss of habitat have resulted in depletion of the populations of these animals often to the limit of extinction. Although many governments have introduced laws to prevent hunting and poaching, legal and illegal trade continues unabated particularly in developing countries. Bioresources from most of the economically important plant species are also exploited beyond their sustenance (Bawa et al. 2011). Legal as well as illegal timber extraction is the main cause for deforestation. Even the sources of non-wood forest products particularly of medicinal value are overexploited. As a majority of medicinal plants are collected from the wild sources, their population is steadily depleting in their natural habitat.

Overexploitation of bioresources is being done not only for local needs but also to cater to the demands of other countries. Lenzen et al. (2012) have linked 25,000

IUCN red-listed animal species to >15,000 commodities produced in 187 countries in terms of their biodiversity impacts; they estimated that about 30% of global species threats are due to international trade (see also Moran and Kanemoto 2017). For example, Himalayan musk deer (*Moschus leucogaster*) occurs in higher ranges of Himalayas. Musk pod in the abdomen of males emits sweet persistent aroma. There is a great demand for musk pods in cosmetic and related industries. According to IUCN, each kg of musk deer pod fetches ca US $45,000. As each pod weighs ca 25 g, the income worksout to be ca $1000/pod (or for each male deer). The animals are extensively poached particularly in Kedarnath Wildlife Sanctuary in the Himalayas of India (Chakraborty 2016).

Large areas of forests are converted into plantations to grow palm oil in Malaysia and Indonesia, and cocoa, coffee and tea in several other countries largely for export. Teak (*Tectona grandis*) is the source of expensive timber. It is not native to Ecuador. However, teak is increasingly grown in the rain forests and agricultural lands of Ecuador in recent years for export of its timber to India (Anonymous 2017b). *Aquilaria malaccensis* (agar wood plant) is native to North-East India and a few other neighboring countries. It produces a resin, the source of an expensive perfume, in the heart wood in response to a fungal infection. The perfume extracted from resinous heart wood is in great demand (1 kg of even crude perfume is reported to cost about US $15,000 (Alibaba.com). Although the cutting of this tree is banned in India, lots of illegal trade is continuing largely for export. Another example of overexploitation is *Commiphora wightii,* the source of guggul, an important oleo-gum-resin secreted as a result of mechanical injury caused to the stem. Apart from its use in a number of medicinal preparations, it is also used in anti-inflammatory and anti-obesity treatments. Guggul products are exported from India to 42 countries. The demand for guggul in India is >900 metric tons. Because of its overexploitation in the country, guggul production has declined in Gujarat, a major production area, to 1.6 mt. Now, 90% of guggul is imported, mostly from Pakistan (https://doi.org/10.1016/j.jep.2018.04.040). In general, developing countries tend to be the net exporters and developed countries tend to be the net importers of products implicated in biodiversity loss (Lenzen et al. 2012).

Grounwater, surface water, snow, and ice are the major sources of terrestrial freshwater. Population growth, unsustainable groundwater extraction combined with climate change has made freshwater availability a major constraint in many parts of the world. Based on satellite data of 14 years (2002–2016) Rodell et al. (2018) have quantified the trends in terrestrial water storage around the world. They have identified a number of areas with declining groundwater availability including northern India as the result of unsustainable groundwater extraction for irrigation. The trend of groundwater depletion persists in several areas despite precipitation being normal. The authors have emphasized the need for implementing effective water conservation and management strategies for sustaining water and food security in the coming decades.

2.3.3 Climate Change

Climate change is one of the main drivers for biodiversity loss. Apart from its effect on biodiversity, climate change has much wider and direct effects on human well-being. Massive combustion of fossil fuels (coal, oil and natural gas) has led to the emission of enormous amount of greenhouse gases (CO_2 and methane) into the atmosphere. Greenhouse gases along with water vapors trap the radiated heat from the earth leading to global warming. Humans have burnt so much fossil fuels since the beginning of the industrial revolution to add 365 billion metric tons of carbon to the atmosphere (Kolbert 2014). Deforestation has added another 180 billion tons.

Impact of climate change on crop productivity has been one of the major concerns. Earlier studies, largely performed in greenhouses or growth chambers, showed the possibilities of increase in crop yield with the doubling of atmospheric CO_2 concentration (see Kimball 1983). However, later studies using elevated CO_2 under fully open airfield conditions have cast serious doubt on such projections (Long et al. 2006). All predictions so far made on the impact of climate change on crop productivity in the coming decades consistently show declines in crop yields. Zhao et al. (2017) reported mean temperature increase by each °C would, on average, reduce global yields of wheat by 6.0%, rice by 3.2%, maize by 7.4%, and soybean by 3.1%. Based on 65 years of weather records and the data on wheat and barley yields in France, Gammans et al (2017) reported yield decline by 17–33% in these crops by the end of the century. Global warming is predicted to cause 17% decrease in the yield of barley (Xie et al. 2018). Apart from a reduction in the yield, Zhu et al. (2018) reported global warming to result in the decline in protein, iron, zinc, and vitamins in rice. Overall reduction in the yield and nutritional quality of crops would affect food and nutritional security of the steadily increasing population around the world.

Insects have been the most successful group of animals and have thrived over millennia. However, biodiversity of insects is also threatened around the world in recent decades. In protected areas of Germany, the biomass of total flying insects has declined by 75% over 27 years (Hallmann et al. 2017). Even in Puerto Rico, the biomass of flying insects has declined by 97% in protected rainforest when compared to 1970s (Lister and Garcia 2018). This has affected the food web of the forest. The decline in the insect biomass has induced parallel declines in other animals such as insectivorous lizards, frogs, and birds that feed on the insects. Based on comprehensive review of 73 historical reports of insect decline and systematic assessment of its drivers, Sanchez-Bayo and Wyckhuys (2019) report that 40% of the global insect species are threatened with extinction over the next few decades. Lepidoptera, Hymenoptera, and Coleoptera seem to be the most affected among the terrestrial ecosystems. Habitat loss, agrochemical pollutants, invasive species, and climate change are the main drivers for this decline.

Some parts of Australia experienced record-breaking temperatures of >42 °C in November 2018. This high temperature killed at least 23,000 spectacled flying foxes, *Pteropus conspicillatus* (fruit-eating bat), amounting to a third of this species present in the country (Akst 2019). Also, the periods of intense drought and excessive warming caused a drop in water level in Australia's Darling River (Nogrady 2019). This created ideal conditions for major blooms of cyanobacteria which depleted the dissolved oxygen in the water leading to the death of hundreds of thousands of fishes in recent months. Biologists are concerned about the fate of many other species sensitive to extreme temperatures.

A part of the atmospheric CO_2 is absorbed by the ocean. Thus, apart from increasing the temperature of the atmosphere, rising CO_2 levels make the ocean warmer and acidic (IPCC 2014). The oceans have become 30% more acidic when compared to 1985 making it harder for sustenance of many marine animals such as corals, oysters, and mussels. Corals are considered as one of the most endangered groups that exceed most of the terrestrial endangered groups. A spike in ocean temperatures in 2016 as a result of global warming killed nearly one-third of the corals of the Great Barrier Reef of Australia (Hughes et al. 2018). This has altered the assemblages of its coral species. If greenhouse gas emissions and global warming continue at the current rate, Hughes et al. (2018) predict that the Great Barrier Reef, the world's largest coral reef, will be unrecognizable in another 50 years. According to the UNESCO World Heritage Centre, 21 of the 29 World Heritage reefs have been subjected to severe heat stress causing some of the worst bleaching ever observed. The analysis predicts, under the current emissions scenario, all 29 world heritage site coral reefs would cease to exist as functioning ecosystems by the end of the century. Coral reefs are comparable to tropical rain forests; they are the homes to thousands of species of marine animals. Breakdown of coral reefs would result in the collapse of the whole ecosystem.

Oceans are warming a faster rate than ever (Cheng et al. 2019). Earlier studies on the effects of climate change on ocean warming concentrated largely in the upper layer (0–700 m). This does not give full information as to how much heat is being trapped by the ocean. Recent investigations have started analyzing the global ocean heat content (OHC) in deeper layers. Gleckler et al. (2016) examined OHC changes in varying depths of the world's oceans, using data and models available, since early industrial-era (1865–2015). Their studies show that over a third of the accumulated heat occurring at depths below 700 m has occurred in recent decades.

Global warming has contributed to the decline in ice sheets and ice caps in the polar regions. Glaciers are receding at a faster rate around the world. Based on the assessment made over five years on the inputs from more than 350 researchers from 22 countries, at least one-third of the Himalayan glaciers are predicted to melt by the end of the century even when ongoing mitigation targets are met; two thirds are expected to melt if global warming continues at the present rate (Wester et al. 2019). The loss of ice caps in Arctic and Antarctic poles and Greenland has been the main cause for the rise in the sea level (Rignot et al. 2019; Trusel et al. 2018). A rise in sea level just by 1 m could lead to submergence of many of the smaller islands and a large part of the coastal areas around the world. When this happens, the biodiversity in such islands and coastal areas become extinct without giving a chance for their conservation. Marked increase in the frequency and intensity of extreme weather conditions in the form of severe cyclones, tsunamis and tornados, snowstorms and thunderstorms, floods and droughts and forest fires in recent years around the world are believed to be the result of climate change. Apart from the impacts of these extreme weather conditions on human sufferings, they may lead to the extinction of many narrow endemics at one go.

Severe storms and tsunamis often cause movement of coastal marine organisms. Recently Carlton (2017) have documented transport of nearly 300 Japanese marine coastal species across Pacific Ocean to the shores of North America and Hawaii in over 6 years after the massive Tsunami generated in 2011 by an earthquake in East Japan. Most of their dispersal has occurred through rafting on non-biodegradable objects. This is the longest recorded transoceanic survival and dispersal of coastal species by rafting (Carlton 2017). Such migration is bound to affect markedly the faunal community of the Pacific coast of the US.

According to the fifth assessment report of the Intergovernmental Panel on Climate Change (IPCC 2014), global warming of the atmosphere and ocean is unequivocal. Concentrations of carbon dioxide (400 ppm), methane (1800 ppb), and nitrous oxide (320 ppb) in the atmosphere have increased to levels unprecedented at least in the last 800,000 years. The report also projects an increase in the mean global temperature of 3.7–4.8 °C relative to preindustrial levels (1850) by 2100 in the absence of new policies to mitigate climate change. The report predicts that climate change not only amplifies existing risks but also creates new risks for natural and human systems. Climate change-induced risks are generally greater for disadvantaged people and communities.

The Paris agreement on climate change adapted in 2015 formally recognized to limit the temperature increase to 1.5 °C above preindustrial level by the end of the 21 century. The possibility of achieving this goal was seen almost impossible on the basis of the present emission scenario. Scientific studies mainly looked at limiting warming to 2 °C. However, the benefit that would incur by achieving this goal of limiting temperature rise to 1.5 °C is enormous. According to recent global-scale analysis of Warren et al. (2018) the proportion of species projected loses (>50% of their geographic range) expected at a rise by 2 °C is reduced by half for vertebrates and plants and by two thirds for insects when the warming is limited to 1.5 °C. New research has shown that this goal, although difficult, is within the reach if strong action is taken by all the countries. The model of Rogelj et al. (2018) shows the conditions under which this can be achieved.

2.4 Impact of Human-Induced Environmental changes on Flowering Plants

Most of the discussions on biodiversity loss deal with animal species. There is very little discussion by both academia and public about the loss of diversity plant species. Plant diversity forms the basis of sustainability of all non-photosynthetic organisms including humans. Apart from providing materials of essential needs of humans (such as food, fiber, fuel, shelter, medicines, gums, and resins), plant diversity regulates the quality of water, soil and climate, and provides educational, recreational and ethical needs. Genetic resources of wild relatives of crop plants are the reservoirs of novel genes not only of nutritional value but also of those imparting resistance to a range of biotic and abiotic stresses needed for genetic improvement of cultivated species. Wild plants also provide raw materials for several industries. The impact of ecological disturbances on the vulnerability of all plant groups is beyond the scope of this chapter; it covers only the flowering plants in their natural habitats. All the three major drivers for species vulnerability—habitat degradation, overexploitation, and climate change—act in synergy and often it becomes difficult to identify the impacts of individual factors.

2.4.1 *Basis of Population/Species Sustainability*

The stability of a population/species of any flowering plant in its natural habitat essentially depends on two factors. Its ability for (i) optimal reproduction resulting in the development of adequate number of fruits and seeds and (ii) efficient recruitment of new individuals to sustain populations. In a stable population, the number of births is equal to or more than the number of deaths. Reproduction and recruitment in flowering plants involve a series of sequential events starting with the flower (Fig. 2.1). Even under ideal conditions, the proportion of seeds that result in seedling establishment is extremely small. According to one estimate, only one seed out of 100,000 may establish into a sapling in tropical forests (Ghazoul and Sheil 2010). A number of hazards operate a teaching step involved in reproduction and recruitment; the hazards reduce this number even further. The most critical steps involved in efficient reproduction and final recruitment are pollination and seed dispersal. Both these steps involve, to a large extent, mutualism between plants and animals, and are the most vulnerable to human-induced environmental changes.

2.4.2 *Environmental Changes and Their Impact On Recruitment*

Habitat loss, overexploitation, and climate change impose major constraints in reproduction and/or recruitment in populations of flowering plants. These constraints operate at different levels in the sequential events shown in Fig. 2.1. Recruitment constraints lead to more deaths than births in the population and result in a gradual decline in its size. When this trend continues for a long time, the population becomes endangered and eventually extinct. Thus, recruitment constraint/failure is the ultimate driver for species vulnerability and eventual extinction.

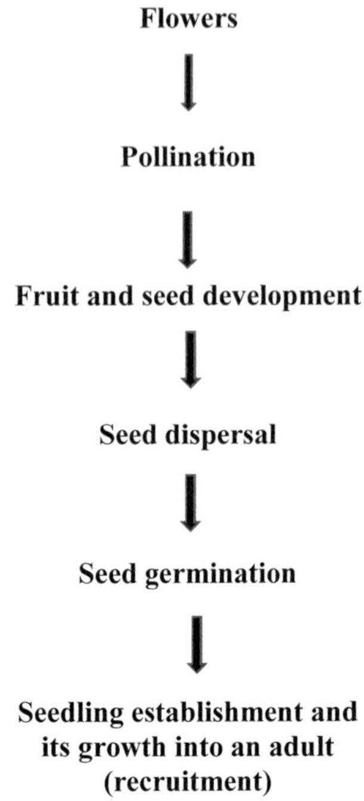

Fig. 2.1 Sequential events involved in recruitment of new individuals in flowering plants

2.4.2.1 'Global Pollinator Crisis'

Pollination is the transfer of pollen grains from the anther to the stigma. This is one of the critical ecoservices and is a prerequisite for fertilization and eventual fruit and seed development. Nearly 90% of all wild flowering plants and 75% of food crops depend fully or partially on animals to achieve pollination (Willmer 2011; Patiny 2012; Shivanna 2014; IPBES 2016). One of the important impacts of human-induced environmental damage is on pollination services both in natural and agricultural habitats. Pollinators play a vital role not only in sustenance of biodiversity but also in world food production and thus in providing food and nutritional security to humans. Since 1990s, a number of publications have appeared highlighting the decline of pollinators in their natural and agricultural habitats and its impact on pollination services (Buchmann and Nabhan 1996; Kearns et al. 1998; Kremen and Ricketts 2000; Wilcock and Neiland 2002; Samejima et al. 2004; Roubik 1995; James and Pitts-Singer 2008; Hegland et al. 2009; Murray et al. 2009; Potts et al. 2010; Obute 2010; Aslan et al. 2013; Tylianakis 2013; Dicks et al. 2016; IPBES 2016).

Based on the analyses of almost one million records from National Entomological databases, Biesmeijer et al. (2006) reported a significant decrease in bee diversity, one of the most important pollinators, in both Britain and Netherlands by about 30% since 1980. This decline was more frequent in habitat and floral specialists, especially univoltine (having one generation per year) and nonmigrant species when compared to generalists (multivoltine and migrant species). More importantly, there was a parallel decline in plants pollinated by declining pollinators. Plants pollinated by abiotic agents (wind or water) showed an increase whereas self-pollinating plant species showed an intermediate response, clearly linking the decline of plant species to their biotic pollinators. The relative abundance of four out of eight species of bumblebees across the US has declined by up to 96% and their surveyed geographic range has contracted by 23–87%, some within the last 20 years (Cameron et al. 2011). Using historical data sets, Burkle et al. (2013) analyzed the changes in plant-pollinator interactions over 120 years in a temperate forest understory community in Illinois, USA and showed the decline of wild pollinators over time leading to the decline in the quantity and quality of pollination services.

Apart from habitat loss and fragmentation, climate change, extensive use of pesticides on agricultural crops and monoculture agriculture, spread of alien species and a range of diseases seem to be responsible for the decline of pollinators. There are a number of publications on the effects of fragmentation on pollination services (see da Silva Melo et al. 2014 and references therein). Meta-analyses carried out by Aguilar et al. (2006) on the susceptibility of pollination and reproduction of plant species to habitat fragmentation during the last two decades revealed large and negative effects of fragmentation on pollination and plant reproduction. They also found a significant positive correlation between the sizes of fragments on pollination and reproductive success of the populations. Similarly, studies carried out on the density of plants and fruit set of *Byrsonima pachyphylla* in relation to fragment size in Brazilian savanna (da Silva Melo et al. 2014) found that plant population densities and extent of fruit set was greater in larger fragments indicating that larger fragments are needed for maintenance of populations of species in the communities.

As the pesticide residues get incorporated in the rewards of the bees (pollen and nectar), pesticide application affects not only managed bees but also wild bees and other species involved in pollination of both wild and crop species. Many pesticides are acutely toxic to bees and kill them. Pesticides at low levels may not kill the bees but have significant effects on their performance such as their ability to forage, learn and reproduce, and thus affect the survival of the hives. In several critical studies, neonicotinoids have been shown to affect negatively wild bee density, solitary bee nesting and bumblebee colony growth and reproduction under field conditions (Rundlof et al. 2015; IPBES 2016), thus indicating substantial risk to wild pollinators. The results of Brittain et al. (2010) also suggest that wild bees are at particular risk from pesticide use. Recently increased genetic diversity, as revealed by a higher electrophoretic banding pattern, in native populations of *Apiscerana* continually exposed to pesticides has been reported (Chakrabarti et al. 2018). This may indicate the development of a more resistant populations in response to threshold pesticide stress as a survival strategy and would have far-reaching implications on the effects of pesticides and this area needs further studies.

Colony collapse disorder (CCD), prevalent in the US and several European countries in which worker bees fail to return to the hives, is one of the main causes for a significant reduction in the number of managed bee colonies. Neonicotinoids are water-soluble, tend to accumulate in soils and leach into waterways. They also reach the main rewards of the pollinators, nectar, and pollen, of treated crops (Goulson 2013). The concentration of neonicotinoids in nectar and pollen are sufficient to affect colony reproduction in bumblebees. Many evidence indicate that CCD is induced by pesticides especially neonicotinoids as well as pathogens (Brittain et al. 2010; Porrinni et al. 2014).

In the light of the importance of pollination services in sustenance of biodiversity and crop productivity, and widespread declines of pollinator diversity and density have led to serious concern about 'global pollinator crisis' (Holden 2006; Murray et al. 2009). This has led to the initiation of many international programmes to study and remedy pollination services. The latest being the establishment of Intergovernmental Science-policy Platform on Biodiversity and Ecosystem Services (IPBES) in 2012 (in collaboration with UNEP, UNESCO, FAO, and UNDP). Its mandate was to "compile scientifically credible and independent up-to-date assessment of available knowledge on pollination and pollinators to make informed decisions at the local, national and international levels." Its report was released in November 2016. It confirms the decline of pollinators, wherever it has been studied, across the world. Globally, 16.5% of vertebrate pollinators and over 40% of invertebrate pollinator species particularly bees and butterflies are facing extinction. The report emphasizes that food and nutritional security depends on sustaining and enhancing pollination services. It also highlights lack of data on wild pollinators and pollination in many developing countries. The report underlines the urgent need for long-term monitoring of pollinators and pollination in these countries to take suitable steps to sustain pollination services (for more details see IPBES 2016).

So far studies on pollinator disruption as a result of anthropogenic activities have concentrated on diurnal pollinators. However, nocturnal pollinators play an important role in several plant species in natural habitats. This aspect has hardly been touched. There has been a rapid global increase in artificial lighting at night which is likely to disrupt nocturnal pollinators (see Callum et al. 2015). Recent studies of Knop et al. (2017) have shown that the visits of nocturnal pollinators to flowers were reduced by 62% in artificially illuminated plant communities when compared to dark areas resulting in 13% reduction in fruit set even when the plants received numerous visits by diurnal pollinators. There is a need to intensify studies on the effects of light pollution on nocturnal pollination services.

Apart from their role in sustaining plant diversity in natural habitats, wild pollinators also play an important role in pollination of crop species even in the presence of managed pollinators (Greenleaf and Kremen 2006; Brittain et al. 2013; Garibaldi et al. 2013). Behavioral interactions between wild bees and honey bees have been shown to increase pollination efficiency of honey bees up to fivefold in hybrid sunflower (Greenleaf and Kremen 2006). Based on 41 crops across 600 field sites around the world, Garibaldi et al. (2013) reported that managed honeybees supplement rather than substitute pollination services rendered by wild insects. Thus, wild pollinators and managed bees promote fruit set of crop species independently. Now the trend in Western countries is to develop integrated pollination services by increasing the density and diversity of native pollinators by making agricultural habitats pollinator-friendly and judicial use of managed pollinators to safeguard pollination services of crop plants. Pollinator-friendly habitats include making the nest sites available around agricultural fields by maintaining uncultivated strips along field margins and/or providing artificial nesting sites and reducing the use of pesticides and other unfriendly agrochemicals.

Many new approaches are being tested to safeguard pollination services, particularly of crop species, in the coming decades. One of them is to vaccinate bees against various diseases (Salmela et al. 2017; Jacobs 2018). Vaccine is mixed with sugar solution that attracts queen bees; the vaccinated queen bee is then introduced to the hive, initiating a new generation of bees with immunity to the disease. Another technology being tried is to develop pollinator drones (Chechetka et al. 2017, Lallensack 2017). The efficacy of remote-controlled drones has been demonstrated under laboratory conditions in Japanese lilies.

In India, there are no direct studies on the density and distribution of pollinators in natural habitats and the effects of human-induced environmental changes on them. There are some indirect evidences to indicate pollinator limitations. Nayak and Davidar (2010) compared pollination efficiency between self-incompatible and self-compatible species and their saplings and adult densities of ten species around Puducherry region of South India. Their studies showed higher levels of pollination limitation in self-incompatible species leading to a lower seed set when compared to self-compatible species. This difference is obviously due to pollinator limitation; unlike self-incompatible species which require interplant movement of pollinators, self-compatible species are less dependent on pollinators and their interplant movement. Also, the sapling and adult densities in fragmented habitats were significantly lower in self-incompatible species when compared to self-compatible species indicating lower recruitment in the former as a result of pollination limitation leading to lower fruit set.

2.4.2.2 Seed Dispersal Constraints

Apart from pollination constraint, another serious limitation of human-induced environmental changes is on seed dispersal. Seed dispersal is an important ecoservice essential for effective recruitment. Seed dispersal provides three important recruitment advantages: (i) escape from the sources of mortality (competition and high level of predation) concentrated around the parents, (ii) increased probability of seedling establishment away from the parents and (iii) ability to colonize new areas. Vertebrates are the major seed dispersers; in tropical forests, seeds of 70% of the flowering plants are dispersed by animals (Howe and Smallwood 1982; Howe 1990; Turner 2001; Norjhauer and Newberg 2010; Tadwalker et al. 2012). Seed dispersers are becoming scarce in natural habitats as they are hunted/poached extensively (see page …) and their habitats are greatly degraded.

There are a number of reports on the impact of dispersers' scarcity in many neo-tropical forests of Brazil, Costa Rica, Mexico, Panama, and Peru (Anonymous 2007). Hunting has resulted in the reduction of large and medium-size primates by >80% when compared to protected areas in south-eastern Peru. Many investigators have compared the seed dispersal efficiency of plant species in hunted and protected areas. These reports invariably show a marked decline in seed dispersal efficiency in hunted areas (Anonymous 2007; Corlett 2007b; Nunez-Iturri and Howe 2007; Wang et al. 2007; Beckman and Muller-Landau 2007; Wright et al. 2007). For example, in *Antrocaryonklaineanum* (Anacardiaceae), a mammal-dispersed species, only 2% of the seeds got dispersed in heavily hunted areas when compared to 40% in protected areas in Cameroon (Wang et al. 2007). In another study in central Panama, similar reduction in dispersal efficiency in hunted areas has been reported in *Oenocarpus mapora* (Arecaceae) (Beckman and Muller-Landau 2007). Seed dispersal constraint has resulted in a marked reduction in the density of individuals in the hunted sites. Such populations with low density and considerable distances between conspecifics exacerbate dispersal limitation as they are unable to attract enough seed dispersers because of low visibility and limited reward they offer to the dispersers. Fragmentation of the habitat also limits seed dispersal efficacy. Interestingly, a marked increase in the density of wind-dispersed lianas in heavily hunted forests has been recorded (Wright et al. 2007).

In India, there are no studies on the impact of persecution of animal dispersers on seed dispersal efficacy. However, Tadwalkar et al. (2012) have reported reduction in the density of animal-dispersed plant species in the disturbed areas of northern Western Ghats when compared to those dispersed by abiotic agents; this reduction was proportional to the extent of disturbances of the habitat. This appears to be the result of dispersal limitation since animal dispersers are affected by forest fragmentation and other anthropogenic disturbances.

Even after the seeds are dispersed, there are constraints for eventual recruitment of new individuals to sustain populations. They operate at different post-dispersal stages. Habitat destruction, fragmentation, and climate change may lead to more and more seeds dispersing on sites not suitable for germination and seedling establishment.

Another major anthropogenic impact on recruitment constraint, particularly in developing countries is collection of seeds and/or fruits, which are economically important, from wild species (Murali et al. 1996; Ganesan and Setty 2004). Such collections hardly leave any seeds in the natural habitat for recruitment. For example, *Vateria indica* (Sinu and Shivanna 2016) is an endangered and economically important tree species endemic to the Western Ghats. Fruits are collected for extraction of oil from seeds which is used in the oil and paint industry. Studies carried out in a Reserve Forest in the Central Part of the Western Ghats, India showed that only 3.8% of the fruits that were naturally-fallen on the forest floor were spared after human collection. Such large-scale fruit collection can hamper local population structure of the species by limiting new recruitments.

2.4.2.3 Shift in Species Range and Phenological Events

The impact of climate change on biodiversity is very complex and often difficult to analyze. Most of the studies so far have been on the most apparent and significant events induced by climate change—shift in distributional range of species (to higher altitudes and latitudes) and phenological events of plants and animals. Phenology is the timing of recurring seasonal events in organisms. In flowering plants, various reproductive events such as the timing of flowering, fruiting, their intensity, and longevity are important phenological events which are responsive to climate change. Shifts in the distributional range and phenology of species can be studied by comparing early records over several decades with the present field studies. In Europe and several other developed countries meticulous records of such details are available, often going back to >100 years, and have been used to estimate species migration and shifts in their phenology. The longest past records of the flowering phenology available in the world (over 1200 years) are the flowering phenology of Cherry blossoms (*Prunus* species) in Japan (see Primack et al. 2009). Another method of analyzing the species migration and shift in their phenology is to compare herbarium specimens (on which the location of collection and the period of flowering, etc. are recorded) deposited in standard herbaria over the past years with the present field studies. Phenological timings in different plant species are generally evolved over thousands of years to achieve optimal reproduction and recruitment under the prevailing environment. Flowering is associated with the availability of pollinators and fruiting is associated with the availability of seed dispersers and optimal conditions for seed germination and seedling establishment. Any change in the phenology of plants or its mutual partners may uncouple these adaptations for pollination, seed dispersal, and seedling establishment, and thus leads to reproductive/recruitment constraints.

Developmental/phenological events in plants and animals are based on environmental cues such as temperature, light, precipitation, and snowmelt. Climate change alters the timing and magnitude of environmental cues. Early warming of the habitat shifts the timing of the phenological events, particularly the timing of flowering and fruiting. There are a number of reports on the shifts in the distributional range and phenology of both plants and animals as a result of climate change (Grabherr et al. 1994; Parmesan et al. 1999; Cleland et al. 2007; Parmesan and Yohe 2003; Beckage et al. 2008; Pimm 2009; Miller-Reshing and Forrest 2010; Miller-Rushing et al. 2010; Lovejoy and Hannah 2005). As the habitat once suitable for the species warms up, many of them extend their range toward higher altitudes (in mountain areas) and poles (in plains) in search of suitable climatic niche. However, many species may not be able to keep pace with the changing conditions and thus lag behind leading to their eventual extinction. Long-term observations over 110 years have shown that many species of North-American and European bumblebees are not keeping up with the changing climate but they are disappearing from the southern portions of their range (Kerr et al. 2015). Bertrand et al. (2011) observed a larger temperature lag between the climate and plant community composition in lowland forests than in highland forests. This has been attributed to greater ability of lowland species for local persistence in response to climate warming, reduced opportunity for short distance escape and greater habitat fragmentation in lowland forests. Thackeray et al (2016) quantified variations in climate sensitivity to temperature and precipitation data, to over 10,000 phenological data sets of terrestrial and aquatic organisms and showed marked variations in the direction, magnitude and timing of climate sensitivity among taxa. Secondary consumers showed consistently lower climate sensitivity when compared to other groups.

There are a number of reports of climate change-induced early flowering. Cherry blossom festival is one of the important cultural events in Japan which is celebrated each year sometime in April coinciding with the peak of the flowering of cherry blossom (Primack et al. 2009). The dates of cherry blossom festival have been documented since 900 s (over 1200 years). During the period from 1971 to 2000, plants flowered an average of 7 days earlier in comparison to the average of all previous records. This is correlated with a rise in temperature. Projections suggest that cherry blossom trees may flower about 30 days earlier by 2100 (Allen et al. 2013). Dunnell and Traverse (2011) have compared first flowering time (2007–2010) of the native prairie species in the Red River Valley, North Dakota, USA, for which historical data (1910–1961) on flowering time was available. Spring temperatures in these plains have increased and the growing season has lengthened significantly over the years. Flowering times of 75% of the 178 species have been reported to be sensitive to at least one variable related to temperature or precipitation. Over the last four years of their study 5–17% of the observed species have significantly shifted their first flowering time earlier or later relative to the previous century.

In India, some information is available based on the interview of elderly people living in the villages in the Himalayan ranges on changes in species composition and their flowering period around their villages (Bharali and Khan 2011). According to an elderly villager, apple trees used to flower only once in a year during February. Now they flower twice a year, once in late March and again in September. The flowers that bloom in March produce fruits, while those that bloom in September do not produce any fruits. Many tree species such as *Rhododendron* and *Abies* were growing in areas of lower altitudes around the village, but now they are restricted to higher ridges of the mountains. The lower ridges are gradually occupied by pine species.

Presently, one of the central Himalayan tree species, *Rhododendron arboretum*, reaches the peak of flowering from early February to mid-March. Long-term temperature data revealed a significant increase in seasonal and annual mean maximum temperatures. Generalized additive model using real-time field observations (2009–2011) and herbarium records (1893–2003) predicted 88–97 days early flowering in this species over the last 100 years (Gaira et al. 2014). This early flowering was correlated with an increase in the temperature. These studies, although limited, clearly show prevalence of species migration to higher altitudes and early flowering in the Himalayan ranges in response to the climate change.

Global warming has also forced the cultivation of many horticultural species to higher altitudes in the Himalayas (Gautam et al. 2014). Since 1981 apple cultivation in lower altitudes has reduced as much as 77% owing to reduced fruit yield due to inadequate chilling. Its cultivation has now shifted to higher altitudes which were earlier considered to be too cold.

2.4.2.4 Uncoupling of Phenological Events and Pollination

Although extensive studies have been carried out on altitudinal and latitudinal shifts of both plant and animal species in response to climate change, the effects of such shifts on pollination efficacy and sustainability of the species are much less studied. The life cycle of plant and animal species in temperate regions is strongly linked to the prevailing climate. Plants initiate leaves when the ice melts and the temperature rises, and flower in the spring. Many of the hibernating insects also hatch when ice melts with a rise in the temperature and feed on newly developed leaves. The insects emerge by the time plants bloom to pollinate their flowers. As pointed out earlier, this synchrony relaying on environmental cues between plants and animals has evolved over millions of years. Global warming may upset these synchronized events. Species may bloom before the emergence of their pollinators or pollinators may emerge from hibernation before the plant's bloom.

Asynchrony in the timing of pollinator emergence from hibernation and blooming of flowers is likely to affect the pollination services as well as the sustenance of pollinators. However, studies documenting such asynchrony are limited (Hegland et al. 2009; Settle et al. 2016; Thomson 2010). As the phenological events of both plant and insect species are equally responsive to air temperatures, there may not be a significant mismatch in many mutualisms between the timing of flowering and emergence of pollinators induced by global warming. Bartomeus et al. (2011) used available data on generalist pollinator, bees, and the blooming time of plant species pollinated by them in the North-Eastern USA. The data suggested that the extent of advancement in the blooming time of the plants and their pollinators kept pace with each other and there was no visible asynchrony. Gordo and Sanz (2005) studied long-term temporal trends in phenological events of 45 plant and 4 insect species in Mediterranean basin in response to climate change. Although most of the phenoevents showed common trends toward the advancement, particularly since mid-1970s, insect phenoevents showed a steeper advance than plant phenoevents during the last decade. These results suggest some asynchrony in the emergence of pollinators and the time of flowering in many instances. There was complete asynchrony between flowering period of *Lathyrus*, an alpine generalist pollinator species, and one of its pollinator *Hoplitis fulgida* during a season in several sites (Forrest and Thomson 2011). However, the plants did not suffer pollination limitation as they were able to use other pollinators.

Memmott et al. (2007) simulated the consequences of phenological shifts that can be expected by doubling of atmospheric CO_2 by using highly resolved empirical pollination networks between 1420 pollinators and 429 plant species. The studies predicted a reduction of 17–50% of floral resources available to pollinators as a result of phenological shifts resulting in possible extinctions of pollinators, plant species and their interactions. Kudo and Ida (2013) investigated phenology and seed set success in *Corydalis ambigua*, a spring ephemeral pollinated by bumblebees, over 10–14 years in three populations and reported that when spring came early, flowering tended to be ahead of pollinator appearance. This resulted in lower seed set because of lower pollination efficiency. In an alpine region of Japan where bumblebees are the major non-specialist pollinators, the availability of flowers in the community and bumble bee abundance was correlated in normal year (Kudo 2014). However, in a warmer year this positive correlation was broken due to phenological mismatch between the onset of flowers and bee emergence; queen bees emerged 8–9 days before the onset of flowering (Kudo 2014). Bumblebee pollinated species, *Erythronium grandiflorum* suffered pollination limitation in early flowering period in the absence of queen bees but set normal fruit through self-pollination (Thomson 2010). Detailed studies of Forrest and Thomson (2011) in the West Elk Mountains of central Colorado, USA, suggested that phenological asynchrony alone is unlikely to threaten the stability of populations.

Asynchrony in flowering period and appearance of pollinators is particularly disadvantageous to specialized pollinator systems (Shivanna 2014) than to generalized systems. Generalist pollinators are able to switch between plant species and generalist plants are able to switch between pollinators. Sexually deceptive orchids represent one of the highly specialized pollination systems. Each orchid species is pollinated by a species-specific pollinator. The flowers not only resemble the female insect of its pollinator but also emit species-specific pheromone to attract males. The orchids generally bloom during a specific time window each spring in synchrony with the emergence of male pollinators from hibernation but before the female bees appear. As there are no females (when the males emerge), males readily visit orchid flowers which resemble the females, try to copulate and bring about pollination. This gap between male and female insect emergence is critical and ensures the visits of the males to the flowers. When once the females arrive, the males hardly visit orchid flowers. Early spring induces female bees to emerge sooner than usual and lure the male bees away from the flowers thus affecting pollination success of sexually deceptive orchids (Robbirt et al. 2014).

Another example of climate change affecting specialized pollination systems is the effects of drought in tropical forests (Harrison 2000). Northern Borneo suffered a very severe drought from January to March 1998 linked to the El Nino Southern Oscillation event. Figs (*Ficus* species) are pollinated by species-specific fig-wasps. The drought caused a substantial break in the production of inflorescences in dioecious figs leading to the local extinction of their pollinators at Lambir Hills National Park, Sarawak (Harrison 2000). Most pollinators had not recolonized the site even six months after the drought. As fig fruits are the keystone resources for many seed disperser vertebrates, this may also have cascading effect on vertebrates involved in fig seed dispersal (Harrison 2000).

There are a number of areas that have so far received limited attention in understanding the impact of human-induced environmental changes on biological diversity. Although the main focus of this review is on the effects of anthropogenic activities on reproductive success and sustenance of pollination and seed dispersal mutualisms of flowering plants, there are a whole range of additional aboveground and belowground mutualistic/antagonistic interactions involving plants with herbivores, pathogens, soil mycorrhizae, parasitoids, and hyperparasitoids (see Carnelissen 2011; Van der Putten 2012; Abhilash and Dubey 2014; Harvey and Malcicka 2014). Many of these relationships are species-specific and sensitive to disruptions. As the species migrate individually and not all species of the community at the same time and rate, range shift of the species would affect markedly the sustainability of the species, particularly those which involve long-distance migration through seed dispersal (Van der Putten 2012).

Desert ecosystems pose extreme conditions for the survival of its biota. We hardly have any information on their strategies to cope with anthropogenic effects (see Franklin et al. 2016). Another effect of global warming that affects the reproductive success of the flowering plants but has not received adequate consideration is on pollen development. Pollen development is sensitive to ambient temperature; a raise of a few degree Celsius beyond optimal temperature may induce pollen sterility (Shivanna 2003; Kartikeya et al. 2012). This would drastically affect reproductive success of the species.

2.5 Concluding Remarks

Now enough evidences have accumulated to show that human-induced environmental changes have catastrophic effects on the sustenance of all levels of biodiversity and ecosystem services. They point to the distinct possibility of the 'sixth mass extinction'. Anthropogenic activities have seriously threatened two of the important plant–animal mutualisms—pollination and seed dispersal. Both these mutualisms are crucial for optimal reproduction and recruitment of plant species. Pollination mutualism also seriously affects crop productivity and thus global food and nutritional security.

In recent years both public and academia have become aware of the importance of biodiversity and the need for its effective conservation for sustenance of human welfare. This reality has compelled many international initiatives to impress upon the Governments to take suitable policies and programmes to reverse this trend. According to IPBES (2018) report on land degradation "Avoiding, reducing and reversing this problem, and restoring degraded land is an urgent priority to protect the biodiversity and ecosystem services vital to all life on Earth and to ensure human well-being." Science Academies of the Commonwealth countries have also highlighted the need for concerted global action to reduce atmospheric carbon to protect ourselves, our children and our planet (Anonymous 2018). It also emphasizes the importance of Paris Agreement on Climate Change and highlights that even if all countries meet their current commitments to greenhouse gas emission reductions, a global temperature rise of >3 °C above preindustrial level is projected by 2100. Some of the steps identified by 17,000 scientists (Ripple et al. 2017) to achieve sustainability transitions include well-funded and well-managed reserves, halting conversion of forests and other native habitats, restoring native plant communities, reducing food wastage, promoting dietary shifts away from meat, encouraging the adoption of renewable energy and limiting human population growth.

UN Report on climate change, released in October 2018, gave another serious warning to the world. Overshooting 1.5 °C rise by the end of the century (compared to preindustrial level) will have devastating effects on ecosystems, communities, and economies. It is expected to result in global food shortages, inundation of coastal cities and a refugee crisis the world has ever seen. A rise of 2 °C, would destroy 99% of the coral reefs. The report warns that immediate action is needed to mitigate this crisis. The report emphasizes that the world cannot afford a 2 °C rise. The coming decade could be one of humanity's last chances to avert this catastrophic impact.

National Research Council (2015) of the National Academies of Sciences, USA has suggested Climate Intervention in the form of removal of atmospheric CO_2 and its reliable sequestration and reflecting sunlight to cool the earth as additional possible approaches although we still do not have the technology to do so economically. However, a Swiss company is already working on a technology to remove carbon dioxide from the air at economical prices (Gertner 2019). This offers hope for mitigating carbon dioxide emission.

What the humanity is doing now is nothing but ecological suicide in the form of deforestation, habitat degradation, unsustainable extraction of freshwater, overexploitation of bioresources, burning of enormous amount of fossil fuels and population increase. Analyses of the collapse of many of the past Societies in different parts of the world clearly show that irreversible ecological suicide was the main cause for their collapse (Diamond 2005). The environmental problems we are facing today are very similar to those which led to the collapse of the past societies. Unless we are able to reverse the present trend soon, as highlighted by many international authorities and organizations, in all probability, present civilization may also start to collapse similar to the past societies. Human society has been able to take drastic steps when it comes to its own survival and hopefully it should be able to overcome this challenge also and survive.

References

Abhilash PC, Dubey RK (2014) Integrating aboveground–belowground responses to climate. Curr Sci 106:1637–1638

Aguilar R et al (2006) Plant reproductive susceptibility to habitat fragmentation: review and synthesis through a meta-analysis. Eco Lett 9:968–980

Akst J (2019) Thousands of australian animals die in unprecedented heatwave. The Scientist, 17 Jan 2019

Allen JM, Terres MA, Katasuki T, Iwamoto K et al (2013) Modeling daily flowering probabilities: expected impact of climate change on Japanese cherry phenology. Global Change Bio. https://doi.org/10.1111/gcb.12364
Anonymous (2007) Special section: pervasive consequences of hunting in tropical forests. Biotropica 39(3):289–392
Anonymous (2014) Special section: vanishing fauna. Science 245(6195):392–412
Anonymous (2017a) State of world’s birds. Birdlife International 2017. https://www.birdlife.org/sites/default/files/attachments/BL_ReportENG_V11_spreads.pdf. Accessed on Apr 2018
Anonymous (2017b) Down to earth, centre for environmental studies, New Delhi, 16–31 Oct 2017, pp 22–23
Anonymous (2018) https://royalsociety.org/~/media/news/2018/commonwealth-academies-consensus-statement-on-climate-change-12-march-2018.pdf. Accessed on Mar 2018
Aslan CE, Zavaleta ES, Tershy B, Croll D (2013) Mutualism disruption threatens global plant biodiversity: a systematic review. PLoS ONE 8(6):e66993. https://doi.org/10.1371/journal.pone.0066993
Barnosky AD et al (2011) Has the Earth’s sixth mass extinction already arrived? Nature 471:51–57
Bartomeus I, Ascher JS, Wagner D et al (2011) Climate-associated phenological advances in bee pollinators and bee-pollinated plants. Proc Natl Acad Sci USA 51:20645–20649
Bawa KS, Primack RB, Oommen MN (2011) Conservation biology: a primer for South Asia. Universities Press, Hyderabad
Beckage B, Osborne B, Gavin DG et al (2008) A rapid upward shift of a forest ecotone during 40 years of warming in the Green Mountains of Vermont. Proc Natl Acad Sci USA 105:4197–4202
Beckman NG, Muller-Landau HC (2007) Differential effects of hunting on pre-dispersal seed predation and primary and secondary seed removal of two Neotropical tree species. Biotopica 39:328–339
Bertrand et al (2011) Changes in plant community composition lag behind climate warming in lowland forests. Nature 479:517–520
Bharali S, Khan ML (2011) Climate change and its impact on biodiversity; some management options for mitigation in Arunachal Pradesh. Curr Sci 110:835–860
Biesmeijer JC, Roberts SPM, Reemer M et al (2006) Parallel declines in pollinators and insect-pollinated plants in Britain and the Netherlands. Science 313:351–354
Beketov M, Kefford B, Schafer R, Liess M (2013) Pesticides reduce regional biodiversity of stream invertebrates. Proc Natl Acad Sci USA 110:11039–11043
Brittain CA et al (2010) Impacts of a pesticide on pollinator species richness at different spatial scales. Basic Appl Eco 11:106–115
Brittain C, Williams N, Kremen C, Klein A-M (2013) Synergistic effects of non-*Apis* bees and honey bees for pollination services. Proc R Soc B 280:20122767. https://doi.org/10.1098/rspb.2012.2767
Buchmann SL, Nabhan GP (1996) The forgotten pollinators. Island Press, Washington, DC
Burkle LA, Marlin JC, Knight TM (2013) Plant-pollinator interactions over 120 years: loss of species, co-occurrence, and function. Science 339:1611–1615
Callum J et al (2015) Pollination by nocturnal Lepidoptera, and the effects of light pollution: a review. Ecol Entomol 40:187–198
Cameron SA, Lozier JD, Strange JP et al (2011) Patterns of widespread decline in North American bumble bees. Proc Natl Acad Sci USA 108:662–667
Carlton JT (2017) Tsunami-driven rafting: transoceanic species dispersal and implications for marine biogeography. Science 357:1402–1406
Carnelissen T (2011) Climate change and its effects on terrestrial insects and herbivory patterns. Neotrop Entomol 40:155–163
Ceballos G et al (2017) Biological annihilation via the ongoing sixth mass extinction signaled by vertebrate population losses and declines. Proc Natl Acad Sci USA 114. https://doi.org/10.1073/pnas.1704949114

Chakrabarti P, Sarkar S, Basu P (2018) Field populations of wild *Apis cerana* honey bees exhibit increased genetic diversity under pesticide stress along an agricultural intensification gradient in eastern India. J Insect Sci 18:1–8
Chakraborty A (2016) Fading scent of musk deer. Down To Earth, 16–30 Sept 2016, pp 37–38
Chechetka SA, Yu Y, Tange M, Miyako E (2017) Materially engineered artificial pollinators. Chem 2:224–239
Cheng L, Abraham J, Hausfather Z et al (2019) How fast are the oceans warming? Science 363:128–129
Cleland EE et al (2007) Shifting plant phenology in response to global change. Trends Ecol Evol 22:357–365
Corlett RT (2007a) The impact of hunting on the mammalian fauna of tropical Asian forests. Biotropica 39:292–303
Corlett RT (2007b) Pollination or seed dispersal: which should worry about most? In: Dennis AJ, Schupp EW, Green RJ, Wescott DA (eds) Seed dispersal: theory and its application in a changing world. CAI International, Wallingford, pp 423–544
Crutzen PJ (2002) Geology of mankind. Nature 415:23
da Silva Melo M, de Oliveira DE, Franceschinelli EV (2014) Density and fertility of *Byrsonima pachyphylla* A. Juss. (Malpighiaceae) in small fragments of the Brazilian Cerrado. Acta Bot Brazilica 28:259–265
DeFries RS, Foley JA, Asner GP (2004) Land-use choices: balancing human needs and ecosystem function. Front Ecol Environ 2:249–257
Diamond J (2005) Collapse: how societies choose to fail or succeed. Penguin Books
Dicks LV et al (2016) Ten policies for pollinators. Science 354:975–976
Dirzo R et al (2014) Defaunation in the anthropocene. Science 245:401–406
Dunnell KL, Travers SE (2011) Shifts in the flowering phenology of the northern Great Plains: patterns over 100 years. Am J Bot 98:935–945
Franklin KA, Sommers PN, Aslan CE (2016) Plant biotic interactions in the Sonoran Desert: current knowledge and future research perspectives. Int J Plant Sci 177:217–234
Forrest JRK, Thomson JD (2011) An examination of synchrony between insect emergence and flowering in the Rocky Mountain meadows. Ecol Monogr 81:469–491
Gaira KS et al (2014) *Rhododendron:* impact of climate change on the flowering of *arboreum*in central Himalaya, India. Curr Sci 106:1735–1738
Gammans M et al (2017) Negative impacts of climate change on cereal yields: statistical evidence from France. Environ Res Lett. https://doi.org/10.1088/1748-9326/aa6b0c
Ganesan R, Setty RS (2004) Regeneration of amla, an important non-timber forest product from Southern India. Conserv Soc 2:365–375
Garibaldi LA, Steffan-Dewenter I, Winfree R et al (2013) Wild pollinators enhance fruit set of crops regardless of honey bee abundance. Science 339:1608–1611
Gaston KJ (2010) Biodiversity. In: Sodhi NS, Ehrlich PR (eds) Conservation biology for all. Oxford University Press, Oxford, pp 27–44
Gautam HR et al (2014) Climate change is affecting apple cultivation in Himachal Pradesh. Curr Sci 106:498–499
Gerland P et al (2014) World population stabilization unlikely this century. Science 346:234–237
Gertner J (2019) The tiny Swiss company that thinks it can help stop climate change. New York Times Magazine, 12 Feb 2019
Ghazoul J, Sheil D (2010) Tropical rain forest ecology, diversity and conservation. Oxford University Press, Oxford, New York
Gleckler PJ et al (2016) Industrial-era global ocean heat uptake doubles in recent decades. Nat Clim Change 6:394–398
Gordo O, Sanz JJ (2005) Phenology and climate change: along-term study in a Mediterranean locality. Ecologia 146:484–495
Goulson D (2013) An overview of the environmental risks posed by neonicotinoid insecticides. J Appl Ecol 50:977–987

Grabherr G, Gottfried M, Pauli H (1994) Climate effects on mountain plants. Nature 369:448
Greenleaf SS, Kremen C (2006) Wild bees enhance honey bees' pollination of hybrid sunflower. Proc Natl Acad Sci USA 103:13890–13895
Hallmann CA et al (2017) More than 75 percent decline over 27 years in total flying insect biomass in protected areas. PLoS ONE 12(10):e0185809. https://doi.org/10.1371/journal.pone.0185809
Harrison RD (2000) Repercussions of El Nino: drought causes extinction and the breakdown of mutualism in Borneo. Proc R Soc B 267:911–915
Harvey JA, Malcicka M (2014) Climate change, range shifts and multitrophic interactions. INTECH. http://creativecommons.org/licenses/by/3.0. Accessed on May 2018
Hegland SJ et al (2009) How does climate warming affect plant-pollinator interactions? Ecol Lett 12:184–195
Holden C (2006) Report warns of looming pollination crisis in North America. Science 314:397
Howe HF (1990) Seed dispersal by birds and mammals: implications for seedling demography. In: Bawa KS, Hadley M (eds) Reproductive ecology of tropical forest plants. UNESCO and The Parthenon Publishing Group, Paris, pp 191–218
Howe HF, Smallwood J (1982) Ecology of seed dispersal. Ann Rev Ecol Syst 13:201–228
Hughes TP et al (2018) Global warming transforms coral reef assemblages. Nature. https://doi.org/10.1038/s41586-018-0041-2
IPBES (2016) Summary for policymakers of the assessment report of the intergovernmental science-policy platform on biodiversity and ecosystem services on pollinators, pollination and food production. In: Potts SG et al (eds). IPBES Secretariat, Bonn, Germany, p 36
IPBES (2018) Summary for policymakers of the thematic assessment report on land degradation and restoration of the Intergovernmental Science-Policy Platform on Biodiversity and Ecosystem Services. In: Scholes R et al (eds). IPBES secretariat, Bonn, Germany, p 31
IPCC (2014) Climate change 2014: synthesis report. Intergovernmental Panel on Climate Change, Geneva, Switzerland
IUCN (2017) The IUCN red list of threatened species. Version 2017-3. International Union for Conservation of Nature and Natural Resources
Jacobs J (2018) Vaccine for honeybees could be a tool to fight population decline. New York Times, Science, 23 Dec 2018
James RR, Pitts-Singer TL (eds) (2008) Bee pollination in agricultural ecosystems. Oxford University Press, Oxford, New York
Kartikeya S et al (2012) Screening of tomato genotypes for reproductive characters under high temperature stress conditions. Sabrao J Breed Genet 44:263–276
Kearns CA, Inouye DW, Waser NM (1998) Endangered mutualisms: the conservation of plant-pollinator interactions. Ann Rev Ecol Syst 29:83–112
Kerr et al (2015) Climate impacts on bumblebees converge across continents. Science 349:177–180
Kimball BA (1983) Carbon dioxide and agricultural yield: an assemblage and analysis of 430 prior observations. Agron J 75:779–788
Knop E et al (2017) Artificial light at night as a new threat to pollination. Nature 548:206–209
Kolbert E (2014) The sixth extinction: an unusual history. Henry Holt and Co, NY
Kremen C, Ricketts T (2000) Global perspectives on pollination disruptions. Conserv Biol 14:1226–1228
Kudo G (2014) Vulnerability of phenological synchrony between plants and pollinators in an alpine ecosystem. Ecol Res 29:571–581
Kudo G, Ida TY (2013) Early onset of spring increases the mismatch between plants and pollinators. Ecology 94:2311–2320
Lallensack RL (2017) Could this pollinating drone replace butterflies and bees? Science, 9 Feb, 2017. https://doi.org/10.1126/science.aal0730
Lenzen M, Moran D, Kanemoto K et al (2012) International trade drives biodiversity threats in developing nations. Nature 486:109–112

Lister BC, Garcia A (2018) Climate-driven declines in arthropod abundance restructure a rainforest food web. PNAS 115(44):E10397–E10406. www.pnas.org/cgi/doi/10.1073/pnas.1722477115

Long SP, Ainsworth EA, Leakey ADB et al (2006) Food for thought: lower-than-expected crop yield stimulation with rising CO_2 concentrations. Science 312:1918–1921

Lovejoy TE, Hannah L (eds) (2005) Climate change and biodiversity. Yale University Press, New Haven, London

Memmott J, Craze PG, Waser NM, Price MV (2007) Global warming and the disruption of plant-pollinator interactions. Ecol Lett 10:710–717

Miller-Rushing AJ, Forrest J (eds) (2010) Theme issue 'The role of phenology in ecology and evolution'. Phil Trans Royal Soc B 365:3101–3260

Miller-Rushing AJ et al (2010) The effects of phenological mismatches on demography. Phil Trans R Soc B 365:3177–3186

Montzka SA, Dutton GS, Yu P, Portmann RW et al (2018) An unexpected and persistent increase in global emissions of ozone-depleting CFC-11. Nature 557:413–417

Moran D, Kanemoto K (2017) Identifying species threat hotspots from global supply chains. Nat Ecol Environ 1. https://doi.org/10.1038/s41559-016-0023

Murali KS et al (1996) Extraction of non-timber forest products in the forests of Biligiri Rangan Hills, India. 2. Impact of NTFP extraction on regeneration, population structure and species composition. Econ Bot 50:252–269

Murray TE et al (2009) Conservation ecology of bees: populations, species and communities. Apidologie 40:211–236

Nayak KG, Davidar P (2010) Pollinator limitation and the effect of breeding systems on plant reproduction in forest fragments. Acta Oecol 36:191–196

Nogrady B (2019) Mass fish deaths in Australian set to continue. Nature News. https://doi.org/10.1038/d41586-019-00146-5

Norjhauer JM, Newberg DM (2010) Recruitment limitation after mast seeding in two African rain forest trees. Ecology 91:2303–2312

Nunez-Iturri G, Howe HF (2007) Bushmeat and the fate of trees with seeds dispersed by large primates in a lowland rain forest in Western Amazonia. Biotropica 39:348–354

Obute GC (2010) Pollination: a threatened vital biodiversity service to humans and the environment. Int J Biodivers Conserv 2:1–13

O'Hanlon SJ et al (2018) Recent Asian origin of chytrid fungi causing global amphibian declines. Science 360:621–627

Parmesan C, Yohe G (2003) A globally coherent fingerprint of climate change impacts across natural systems. Nature 421:37–42

Parmesan C et al (1999) Poleward shifts in geographical ranges of butterfly species associated with regional warming. Nature 399:579–583

Patiny S (ed) (2012) Evolution of plant-pollinator relationships. The Systematic Association Special, vol 81. Cambridge University Press, New York

Pimm SL (2009) Climate disruption and biodiversity. Curr Biol 19:R595–R601

Porrinni C et al (2014) Using honey bee as bioindicator of chemicals in Campanian agroecosystems (South Italy). Bull Insectology 67:137–146

Potts SG, Biesmeijer JC, Kremen C et al (2010) Global pollinator declines: trends, impacts and drivers. Trends Ecol Evol 25:345–353

Primack RB, Higuchi H, Miller-Rushing AJ (2009) The impact of climate change on cherry trees and other species in Japan. Biol Conserv 142:1943–1949

Rignot E, Mouginot J, Scheuchl B et al (2019) Four decades of Antarctic ice sheet mass balance from 1979–2017. Proc Natl Acad Sci USA 116:1095–1103

Ripple WJ et al and 15,364 scientist signatories from 184 countries (2017) World scientists' warning to humanity: a second notice. BioSci 67:1026–1028

Robbirt KM et al (2014) Potential disruption of pollination in a sexually deceptive orchid by climatic change. Curr Biol 24:2845–2849

Rodell M, Famiglietti JS, Wiese DN et al (2018) Emerging trends in global fresh water availability. Nature. https://doi.org/10.1038/s41586-018-0123-1

Rogelj J et al (2018) Scenarios towards limiting global mean temperature increase below 1.5 C. Nat Clim Change. https://doi.org/10.1038/s41558-018-0091-3

Rothman DH (2017) Thresholds of catastrophe in the Earth system. Sci Adv 3:e1700906

Roubik DW (ed) (1995) Pollination of cultivated plants in the Tropics. FAO Agriculture Service Bulletin 118. FAO, UN, Rome

Rundolf M et al (2015) Seed coating with a neonicotinoid insecticide negatively affects wild bees. Nature 521:77–80

Salmela H, Amdam GV, Freitak D (2017) Transfer of immunity from mother to offspring is mediated via egg-yolk protein vitellogenin. PLoS Pathog 11(7):e1005015. https://doi.org/10.1371/journal.ppat.1005015

Samejima H et al (2004) The effects of human disturbance on a stingless bee community in a tropical rainforest. Biol Conserv 120:577–587

Sanchez-Bayo F, Wyckhuys KAG (2019) Worldwide decline of the entomofauna: a review of its drivers. Biol Conserv 232:8–27

Sax DF, Gaines SD (2008) Species invasions and extinction: the future of native biodiversity on islands 105:11490–11497

Settele J, Bishop J, Potts SG (2016) Climate change impacts on pollination. Nat Plants 2:16092

Shivanna KR (2003) Pollen biology and biotechnology. Science Publishers Inc, Enfield/Plymouth

Shivanna KR (2014) Biotic pollination: how do plants achieve conflicting demands of attraction and restriction of potential pollinators. In: Ramawat KG, Merillon J-M, Shivanna KR (eds) Reproductive biology of plants. CRC Press, pp 218–267

Sinu PA, Shivanna KR (2016) Factors affecting recruitment of a critically-endangered Dipterocarp species, *Vateria indica* in the Western Ghats, India. Proc Natl Acad Sci, India B 86:857–862

Skerratt LF et al (2007) Spread of chytridiomycosis has caused the rapid global decline and extinction of frogs. EcoHealth 4:125–134

Sodhi NS, Ehrlich PR (eds) (2010) Conservation biology for all. Oxford Univ Press, Oxford

Tadwalkar MD et al (2012) Dispersal modes of woody species from the northern Western Ghats, India. Trop Ecol 53:53–67

Thackeray et al (2016) Phenological sensitivity to climate across taxa and trophic levels. Nature 535:241–245

Thomas CD, Cameron A, Green RE et al (2004) Extinction risk from climate change. Nature 427:145–148

Thomson JD (2010) Flowering phenology, fruiting success and progressive deterioration of pollination in an early-flowering geophyte. Phil Trans R Soc B 365:3187–3199

Tilman D, Reich PB, Knops J, Wedin D, Mielke T (2001) Diversity and productivity in a long-term grassland experiment. Science 294:843–845

Trusel LD, Das SB, Osman MB et al (2018) Nonlinear rise in Greenland runoff in response to post-industrial Arctic warming. Nature 564:104–108

Turner IM (2001) The ecology of seeds in tropical rain forests. Cambridge University Press, Cambridge

Tylianakis JM (2013) The global plight of pollinators. Science 339:1532–1533

Urban MC (2015) Accelerating extinction risk from climate change. Science 348:571–573

Van der Putten WH (2012) Climate change, aboveground-belowground interactions, and species' range shifts. Ann Rev Ecol Evol Syst 43:365–383

Wang BC et al (2007) Hunting of mammals reduces seed removal and dispersal of the Afrotropical tree *Antrocaryon klaineanum* (Anacardiaceae). Biotropica 39:340–347

Warren et al (2018) The projected effect on insects, vertebrates, and plants of limiting global warming to 1.5 °C rather than 2 °C. Science 360:791–795

Wester P et al (eds) (2019) The Hindu Kush Himalaya assessment: mountains, climate change, sustainability and people. Springer, Berlin

Wilcock C, Neiland R (2002) Pollination failure in plants: why it happens and when it matters. Trends Plant Sci 7:270–277

Willmer P (2011) Pollination and floral ecology. Princeton University Press, Princeton, NJ

Wilson EO (1997) Introduction. In: Wilson DE, Wilson EO, Reaka-Kudla ML (eds) Biodiversity II. Understanding and protecting our biological resources. Joseph Henry Press, Washington DC

Wright SJ et al (2007) The plight of large animals in tropical forests and the consequences for plant regeneration. Biotropica 39:289–291

Xie W et al (2018) Decreases in global beer supply due to extreme drought and heat. Nat Plants 4. https://doi.org/10.1038/s41477-018-0263-1

Zhao C et al (2017) Temperature increase reduces global yields of major crops in four independent estimates. Proc Natl Acad Sci USA 114:9326–9331

Zhu C et al (2018) Carbon dioxide (CO2) levels this century will alter the protein, micronutrients, and vitamin content of rice grains with potential health consequences for the poorest rice-dependent countries. Sci Adv 4:eaaq1012. https://doi.org/10.1126/sciadv.aaq1012

Chapter 3
Molecular Approaches to Explore Coastal Benthic Metazoan Diversity—Success and Constraints

Punyasloke Bhadury

Abstract Coastal environments are represented by rich biotopes and harbour diverse organismal groups, many of which are yet to be explored. The metazoan phyla found in the sediment of coastal environments are critical to ecosystem functioning. The abundance and diversity of benthic metazoan phyla such as the free-living marine nematodes in various coastal biotopes are not fully understood from the viewpoint of biodiversity. Molecular tools such as next-generation sequencing (NGS) approach offer a way to develop robust metabarcodes. Generation and processing of NGS data including metabarcode sequences involve computational understanding. Metabarcodes obtained using NGS platforms are providing improved understanding of biodiversity-rich sedimentary metazoan groups such as free-living marine nematodes. Some of these aspects in terms of NGS platforms, data processing and examples of application of NGS to explore benthic metazoan diversity with focus on free-living nematode communities have been discussed.

Keywords Metazoa · Free-living marine nematode · Next-generation sequencing · Metabarcoding · Biodiversity

3.1 Introduction

Since the origin of life, biological communities have shaped past and present of Earth including habitat diversity, ecosystem functioning and geochemistry. These have also led to evolution of biota spanning across millions of years resulting in rich biodiversity as represented in the Tree of Life. The magnitude and relative diversity of species richness globally across archaeal, prokaryotic and eukaryotic

P. Bhadury (✉)
Integrative Taxonomy and Microbial Ecology Research Group, Department of Biological Sciences & Centre for Climate and Environmental Studies, Indian Institute of Science Education and Research Kolkata, Mohanpur, Nadia 741246, West Bengal, India
e-mail: pbhadury@iiserkol.ac.in

K. G. Ramawat (ed.), *Biodiversity and Chemotaxonomy*, Sustainable Development and Biodiversity 24, https://doi.org/10.1007/978-3-030-30746-2_3

domains remain unknown. The Animal Kingdom represents an important biological domain of the Tree of Life, and more than 1.5 million species encompassing 35 different phyla have been reported (e.g. del Campo et al. 2014). It has been speculated that the biodiversity in Animal Kingdom can be represented by more than 10 million species, many of which are unknown to modern biology. Interestingly, unexplored or hidden diversity could be within the metazoan groups some of which are microscopic and represent size below 2 mm (Blaxter et al. 2005). The metazoan groups have a wide ecological distribution ranging from soil to deep-sea sedimentary environments.

The marine environment represents one of the richest ecosystems on our planet. It has been estimated that there are millions of marine eukaryotic species (e.g. Leray and Knowlton 2016), many of which belong to diverse metazoan groups. In land–ocean boundary, rich coastal biomes including estuaries, mangroves and lagoons are home to metazoan communities that inhabit sedimentary layers and contribute to critical ecosystem processes including coastal carbon and nitrogen cycling (e.g. Danovaro and Fraschetti 2002; Giere 2009). The biodiversity estimation of benthic metazoan groups in coastal ecosystems is largely dependent on taxonomy. Identification of many benthic metazoan groups is mostly undertaken on morphological attributes which require considerable taxonomic expertise (e.g. Boufahja et al. 2015). Thus, identification of metazoan groups such as meiofauna or micro- and mesozooplanktons is challenging and their biodiversity is often underestimated at the community level (Fonseca et al. 2010; Hirai and Tsuda 2015; Hirai et al. 2015). For example, free-living marine nematodes which constitute a key component in coastal sedimentary domain in terms of abundance and diversity are identified based on morpho-taxonomy. However, in many juvenile free-living marine nematodes, fine morphological structures are not clearly developed resulting in underestimation of their biodiversity. Morphological identification of 10% of nematodes encountered in a sample requires an effort 120 times higher than that required to identify vertebrate morphospecies in tropical forests (e.g. Lawton et al. 1998). Such underestimation could restrict our understanding of the critical roles played by many benthic metazoan groups in coastal ecosystems. This in turn could hamper understanding of rates and fluxes linked to cycling of major elements such as carbon in coastlines globally. Besides, the extent of hidden biodiversity or phenotypic plasticity in coastal metazoan groups is largely unknown (e.g. Knowlton 1993; Leray and Knowlton 2015; Thomsen and Willerslev 2015) and their biogeographic patterns are not clearly well understood. This is particularly the case for metazoan groups that are found in tropical and subtropical coastal realms. Molecular tools can offer a potential aid to improve understanding of biodiversity of coastal metazoan groups such as free-living marine nematodes. The application of molecular techniques and development of DNA sequence-based approaches have revolutionized the discipline of life sciences. Molecular techniques such as polymerase chain reaction-dependent denaturing gradient gel electrophoresis (PCR-DGGE), clone library, sequencing approach and terminal restriction fragment length polymorphism (TRFLP) have provided deeper understanding of microbial diversity.

In this chapter, an overview has been provided on next-generation sequencing (NGS) platforms, data processing along with examples of the application of NGS data to unravel benthic metazoan diversity from coastal habitats. The concept of metabarcoding based on NGS data has been also discussed. Moreover, as part of this chapter as an example, the applicability of NGS data for exploring biodiversity of free-living marine nematodes has been thoroughly discussed.

3.2 Next-Generation Sequencing

Over the last three decades, sequencing technology has been based on the Sanger method, and even in present time, it is still used for regular molecular studies. The automation of Sanger sequencing with coupling of polymerase chain reaction (PCR) method was subsequently followed by shotgun sequencing strategy. The shotgun strategy has provided functional understanding of biological communities at the genome scale and most widely used approach for large-scale biodiversity studies until recently (e.g. Adams et al. 2000). Since the sequencing reaction has to be undertaken separately in shotgun strategy, this leads to a limited number of base pair data generation per day using Sanger sequencing as well as the cost per base pair is higher at the genomic level resulting in limitations when looking investigating functional biodiversity of biological communities across ecosystem scales. The advent of next-generation sequencing technology over the last decade or so has revolutionized the domain of Biological Sciences.

Next-generation sequencing (NGS) is based on sequencing of nucleotides faster and cheaper compared to that of Sanger sequencing. Compared to Sanger sequencing, NGS technologies can provide large amount of data, generation and separation of millions of individual fragments based on two-level parallelization at a lower cost (e.g. Mardis 2008; Thompson and Milos 2011). A critical component of NGS technology is the incorporation of thin layers, nanopores, nanoscale emulsions or microspheres as physical way to adhere to DNA fragments and probes. This is a key component of the parallelization of processes resulting in generation of millions of fragment sequence data within a short time. Thus, NGS approach can be very effective for addressing questions related to biodiversity such as at the population scales. Importantly, the field of biodiversity and conservation has flourished with the advent of NGS as broad questions pertaining to ecology and evolution of organismal groups across different ecosystems can be posed and answered (e.g. Bonilla-Rosso et al. 2008; Eguiarte et al. 2013). NGS can therefore be applied to investigate genetic variation or sequence variation within a population or across community scale and understand the functional consequences of such variations as part of ecosystem processes (Eguiarte et al. 2013). Thus, it is possible to address arrangement of genes and evolutionary conservation, dynamics of species at the genome scale or the evolutionary history of species at the phylogenomics based on NGS approaches (e.g. Mathee et al. 2008; Yi et al. 2010; Ibarra-Laclette et al. 2013). The following section provides a brief overview about the platforms used for generation of sequence data using NGS approach.

3.3 Roche 454 GS FLX Platform

The Roche 454 platform is also referred to as pyrosequencer. In this platform, the fragments for sequencing are being selected by size (up to 800 bp) and then processed to ligate adapters key for library construction. In pyrosequencing, PCR emulsion consisting of synthesis of the complementary strand of DNA molecules attached to the beads, followed by incorporation of enzymatic activity of luciferase and sulphurylase, is a critical step for downstream processing and data acquisition. The length of reads obtained by pyrosequencing has increased (up to 800 bp), but relatively high error rate and low coverage (1 million reads per run) compared to currently available other NGS platforms have resulted in limited application (Shokralla et al. 2012; Liu et al. 2012).

3.4 Illumina MiSeq and HiSeq Platforms

At present, the most widely used platform for biological studies involves the use of MiSeq or HiSeq platforms based on Illumina sequencing. One of the primary advantages of Illumina over pyrosequencing involves generation of high-quality longer read lengths. The library construction is based on the synthesis of complementary strands of DNA using PCR. But, the synthesis is significantly different from emulsion PCR used in pyrosequencing. Based on the selection of desired fragment size, platform-specific adaptors are added for attachment to the sequencing matrix which is a key to subsequent steps of amplification and data acquisition. In Illumina platforms, the sequences are generated in the form of a forward R1 and a reverse R2 file which can be stitched together to form a contig of longer stretch of sequence. The stitched file can be subsequently used for downstream bioinformatics analyses. Presently, the platforms based on Illumina sequencing can generate 10 million–4000 million reads per run.

3.5 Other Sequencing Platforms

Besides pyrosequencing and Illumina sequencing platforms, other platforms such as SOLiD™ sequencer based on oligo ligation detection provide highest quality data with lowest error. The read length in SOLiD™ platform was 35 bp but has increased to 75 bp with a final output of 10 Gb data per run (Shokralla et al. 2012). In recent times, third-generation sequencing platforms such as Ion Torrent, PacBio and Nanopore have become more popular (Liu et al. 2012; Ghosh and Bhadury 2019). The sequencing principles range from emulsion PCR to single-molecule real-time sequencing (SMRT) according to individual third-generation platform, while the length (100–1050 bp), depth (up to 6075 Mb per run) and quality of data also vary platform-wise and while they are cost-effective (e.g. Rothberg et al. 2011; Shokralla et al. 2012).

3.6 Sequence Assembly of Data Generated from NGS Platforms

The analysis of NGS data requires significant computational capabilities. As the data generated from NGS platforms can vary, therefore assembly of data while filtering poor-quality reads require extensive computational skills. Proper filtering of raw data, removal of ambiguous bases, determining the desired read length of sequences and trimming of primer and adapter sequences are critical steps before downstream analyses. For undertaking biodiversity or ecology-related studies, generated raw data such as that from Illumina platforms can be subjected to filtering steps using programs such as UPARSE integrated into many pipelines including Mothur and QIIME (Caporaso et al. 2010). Subsequently, the filtered 'clean data' can be further processed through chimera check programs such as ChimeraSlayer to remove chimeric sequences. The presence of chimeric sequences could lead to erroneous output and ultimately can overestimate biodiversity with an ecosystem. Once the chimera sequences are removed, sequences are aligned using any of the available alignment methods. This initial alignment could help in correct grouping and determination of operational taxonomic units (OTUs). The determination of OTUs is an important step in biodiversity studies. Often, the presence of gaps or missing bases could lead identical sequences to be binned into separate OTUs (Ghosh and Bhadury 2019). Subsequently, taxonomic assignment or phylogeny can be undertaken using the processed 'clean data'. Moreover, OTUs can be compared with published sequences or barcode sequences available in published databases including GenBank/ENA/DDBJ. This is the basis of metabarcoding which is extensively used in biodiversity assessment. For NGS data, data storage and implementation of statistical models pose a major challenge (e.g. Yoccoz 2012).

3.7 Metabarcoding

The term DNA metabarcoding was first introduced by Taberlet et al. (2012) to designate high-throughput multispecies identification using the total or typically degraded DNA extracted from an environment sample or from bulk samples of entire organisms. Metabarcoding differs from metagenomics in several ways as metagenomics describes the functional and sequence-based analysis of the collective genomes contained in an environmental sample (Riesenfeld et al. 2004), whereas metabarcoding aims to study a subset of genes or gene. For example, metabarcoding approach has been successfully used for characterizing microbial populations from different environments (e.g. Chariton et al. 2010; Epp et al. 2012, Pearman and Irigoien 2015). The NGS data provides an opportunity to explore biodiversity within an ecosystem based on metabarcoding. However, the NGS data has to undergo rigorous quality control in terms of bioinformatics ultimately to be used for metabarcoding. Briefly, raw NGS data following filtering is corrected by clustering using frequency-based heuristics or approximate likelihood based on empirically derived error distributions. The FASTA formatted files can be corrected based also on alignments (Reeder and Knight 2010). The result is that all sequences with putative random errors get removed from downstream analyses. Denoising procedure can work effectively when using short metabarcoding markers with low natural diversity (e.g. nuclear small subunit ribosomal RNA sequence). However, they are not very useful when using longer markers with high natural variability (mitochondrial cytochrome oxidase I sequence), because a high proportion of the initial reads will be removed by the denoising algorithm. In that case, clustering the sequences into OTUs can be useful. Algorithms such as CROP and SWARM can be used for clustering into OTUs (Hao et al. 2011; Mahé et al. 2014). Subsequently, from among each cluster, the longest read can serve as a reference sequence which can then be taxonomically affiliated to known eukaryotic sequence by searching against databases including Barcode of Life Database System (BOLD; Ratnasingham and Hebert 2007). Figure 3.1 provides an overview of the broad steps undertaken in generation of metabarcode data followed by taxonomic assignment. However, it is important to note that the published databases should have sequences from biodiversity-rich regions such as coastal ecosystems located in subtropics and tropics.

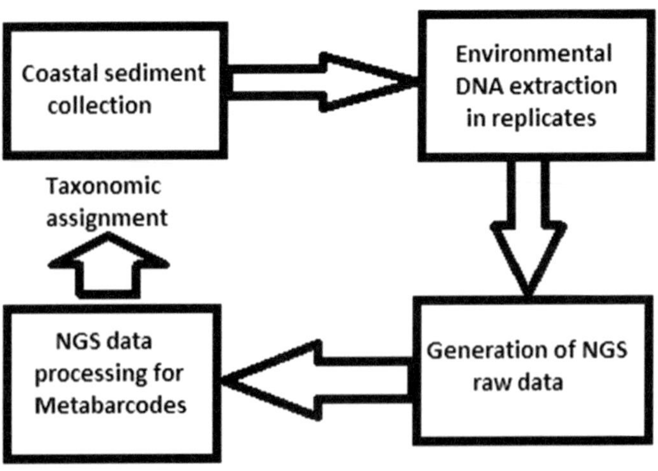

Fig. 3.1 Schematic representation of steps to be undertaken for assessment of coastal sedimentary biodiversity using metabarcoding approach

3.8 Examples of Coastal Benthic Metazoan Biodiversity with Emphasis on Free-Living Marine Nematodes

The application of NGS technology has improved our understanding of coastal benthic metazoan biodiversity globally. The coastlines represent land–ocean boundary system and are characterized by dynamic physical, chemical and hydrological gradients which results in rich biotopes such as estuaries, lagoons and mangroves with associated biodiversity. In particular, biological communities such as free-living marine nematodes that inhabit benthic sediments play critical roles in influencing rates and fluxes of elemental cycling. A biodiversity assessment study was undertaken in coastal two locations—Virginia (temperate) and Florida (sub-temperate) in the USA based on metabarcoding approach (Leray and Knowlton 2015). The metabarcode mitochondrial cytochrome oxidase I sequences were generated based on Ion Torrent technology. Based on the analysis of 983,056 sequences, 2179 operational taxonomic units (OTUs) were recorded. However, it was found that 10.9% could be matched to reference barcodes in public sequence databases, with only 8.2% matching barcodes with both genus and species names. The authors reported broad taxonomic coverage particularly for animals (22 phyla recorded), but 35.6% of OTUs detected via metabarcoding could not be confidently assigned to a taxonomic group. The authors also found that in Florida sample sequences representing the phylum Nematoda were present, in addition to other metazoan groups. This particular study stressed the importance of undertaking robust biodiversity assessment of metazoan communities along the coastal regions in order to understand structure and function of biological communities. Kim et al. (2017) undertook an assessment of benthic metazoan diversity from intertidal sediments representing from the coastlines of Korea. Based on 436,154 high-quality reads of 18S rRNA generated using Illumina MiSeq platform, authors were able to assign the reads into 334 OTUs. They found that sequences representing members of Annelida were the most common, followed by Arthropoda, Mollusca, Nematoda and Platyhelminthes. The study revealed that the distribution of metazoan groups was correlated with existing physical attributes of the coastlines of Korea. The authors also reported that several taxa encountered in their study based on the analysis of NGS data were not reported before from the coastlines of Korea highlighting the importance of NGS technology to unravel metazoan diversity such as those from sedimentary environment (Kim et al. 2017). In a recent study, diversity of metazoan phyla was evaluated across several coastal locations in Europe based on nuclear small subunit ribosomal RNA (18S rRNA) sequences (López-Escardó et al. 2018). The NGS data generated using 454 Roche platform was targeted from DNA and RNA samples representing water column and sediments, and oxic and anoxic environments representing six European coastal locations. The total number of reads after quality control for sediment was 50,771 out of which 25,438 reads were from DNA and the remaining from RNA (López-Escardó et al. 2018). The authors reported high percentage of novel 18S rRNA sequences in most phyla. In particular, putative novelty of generated sequences was found to be high among Platyhelminthes, Acoelomorphs and Nematodes and majority of their OTUs (75%) showed BLAST identity lower than 97%. This clearly indicated that coastal environments are yet to be sampled fully at a molecular level in particular to understanding biodiversity of benthic fauna such as free-living marine nematodes.

Nematodes are diverse, present in soil to deep sea as parasite or free living at a population up to 10^8 per m^2 (Lambshead 2004). However, free-living marine nematodes are lesser studied than soil nematodes from the perspectives of community composition, diversity and functional community structure. Given the importance of these organisms as reliable bioindicator for tracking disturbance (natural or anthropogenic) in marine environments (Semprucci et al. 2015), therefore it is very important to understand the biodiversity and patterns of them across marine environments. Creer et al. (2010) undertook an attempt to characterize organismal biodiversity from sediment samples collected from an intertidal coastal area in South of England, United Kingdom as well as soil, litter and understory habitats sampled from La Selva Biological Station, Costa Rica. Using 454 sequencing, 29756 high-quality 18S rDNA sequences (length over 200 bases) from coastal samples and 40334 high-quality sequences representing tropical rain forest samples were generated as part of their study. In case of terrestrial data, studied soil habitat was represented by fewer nematode operational cluster taxonomic units (OCTUs) (35) than either the coastal habitat (149) or canopy (97).

Fonseca et al. (2010) assessed relative levels of richness and patterns of diversity of multiple metazoan phyla including free-living marine nematodes using 454 sequencing of the 18S nuclear small subunit ribosomal RNA (rRNA). The authors looked at eight coastal sediment samples collected from low-tide zone of an estuarine beach located on the West coast of Scotland, and from one sample from a beach located in the South of England. Based on the analysis of filtered 305,702 sequences, the authors were able to assign 182 OTUs representing free-living marine nematodes. The authors also found that majority (95%) of Nematoda OCTUs have never been sequenced before reflecting the enormous diversity that needs to be explored and documented. For sediment samples representing the West coast of Scotland, authors detected 182 Nematoda OCTUs, compared to 450 species of free-living marine nematodes that have been described from around entire British Isles using morpho-taxonomy. Bik et al. (2012), using 454 sequencing, assessed microbial eukaryotic communities across depth (shallow water to abyssal) and ocean basins (deep-sea Pacific and Atlantic). Within the 12 sites examined, they found that some taxa exhibited eurybathic ranges and cosmopolitan deep-sea distributions, but majority of species appear to be regionally restricted. Contrary to previous observation of free-living marine nematodes being the most abundant members in sediment, authors found equal or more dominant role for other taxonomic groups in some of the deep-sea sites (e.g. unicellular eukaryotes in the Pacific) based on 454 sequencing. In another study representing the deep-sea sites from Southern Ocean, Bhadury et al. (2011) investigated association between deep-sea nematodes and marine fungi and potential implications including food preference of nematodes based on capillary and 454 sequencing approaches. In more recent studies such as the one undertaken by Lallias et al. (2015) based on 454

Roche sequencing of 18S rRNA barcodes, authors showed that free-living marine nematode species richness accounted for almost 22% and 31% in Thames and Mersey estuaries of UK and more than 55% of the generated sequence reads (out of 957216 reads) represented free-living marine nematode signatures.

3.9 Way Forward

The NGS approaches based on metabarcoding will continue to improve understanding of benthic metazoan diversity including extent of their hidden diversity. In addition, potential new DNA barcode regions need to be identified in mitochondrial COI DNA metabarcodes, which should be expanded and tested across biodiversity-rich coastal regions including tropics.

The long-term success of metabarcoding as a tool based on NGS is also dependent on further development of bioinformatics pipelines and the ability to augment databases with more sequences of benthic metazoan phyla such as free-living marine nematodes from diverse regions, in particular tropical coastal ecosystems. To conclude, morpho-taxonomy coupled with metabarcoding could become an essential tool for undertaking large-scale biodiversity assessment studies in order to address critical questions pertaining to biogeography and evolution of animal phyla on a global scale.

References

Adams MD, Celniker SE, Holt RA, Evans CA et al (2000) The genome sequence of Drosophila melanogaster. Science 287:2185–2195

Bhadury P, Bik H, Lambshead JD, Austen MC, Smerdon GR, Rogers AD (2011) Molecular diversity of fungal phylotypes co-amplified alongside nematodes from coastal and deep-sea marine environments. PLoS ONE 6:e2644

Bik HM, Porazinska DL, Creer S, Caporaso JG, Knight R, Thomas WK (2012) Sequencing our way towards understanding global eukaryotic biodiversity. Trends Ecol Evol 27:233–243

Blaxter ML et al (2005) Defining operational taxonomic units using DNA barcode data. Philos Trans R Soc B Biol Sci 360:1935–1943

Bonilla-Rosso G, Souza V, Eguiarte LE (2008) Metagenómica, genómica y ecología molecular: la nueva ecología en el bicentenario de Darwin. TIP Revista Especializada en Ciencias Químico-Biológicas 11:41–51

Boufahja F, Semprucci F, Beyrem H, Bhadury P (2015) Marine nematode taxonomy in Africa: promising prospects against scarcity of information. J Nematology 47:198–206

Caporaso JG, Kuczynski J, Stombaugh J, Bittinger K, Bushman FD et al (2010) QIIME allows analysis of high throughput community sequencing data. Nat Methods 7:335–336

Chariton AA, Court LN, Hartley DM, Colloff MJ, Hardy CM (2010) Ecological assessment of estuarine sediments by pyrosequencing eukaryotic ribosomal DNA. Front Ecol Environ 8:233–238

Creer S, Fonseca VG, Porazinska DL, Giblin-Davis RM, Sung W, Power DM, Packer M, Carvalho GR, Blaxter ML, Lambshead PJD, Thomas WK (2010) Ultrasequencing of the meiofaunal biosphere: practice, pitfalls and promises. Mol Ecol 19:4–20

Danovaro R, Fraschetti S (2002) Meiofaunal vertical zonation on hard bottoms: comparison with soft-bottom meiofauna. Mar Ecol Progr Ser 230:159–169

del Campo J et al (2014) The others: our biased perspective of eukaryotic genomes. Trends Ecol Evol 29:252–259

Eguiarte LE, Aguirre-Liguori JA, Jardón-Barbolla L, Aguirre-Planter E, Souza V (2013) Genómica de poblaciones: nada en Evolución va a tener sentido si no es a la luz de la genómica, y nada en genómica tendrá sentido si no es a la luz de la evolución. TIP Revista Especializada En Ciencias Químico-Biológicas 16:42–56

Epp LS, Boessenkool S, Bellemain EP, Haile J, Esposito A et al (2012) New environmental metabarcodes for analysing soil DNA: potential for studying past and present ecosystems. Mol Ecol 21:1821–1833

Fonseca VG, Carvalho GR, Sung W, Johnson HF, Power DM, Neill SP, Packer M, Blaxter ML, Lambshead PJD, Thomas WK, Creer S (2010) Second-generation environmental sequencing unmasks marine metazoan biodiversity. Nat Comm. https://doi.org/10.1038/ncomms1095

Ghosh A, Bhadury P (2019) Methods of assessment of microbial diversity in natural environments. In: Das S, Dash H (eds) Microbial diversity in the genomic era, Academic Press, p 770

Giere O (2009) Meiobenthology: the microscopic motile fauna of aquatic sediments. Springer, Heidelberg

Hao X, Jiang R, Chen T (2011) Clustering 16S rRNA for OTU prediction: a method of unsupervised Bayesian clustering. Bioinformatics 27:611–618

Hirai J, Tsuda A (2015) Metagenetic community analysis of epipelagic planktonic copepods in the tropical and subtropical Pacific. Mar Ecol Prog Ser 534:65–78. https://doi.org/10.3354/meps11404

Hirai J, Yasuike M, Fujiwara A, Nakamura Y, Hamaoka S, Katakura S et al (2015) Effects of plankton net characteristics on metagenetic community analysis of metazoan zooplankton in a coastal marine ecosystem. J Exp Mar Biol Ecol 469:36–43

Ibarra-Laclette E, Lyons E, Hernández-Guzmán G, Pérez-Torres CA, Carretero-Paulet L et al (2013) Architecture and evolution of a minute plant genome. Nature 498:94–98

Knowlton N (1993) Sibling species in the sea. Ann Rev Ecol Syst 24:189–216

Kim H, Kim H, Hwang H, Kim W (2017) Metagenomic analysis of the marine coastal invertebrates of South Korea as assessed by Ilumina MiSeq. Animal Cells Sys 21

Lambshead PJD (2004) Marine nematode biodiversity. In: Chen ZX, Chen SY, Dickson DW (eds) Nematology: advances and perspectives, vol 1. CABI Publishing, Wallingford, pp 439–468

Lallias D, Hiddink JG, Fonseca VG, Gaspar JM, Sung W, Neill SP, Barnes N, Ferrero T, Hall N, Lambshead PJD, Packer M, Kelley Thomas W, Creer S (2015) Environmental metabarcoding reveals heterogeneous drivers of microbial eukaryote diversity in contrasting estuarine ecosystems. ISME J 9:1208–1221

Lawton JH, Bignell DE, Bolton B et al (1998) Biodiversity inventories, indicator taxa and effects of habitat modification in tropical forest. Nature 391:72–76

Leray M, Knowlton N (2015) DNA barcoding and metabarcoding of standardized samples reveal patterns of marine benthic diversity. Proc Natl Acad Sci USA 112:2076–2081

Leray M, Knowlton N (2016) Censusing marine eukaryotic diversity in the twenty-first century. Phil Trans R Soc B 371:20150331

Liu L, Li Y, Li S, Hu N, He Y, Pong R, Lin D, Lu L, Law M (2012) Comparison of next-generation sequencing systems. J Biomed Biotechnol 2012:251364

López-Escardó D, Paps J, de Vargas C, Massana R, Ruiz-Trillo I, del Campo J (2018) Metabarcoding analysis on European coastal samples reveals new molecular metazoan diversity. Sci Rep 8:9106

Mahé F, Rognes T, Quince C, de Vargas C, Dunthorn M (2014) Swarm: robust and fast clustering method for amplicon-based studies. Peer J 2:e593

Mardis ER (2008) Next-generation DNA sequencing methods. Annu Rev Genomics Hum Genet 9:387–402
Mathee K, Narasimhan G, Valdes C, Qiu X, Matewish JM et al (2008) Dynamics of *Pseudomonas aeruginosa* genome evolution. Proc Natl Acad Sci USA 105:3100–3105
Pearman JK, Irigoien X (2015) Assessment of Zooplankton Community Composition along a Depth Profile in the Central Red Sea. PLoS ONE 10(7): e0133487.https://doi.org/10.1371/journal.pone.0133487
Ratnasingham S, Hebert PDN (2007) BOLD: The Barcode of Life Data System. Mol Ecol Notes 7:355–364. http://www.barcodinglife.org
Reeder J, Knight R (2010) Rapidly denoising pyrosequencing amplicon reads by exploiting rank-abundance distributions. Nat Methods 7:668–669
Riesenfeld CS, Schloss PD, Handelsman J (2004) Metagenomics: genomic analysis of microbial communities. Annu Rev Genet 38:525–552
Rothberg JM, Hinz W, Rearick TM, Schultz J, Mileski W et al (2011) An integrated semiconductor device enabling non-optical genome sequencing. Nature 475:348–352
Semprucci F, Losi V, Moreno M (2015) A review of Italian research on free-living marine nematodes and the future perspectives in their use as Ecological Indicators (EcoInd). Mediterr Mar Sci 16:352–365
Shokralla S, Spall JL, Gibson JF, Hajibabaei M (2012) Next-generation sequencing technologies for environmental DNA research. Mol Ecol 21:1794–1805
Taberlet P, Coissac E, Pompanon F, Brochmann C, Willerslev E (2012) Towards next generation biodiversity assessment using DNA metabarcoding. Mol Ecol 21:2045–2050
Thompson JF, Milos P (2011) The properties and applications of single-molecule DNA sequencing. Genome Biol 12:217–226
Thomsen PF, Willerslev E (2015) Environmental DNA—an emerging tool in conservation for monitoring past and present biodiversity. Biol Conserv 183:4–18
Yi X, Liang Y, Huerta-Sánchez E, Jin X, Xi Z et al (2010) Sequencing of 50 human exomes reveals adaptation to high altitude. Science 329:75–78
Yoccoz NG (2012) The future of environmental DNA in ecology. Mol Ecol 21:2031–2038

Chapter 4
Soft Coral Biodiversity in the Red Sea Family Alcyoniidae: A Biopharmaceutical and Ecological Perspective

Erick E. Dokalahy, H. R. El-Seedi and Mohamed Ali Farag

Abstract Seas cover over 70% of the Earth surface, and its total global biodiversity is estimated to have some 500 × 10^6 species of prokaryote and eukaryote organisms. Moreover, the Red Sea with a high percentage of endemic biota is an epicenter for marine biodiversity. Indeed, of the 180 soft coral species identified worldwide, approximately 40% are native to the Red Sea area. Such coral reef ecosystems support enormous biological diversity, including structural and functional complex benthic communities. The marine metabolome is quite complex, and its diversity exceeds that of mammals because the selection and retention of chemical diversity is a critical factor in an organism's adaptation and fitness and a primary reason for the large number of natural products. Only a few thousand compounds have been reported from the Red Sea of marine origin, and hence, it is believed to have an enormous potential as a provider for new bioactive metabolites. Marine natural products display an extraordinary chemical and pharmacological scope. This could be attributed to their necessity to release secondary metabolites as their own chemical defense tools to survive in extreme environment, to resist their predators, or to provide chemical communication in symbiotic relationships. The growing interest in marine natural products, particularly in the area of anticancer compounds, is attributed to the urgent therapeutic need in this area. The biological and chemical research of the coral reefs has made a remarkable progress as reviewed herein yet the support information of the biodiversity, functions profile and ecological landscapes still to be acquired. This chapter overviews current

E. E. Dokalahy · M. A. Farag (✉)
Pharmacognosy Department, College of Pharmacy, Cairo University,
Kasr el Aini St., P.B. 11562 Cairo, Egypt
e-mail: Mohamed.farag@pharma.cu.edu.eg

H. R. El-Seedi
Pharmacognosy Group, Department of Medicinal Chemistry,
Uppsala University, Biomedical Centre, Box 574, 751 23 Uppsala, Sweden

H. R. El-Seedi
International Research Center for Food Nutrition and Safety,
Jiangsu University, Zhenjiang 212013, China

K. G. Ramawat (ed.), *Biodiversity and Chemotaxonomy*, Sustainable
Development and Biodiversity 24, https://doi.org/10.1007/978-3-030-30746-2_4

research in octocoral order: *Alcyonacea* growing in the Red Sea area with focus on its medicinal potential within its chemical rich niche as well as their ecological functions. The chapter emphasizes also on the potential research areas for the marine natural products that are yet to be investigated.

Keywords Red sea · Alcyoniidae · Biopharmaceutical · Ecological functions

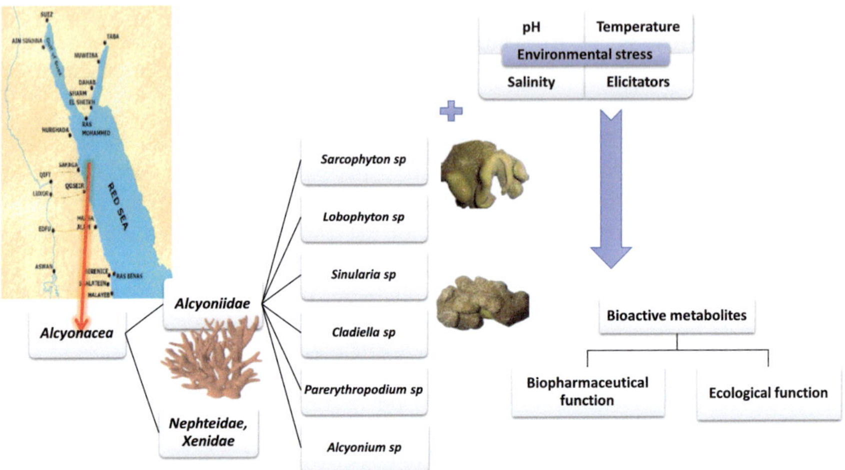

4.1 Introduction

The Red Sea has long been considered as an epicenter for marine biodiversity whether regarding its fish population (DiBattista et al. 2015), microbiome (Mustafa et al. 2016) or more for its hermaphytic corals. Among the animal phylum, octocorals or soft corals (Phylum *Cnidarian*, class *Anthozoa*, subclass Octocorallia or Alcyonaria) have emerged as potential animal source of vast research interest. Unlike the stony corals which are protected by a calcium carbonate skeleton, soft corals have a soft bodies supported by a spiny, minutes skeletal elements called sclerites. As a means to survive through the various stressors in marine life, protect themselves against predators, or communicate among symbiotic organisms (Farag et al. 2017b), octocorals have developed the machinery (Sammarco and Coll 1992) to produce a myriad of secondary metabolites such as ceramides (Cheng et al. 2009), sterols (Santalova et al. 2004), and predominantly terpenoid compounds and its derivatives (Hegazy et al. 2015) with interesting medicinal and ecological properties. As a matter of fact, the marine environment presents a great wealth of untapped natural sources of complex chemicals with potential biopharmaceutical usage such as cytotoxicity (Ellithey et al. 2014), antitumor (Sarmiento-Vizcaíno et al. 2017), antimicrobial (Mariottini and Grice 2016) as well as their antifouling (Soliman et al. 2017), ichthyotoxic (Sammarco and Coll 1992), feeding deterrence (Kelman et al. 1999), and antimicrobial (Kelman et al. 2006) activities in the context of marine ecology.

Up to 40% of the 180 soft corals species identified worldwide are known to be endemic to the Red Sea area (Al-Lihaibi et al. 2014); however, the status of their research is still in its infancy with potential bioactive chemicals yet to be revealed. Berumen et al. (2013) conducted a comparative listing on the existing literature on the soft coral ecology performed in the Great Barrier Reef (GBR) in Australia, Caribbean Sea, and the Red Sea, suggesting that the majority focuses only on 2% of the Red Sea region. A guideline to pinpoint the future horizon of the soft coral research status in the Red Sea is needed to avoid replication of previously investigated corals and highlight the need of future unexplored research areas. Understanding of the Red Sea corals chemistry diversity and ecological function shall be of value to understand the influence of ecological relationships on corals and their change upon (man-made) environmental impact, i.e., global warming or pollution (Riegl et al. 2009). The main interaction between organisms but also within them is indeed of chemical nature. With regard to the Red Sea, Nature Middle East reports that the Red Sea is getting hotter at a rising temperature rate exceeding that of the global average. The Red Sea is already roughly 0.2 °C higher in temperature than the global average (Laylin 2011), which might seem like an insignificant number, but even small changes in temperature can have wide-ranging impacts on the overall ecosystem and marine life, and averages also conceal often more relevant hot spells. Corals, when exposed to seawater temperatures above normal levels for their region, will exhibit "bleaching"; i.e., they lose their zooxanthellae, which provide color to the host coral tissue, leaving the tissue transparent and ultimately leading to coral death (Wooldridge 2010; Sammarco and Strychar 2013). To understand how bleaching occurs in the host coral in terms of chemistry is also just at the beginning, with membrane lipids of symbiotic algae identified as a diagnostic parameter for the sensitivity to thermal bleaching in corals (Tchernov et al. 2004).

Searching literature on the Red Sea soft corals, family Alcyoniidae appeared as the most examined with a total of 39 soft coral species being studied belonging to the genus *Sarcophyton, Sinularia, Lobophyton, Alcyonium, Cladiella, and Parerythropodium* and suggesting that Alcyoniidae has received the most chemical investigation. Among these genotypes, most of the bioactive chemicals appeared to be associated with the genus *Sarcophyton* and in agreement with the review by Liang and Guo (2013). A comparative study on the bioactive terpenes from Red Sea marine organism starting from 1980 till 2014 has been reported by Hegazy et al. (2015). Nevertheless, several new coral species and novel chemicals are continuously being reported which warrants a more comprehensive review of the Red Sea octocorals.

To the best of our knowledge, this work represents not only the most comprehensive study of the Red Sea Alcyoniidae corals chemistry but also its biological and ecological functions. It compiles the reported secondary metabolites of the Red Sea Alcyoniidae family and is subdivided into two main sections including:

(1) reported bioactive components of each Alcyoniidae with a focus on their source followed by (2) ecological function. The chapter ends with a review of reported coral chemicals of yet no biological effects that are presented for researchers to consider in their future work.

4.2 Bioactive Secondary Metabolites of Red Sea *Alcyoniidae*

During the past 20 years, thousands of novel marine metabolites have been reported and assayed for anticancer activity based on their ability to either inhibit the proliferation, migration, tumor formation, the metastasis or even completely kill cancer cell. Extract and fractions from the Red Sea Alcyoniidae family has been assayed in the past decade for their potential source of anticancer and other biological properties (Hegazy et al. 2012). Terpenoids were the most studied among reported metabolites, particularly the cembranoid diterpenes found in abundance in the *Sarcophyton* (family Alcyoniidae). Cembranoids contain a 14-membered macrocyclic skeleton fused to a five-membered unsaturated lactone ring (Fig. 4.1, compound **10**) and exhibit a wide range of biological activities including most prominently antitumor activity (Hegazy et al. 2012). The furanocembranoid diterpene sarcophine (**10**) has been investigated since 1998 for its potential as a chemo-preventive agent and cytotoxic agent. The upcoming section highlights the anticancer activities reported of each Alcyoniidae genus, viz. *Sarcophyton, Sinularia, Cladiella, and Lobophyton* considering the wealth of cytotoxic activity reports for each genotype.

4.2.1 *Sarcophyton Genus Cytotoxic Effect*

Sarcophyton (phylum, Cnidaria; class, Anthozoa; subclass, octocoralia; order Alcyonaceae; family, Alcyoniidae) is one of the genus that has been extensively

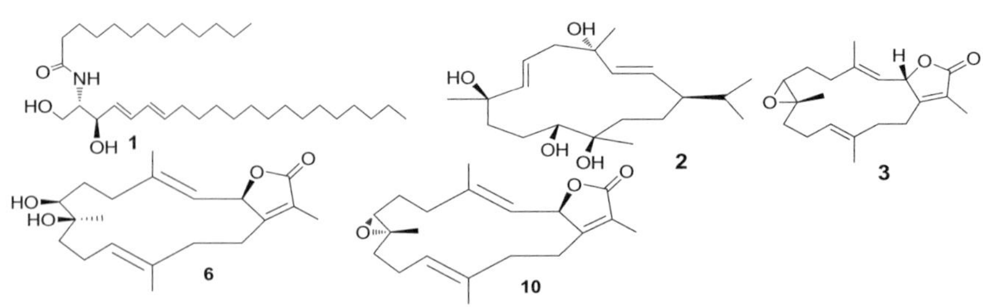

Fig. 4.1 Ceramide (**1**) and Cembranoid diterpenes (**2**, **3**, **6**, and **10**) from *S. auritum*

studied during the past three decades (Zubair et al. 2016) to include *Sarcophyton glaucum* (Abdel-Lateff et al. 2015; Eltahawy et al. 2014; Ne'eman et al. 1974), *Sarcophyton ehrenbergi* (Abou El-Ezz et al. 2013; Eltahawy et al. 2014; Hegazy et al. 2017), *S. trocheliophorum* (Řezanka and Dembitsky 2001; Shaaban et al. 2015), *S. auritum* (Eltahawy et al. 2014) for their antitumor and or cytotoxic activities.

Eltahawy et al. (2014, 2015) investigated the bioactive compounds of *Sarcophyton auritum* leading to the discovery of a ceramide (**1**) and four cembranoid diterpenes **2–3, 6, 10** found against two cancer lineHepG2 (liver cancer cell line) and MCF-7 (breast cancer cell line), Fig. 4.1. While **2 and 3**, reported for the first time, showed a moderate toxicity with an IC_{50} ranging from 19.7 to 21.1 μg/ml, whereas compound **6 and 10** were found the most and least cytotoxic.

The polar and non-polar extracts of *S. ehrenbergi* encompass a myriad of 10 or more ring-structured terpenoids such as cembranes (**4–10**), cembrenes (**11–13, 135–136**) diterpenoid, sesquiterpenes (**14**), bicyclic cembranolide (**15**), steroid (**16**), and tetraterpene (**17**), Fig. 4.2 with potential antiproliferative and cytotoxic activities toward selected cell lines (Hegazy et al. 2017). The antitumor activities of compounds **7–9** were assessed against breast carcinoma (MC-7), with IC_{50} values of 192.87, 68.57, and 114.41 μmol/mL, respectively. It is important to note that these compounds are novel to that species (Elkhateeb et al. 2014). A study by Hegazy et al. (2017) reported a variable antiproliferative activity of the cembrenes (**11–13, 135–136**), cembrane (**6**) diterpenoids, and the steroid (**16**) against three human tumor cell lines, viz. lung (A549), colon (Caco-2), and HepG2 coupled with a molecular docking technique study as the ring structure density plays a substantial role in the receptors/metabolites interaction and or binding. The first reported antiproliferative activity of **13** showed strong cytotoxicity with IC_{50} of 27.3 μM against A549 cell, whereas **13** and **16** showed a moderate inhibition against HepG2 cell line at IC_{50} of 53.8 and 56.8 μM, respectively.

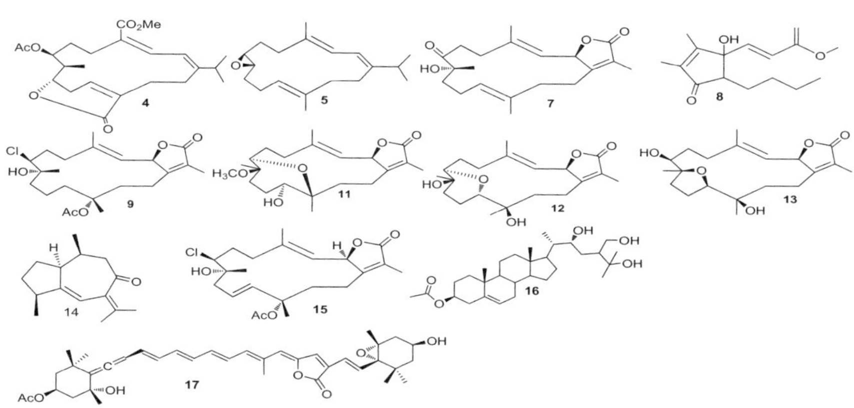

Fig. 4.2 Structure of metabolites from *S. ehrenbergi* **4–17**

Isolation attempt from the non-polar extract of *Sarcophyton glaucum* also provided different secondary metabolites class, namely cembranoids (**10, 18–28, 137–140**), a cembrene **29**, cembranolides **30–32**, sesquiterpenes **33–36, 135** and other miscellaneous classes **37–38**. Abou El-Ezz et al. (2013) assayed the cembranoid diterpenes (**10, 19, 22–25**) against several cancer cell lines. Among the most repeatedly isolated compounds from *Sarcophyton*, the furanocembranoid diterpene **10** has been massively reported by several researchers (Abou El-Ezz et al. 2013; Al-Footy et al. 2015; Eltahawy et al. 2014; Hegazy et al. 2015; Ne'eman et al. 1974; Shaaban et al. 2015; Shaker et al. 2010) to have a dry weight yields of 3% in *S. glaucum* and found most active against cholinesterase in vitro which warrants further development of its activity using combinatorial chemistry or structure active relationships SAR studies. Bioactive cembranoids **18, 22, and 23** were reported to exhibit a similar antitumor and cytotoxic activity toward the mouse melanoma B16F10 and monkey kidney CV-1 cell line. From an EtOAc extraction, bioactive sesquiterpene **35** was reported by Sawant et al. (2007) to exhibit a potent antiproliferative effect against the highly malignant mouse tumorous cell line (+SA mammary epithelial cells) at a dose 20 μM, while the dioxolane sesquiterpene alcohol **136** expressed no activity toward the tested cell line. Al-Lihaibi et al. (2014) reported also the isolation of cembranoids **19, 21, 26, 27,** and the sesquiterpene **34** from an organic (diethyl ether) extract of *S. glaucum* and tested their activity against five cancer lines, *viz*. MCF-7, HepG2, A549, PC-3, and VERO compared to the standard anticancer drug (Doxorubicin). Compounds **21, 26, 34** whose antiproliferative activity were related to their ability to induce cellular apoptosis were cytotoxic against HepG2 (IC_{50}20 μM), whereas metabolites **20, 27** exhibited a less potent against MCF-7 with an IC_{50} of 25 and 29 μM, respectively, as manifested by its much higher IC_{50} value in the micromolar range. Report on the ethyl acetate extract of the same *Sarcophyton* species let to the identification of five compounds including two new peroxide cembranoids **31–32**, two previously reported cembranoids **10, 30,** and a new cembrene derivative **29** (Hegazy et al. 2012). *In vitro* assay revealed that **29 and 31** exhibit a promising inhibitory effect on the cytochrome P450 1A activity (IC_{50} values: 2.7 and 3.7 nM) as well as gluthatione-*S*-transferase (GST) and quinone reductase (QR) inducing potential, which demonstrate their ability to affect the carcinogen metabolizing enzymes in Murine hepatoma cells (Hepa1c1c7).

The chloroform-methanol (1:1) extract and fraction of *Sarcophyton trocheliophorum* has been investigated by Hegazy et al. (2013), Al-Footy et al. (2015), Shaaban et al. (2016), El-Seedi et al. (2016) which has led to the isolation of metabolites (**10, 18, 29–30, 39- 46, 98, 141–148**) belonging to several chemical classes, namely pyrane-based cembranoids (**18, 39, 98**), bicyclo-(5,7)sesquiterpenes (**44, 143–146**), and cembranoid diterpenes (**10, 41, 98**). While their cytotoxicity preliminary testing in a brine shrimp assay showed no response except for **41** with a weak toxicity of 22.5% mortality rate, antitumor activity against two cell lines (Lymphoma and Erlish cell line) of the two diterpenes **10 and 41** exhibited strong effects with IC_{50} of 2.5–3.5 μM (Al-Footy et al. 2015). These results suggest that the application of general cytotoxic assays, i.e., shrimp bioassay, might evade the

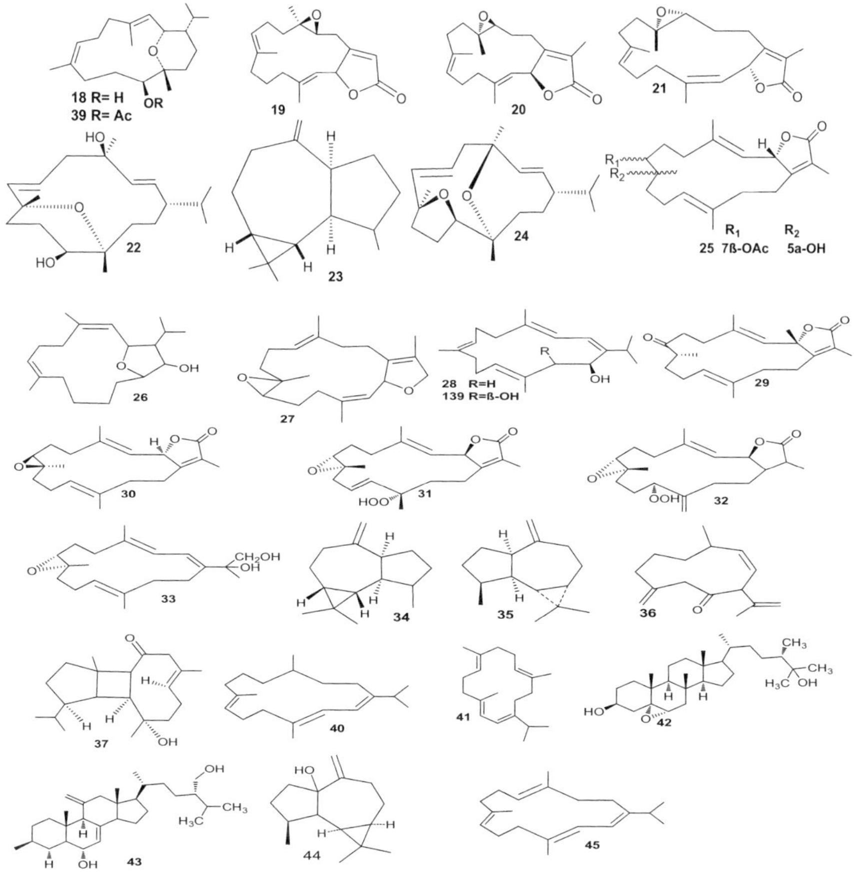

Fig. 4.3 Bioactive metabolites (**18–45**) reported from *S. trocheliophorum* and *S. glaucum*

detection of potential cytotoxic drugs in typical laboratory assays. While **29** and **30** were reported from *S. trocheliophorum* species for the first time, the cembranolide **151** has been isolated by Hegazy et al. (2015) for the first time in nature. Further analysis of its potential pharmacokinetic properties should be investigated (Fig. 4.3).

4.2.2 *Sinularia* Genus Cytotoxic Effect

Sinularia genus reported anticancer effects include firstly that if (El Sayed and Hamann 1996) from *Sinularia gardineri*'s EtOH extract leading to the isolation of a sesquiterpene (**47**), a new heptacyclic norcembranoid dimer (**48**), a known norcembranoid **152** lacking a methyl group at C_4 and a C_4 norcembranoid **153**. Cytotoxicity assay of **47** and **48** were found effective against four cancer cell lines, i.e., murine leukemia (P-388), A549, human colon carcinoma (HT-29), and human melanoma cells (MEL-28) with an IC_{50} ranging from 1 to 5 μg/mL. Investigation on *Sinularia polydactyla* extract uncovered five steroids (**49–53**), two sesquiterpenes (**54, 55**), and a cembranoid diterpene (**56**) with varied cytotoxic efficacies. Metabolites (**41, 49–51, 54–55**) (Shaaban et al. 2013b) exhibited marginal cytotoxic effect toward brine shrimp assay with 4–7% mortality rate at 10 μg/mL except for (**49**), displaying 24%. (Aboutabl et al. 2013) unraveled the antitumor and cytotoxic effects of (**49–54**) against three human cancer lines, viz. liver ($HepG_2$), colon (HCT-116), and epidermoid larynx carcinoma (Hep2). While the crude extract **57** of the same species *Sinularia polydactyla* was active on all cell lines, **56** showed a strong and selective toxicity toward Hep2 (IC_{50} 1.0 μg/mL) suggestive for a synergized effect for all chemicals in crude extract to function independently against several cell lines as commonly reported in plant extracts. The bioactive sterols isolated from *Sinularia terspilli* (Mohammed et al. 2017) included eight compounds (**58–62, 159–161**). Strong cytotoxicity with more than 80% inhibition was observed for **58, 60–62** when tested against K-562 versus **58 and 62** found active against human leukemia cell lines HL60 and K562, with IC_{50} values of 0.002–0.025 μM, comparable to that of taxol drug. The first report on the cytotoxicity of a *Cnidarian* species extract of *Sinularia maxima* (**63**) was made by Ellithey et al. (2014) with a potent cytotoxic effect against leukemia (U937) and cervical cancer (HeLa) cells lines (Fig. 4.4).

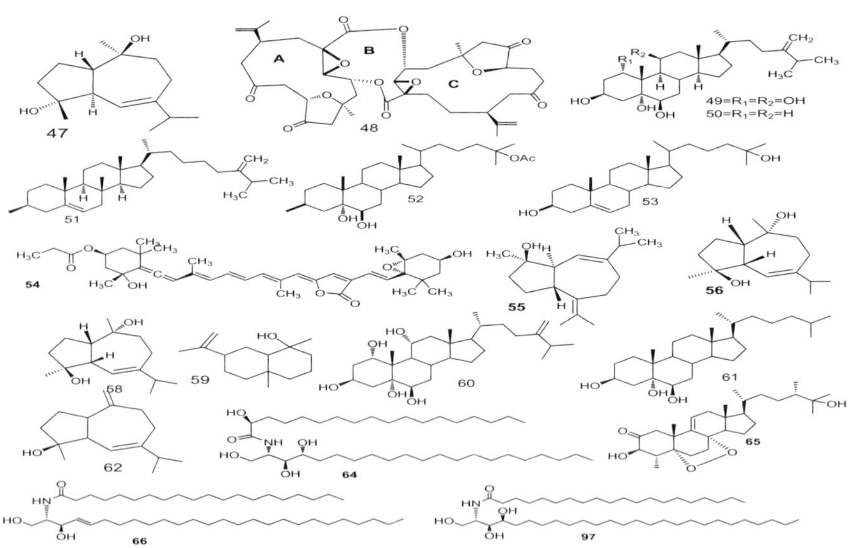

Fig. 4.4 Bioactive metabolites reported from *Sinularia* genus

4.2.3 *Cladiella* Genus Cytotoxic Effect

Study on *Cladiella pachyclados* (Hassan et al. 2010) suggested that next to *Sarcophyton*, *Cladiella* genus is a rich source of a diterpenes with five isolated new (**71–85**) and 11 known (**76–85, 106**) eunicellin-based diterpenoid characterized by the presence of a cladiellane-based skeleton which contains a C_2 and C_9 ether bridge (Fig. 4.5). Three biological assays, namely the MTT, wound-healing, and Cultrex Basement Membrane extract cell invasion assays were performed to assess their cytotoxic potential. Antimigration and antimetastatic assays of the tested compounds suggested the anti-evasive potential of **73, 76, 83,** and **85** compared to the 200 μM dose of the positive drug 4-hydroxyphenylmethylene hydantoin (PMH). This report was the first to reveal for the effect of the eunicellin-based diterpenoid class as anti-invasive or acting as an antimigration of different cancer lines and suggest for searching of other more active agents of that class.

4.2.4 *Lobophyton* Genus Cytotoxic Effect

A study of the methanol extract of *L. crassum* by Aboutabl et al. in (2017) revealed for five polyhydroxysterols **98–100, 161–162** and a sesquiterpene **86**. Biological evaluation of compounds against three human cancer lines, viz. $HepG_2$, Hep-2, and HCT-116, revealed a strong cytotoxic effect of **99** toward $HepG_2$, Hep-2, and HCT-116, with IC_{50} values of 1.9, 5.8, and 6.4 μM, respectively. On the other hand, compounds **86, 98, and 100** expressed a selective affinity to HepG2 compared to the other cell lines with a respective IC_{50}values of 1.9, 3.0 and 3.7 μM. Reports on the organic extract of *Lobophyton* species are also reported in

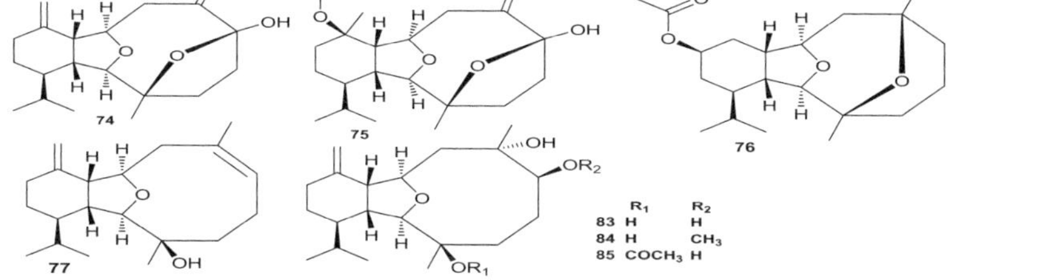

Fig. 4.5 Eunicellin-based diterpenoid from *Cladiella pachyclados*

Table 4.2 (**107–133, 161–162**) for which no reported bioactivity made. Research on *Lobophyton* species producing antiproliferatives cembranolides (Peng et al. 2018) and antiviral seco-cembranoids (Cheng et al. 2014) in other geographical location such as Dongsha Atoll in Taiwan, further investigation on the biopharmaceutical potential of the Red Sea *Lobophyton* are suggested especially considering that coral metabolism varies according to its environmental conditions (Farag et al. 2016, 2018). Other secondary metabolites isolated from other Alcyoniidae Family such as *A. flaccidum, A. utinomii,* and *Sarcophyton* sp are reported in Table 4.2 (Fig. 4.6).

4.2.5 Red Sea Alcyoniidae Corals Antimicrobial Effect

Octocorals, as a response to the marine extreme environment, have adapted its metabolism in a distinct way compared to that of hard corals, especially considering their anatomical structure lacking a hard protective shell. Moreover, like other coelenterate, their unique cavity opening used as food ingestion and waste disposal made them more vulnerable to microbial contamination. Consequently, octocorals produce a plethora of secondary metabolites as their own chemical antimicrobial defense tools compared to the stony corals found to exhibit less antimicrobial activity against marine bacteria (Kelman et al. 2006). Those metabolites present an untapped potential to combat the emerging antibiotic resistance by bacteria due to the abusive use of antibiotics over the past 60 years (Al-Footy et al. 2015). Marine scientists have investigated the antibacterial, antiviral, and antifungal of crude extract or metabolites of the Red Sea soft corals (Kelman et al. 2006) and the next section outlines corals effect against different microorganisms.

4.2.5.1 Antibacterial Activity

Early investigation of the Red Sea soft corals antibacterial effect dates back to the 1990 to encompass mostly extracts of coral species from three different genotypes (*Sinularia, Sarcophyton, and Lobophyton*) that has been tested against marine bacteria and human pathogens. Compound (**45**) isolated by Gomaa et al. (2016) from n-hexane extract of *Sarcophyton trocheliophorum* was found active against several pathogens, namely *Bacillus cereus, Salmonella typhi, Escherichia coli, Staphylococcus aureus,* and *Pseudomonas aeruginosa*. Biological octocorals' antibacterial agents mostly belonged to diterpenoids (**39**) (Al-Footy et al. 2015), sesquiterpenoids (**62, 86–88**), and steroids (**89, 90**) (Al-Footy et al. 2016). In (2006), Kelman compared the antibacterial activity of two different cnidarian orders, namely scleractinian and alcyonacean. Results revealed that the majority of the Red Sea soft corals (*Litophyton arboreum, Rythisma fulvum, Heteroxenia fuscescens, Sarcophyton glaucum, Dendronephthya hemprichi, Xenia macrospiculata*) were 83% more active against marine bacteria *Arthrobacter* sp. (two strains)

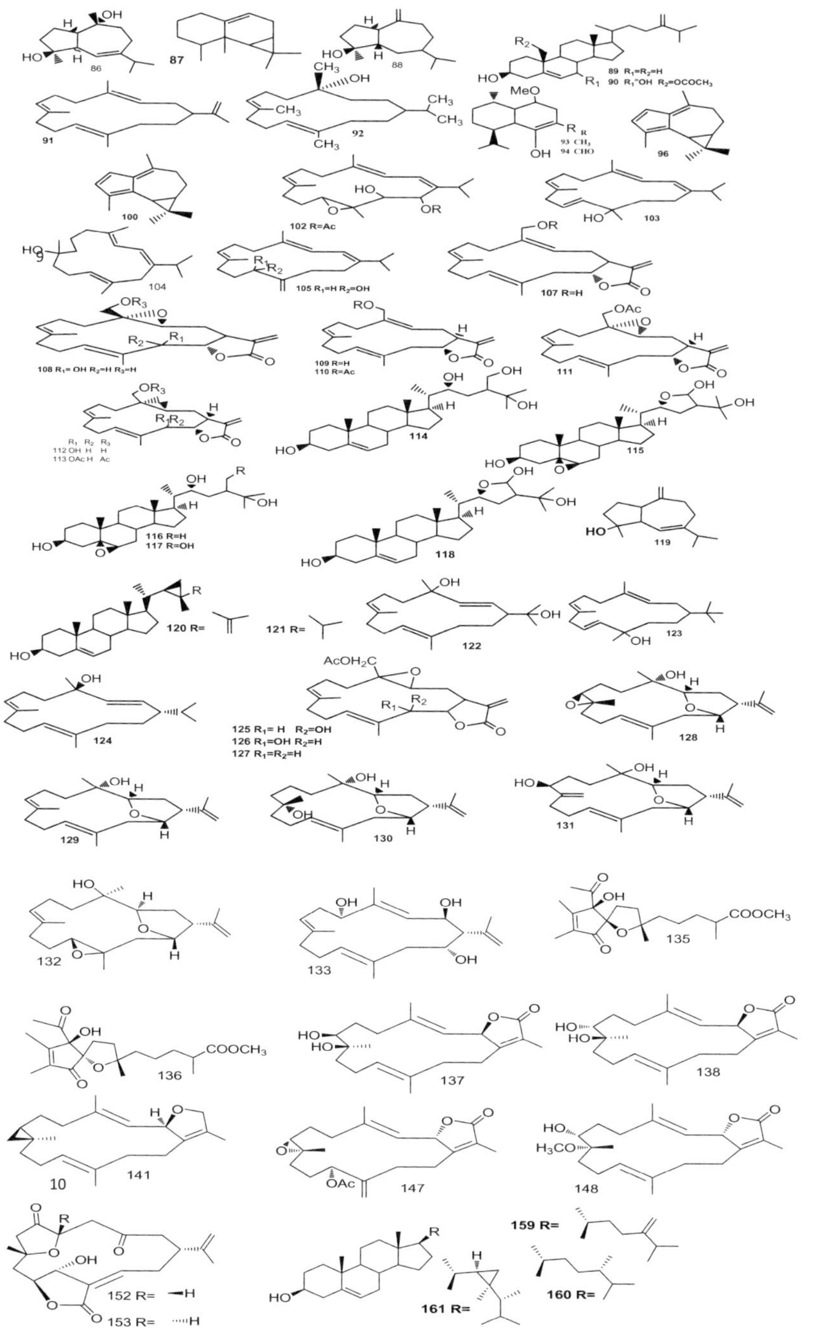

Fig. 4.6 Bioactive metabolites from *Lobophyton* (**87–92**, **103–133**) and *Sarcophyton* (**135–141**) and *Sinularia* metabolites with non-reported activity

compared to hard corals which have mainly developed other strategies to combat microbial invasion in the marine environment (Kelman et al. 2006).

Collected off the Saudi Arabia Red Sea coast, sesquiterpenes (**62, 86–88**) and steroids (**89, 90**) isolated from *Lobophyton* sp. extract exhibited a weaker antibacterial activity compared to sesquiterpene (**91**) and the steroid (**42–43**) as assessed against some gram-positive bacteria (*S. aureus, S. epidermis, and S. pneumonia*) and gram-negative bacteria (*P. aeruginosa*). A further study on *Sarcophyton trocheliophorum* extract showcased the isolation of two cembranoids (**10, 39**) and two pyrane-based cembranoid (**18, 40**) exhibiting significant antibacterial activity, especially against *S. aureus, Acinetobacter spp.*, and MRSA with (MICs) ranging from 1.5 to 4.3 μM (Al-Footy et al. 2015). Aboutabl et al. (2013) assessed the effect of *S. polydactyla* extract against gram-positive bacteria, reporting that compound **53** displayed a moderate antimicrobial activity. Crude extract and fractions including the steroids (**42, 43**) were reported to possess different responses toward *S. aureus* by Shaaban et al. (2013a). While the keto-hydroxysterol **43** showed high antimicrobial activity, oxirane containing compound **42** and the crude extract were found inactive.

Kelman et al. (1998) assessed the extract of several developmental stages of the soft coral *P. fulvum fulvum* against *Vibrio* sp strain from a necrotic coral and other coral associated bacterial strains. The sensitivity toward the *Vibrio* sp with an MIC of 1.25 mg/mL suggested that in general corals do not have a broad-spectrum antibacterial activity against the growth of common, co-occurring, and potentially harmful bacteria, but has a specific activity. Nevertheless, such hypothesis needs to be proved by assaying larger specimens of coral species against several pathogens.

Recently, investigation in the Egyptian Red Sea was done by Soliman et al. (2017) on the methanol extract derived from soft coral species, viz. *S. glaucum, S. cruciate, H. fuscescens,* and *S. compressa* against selected marine bacteria (*P. aeruginosa* ATCC653, *Staphylococcus aureus* ATCC6538, and *Escherichia coli, P. aeruginosa* ATCC6539) using a microdilution broth susceptibility test. The *S. compressa* extract (**67**) showed the strongest effect as revealed from its low MIC/MBC values ranging from 0.5 to 1/1 to 10 mg/mL followed by *S. cruciate* (**68**) and *S. glaucum* (**38**) with a MIC/MBC of 5–50/10–100 mg/mL, whereas *H. fuscescens* exhibited no activity.

4.2.5.2 Antiviral Activity

As a potential marine drug source of antimicrobials, the Red Sea Alcyoniidae coral extracts also demonstrated their capacity to encompass strong antiviral metabolites, though with much less evidence-based assays compared to that derived from antibacterial. Ahmed et al. (2013) reported two classes of compounds, ceramides and sterols to exhibit an antiviral effect against H5N1 virus from *S. candidula*. Bioactivity guided fractionation of the EtOAc extract against influenza H5N1 led to the isolation of three bioactive ceramides (**64, 66** and **97**) and a polyhydroxylated sterol (**65**). All compounds managed to reduce the virus titer by 55.16%, 48.81%,

10.43%, and 15.76% at a dose of 1 ng/mL comparable to the standard drug Zanamivir, while the crude extract expressed a complete inhibition at 12 μg/mL.

In search of anti-HIV/AIDS drugs from marine resources, Ellithey et al. (2014) assessed the response of Red Sea organisms' extracts (**63**) toward the inhibition of HIV-1 reverse transcriptase (RT) and protease (PR) enzymes. Although extracts exhibited no significant effect against HIV-1 RT enzymes, inhibitory activities against HIV-1 PR were shown from the soft corals *S. heterospiculata* (8.6 μg/mL), *L. arboreum* (12 μg/mL), and *S. maxima* (13.1 μg/mL). These results highlight the bioactivity of the crude extract of the tested marine organism, and calling for a further bioassay guided fractionation to identify their active site as potential anti-HIV agents.

4.2.5.3 Antifungal Activity

To evade the predation and infection of soft corals from fungi that may affect their marine life, Alcyoniidae family produce various antifungal biomolecules, viz. terpenes and steroids that can be exploited as therapeutic agents in the pharmaceutical industry for humans (Mohammed 2012). Fungi may be opportunistic pathogens in corals under environmental stress. Abou El-Ezz et al. (2013) evaluated the methanol extract of *Sarcophyton glaucum* for its antifungal activity. Compared to isolated metabolites, crude extract showed no obvious biological activity. In contrast, cembrane-based diterpene **40** and **10** with a MIC of 0.68 μM were found active against fungal pathogens *Candida albicans, Aspergillus flavus* and with an IC_{50} of 20 μg/mL against *Cryptococcus neoformans*, respectively. Al-Footy et al. (2015) assessed compounds isolated from the lipophilic extract of *Sarcophyton trocheliophorum* against *Aspergillus flavus* and *C. Albicans*. While compound **40** exhibited a low antifungal activity with an MIC of 0.68 μM, other pyrane-based cembranoid (**18 and 39**), cembranoid **10,** and a sesquiterpene **44** did not exhibit any response. Analyzing biological structure–activity relationships (SAR) among coral terpenoids analogues also may help identify more biologically active antifungal drugs and reveal crucial structural motifs that promote an antimicrobial effect.

In a similar location, another species *S. gardineri* was investigated (El Sayed and Hamann 1996) for its effect against *C. albicans* B311 and *C. neoformans* (El Sayed and Hamann 1996). The heptacyclic norcembranoid dimer **48** showed a growth inhibition results comparable to amphotericin B. In contrast, sesquiterpene **47** and cembranolides (**159–160**) were not able to inhibit fungal growth. Investigating the lipophilic extract of *S. terspili*, Mohammed et al. (2017) reported that the sterol (**61**) exhibited a moderate activity on the fungus *C. neoformans* with an IC_{50} value of 9.6 μg/mL. The assessment of lypophilic extract in all studies seems rational as to identify positive antifungal hits considering the lypophilicity of the fungal cell wall, a prerequisite type for a chemical to penetrate first prior to exerting a killing effect (Georgopapadakou 1995).

4.2.5.4 Antileishmanial Effect

Leishmaniasis is a disease that is more peculiar to the Third World countries, with an increasing mortality and morbidity rate in Africa, Asia, and Latin America caused by the protozoan *Leishmania* which uses mosquito as its principal vector to human from rodents (Rocha et al. 2005). The need for an effective drug to cure the disease aside from vaccine treatment prompted the search of bioactive chemicals of marine origin. *S. terspilli*, endemic to the Red Sea in Hurgada Egypt, has been reported by Mohammed et al. (2017) as a source of three bioactive sesquiterpenes [**58, 59,** and **62**] and sterols [**60, 61, 146, 159** and **161**]. In vitro culture of *Leishmania donovani* promastigotes was tested against isolated compounds, with sesquiterpenes [**58, 59**] and sterols [**60, 61**] exhibiting antileishmanial activity with IC_{50} values ranging from 10 to 30 μg/mL (Mohammed et al. 2017).

4.2.6 Red Sea Alcyoniidae Miscellaneous Biological Effects

Aside from the previously reported cytotoxic and antimicrobial activities for Red Sea Alcyoniidae family, other activities such as antiepileptic, anxiolytic, anti-inflammatory, and antimalarial activities were reported but to less extent. Eltahawy et al. (2014) examined the effect of crude polar extract and fractions of *S. auritum* enriched in ceramides. In vivo screening of the antiepileptic properties using PZP-induced seizure model, ceramide [**1**] was found successful to antagonize the lethality of pentylenetetrazole in mice. Moreover, the same compound was assessed in vitro of the light–dark transition box and elevated plus maze and confirming its anxiolytic activity.

A semi-synthesized products from furanocembranoid diterpenes (**10**) of *S. glaucum* were reported to exhibit potential anti-inflammatory effects. Sawant et al. (2006) reported the first anti-inflammatory activity of **10**, in addition to its hydroxylated semi-synthetic derivatives, found effective to release inflammatory mediators such as thromboxane B2 and superoxide anion in activated rat neonatal microglia. A promising result with an improved bioactivity of the sulfur derivative of **10** was found to also exhibit strong anti-inflammatory effect on highly malignant +SA mammary epithelial cell proliferation. In general and although in some cases, drug leads from marine resources fail to exhibit a prominent effect, semi-synthesis or biotransformation of these chemicals could present more efficacious drugs and with less side effects. As an example, the bioconversion of the *S. glaucum* fractions (**10, 40, 136**) using preparative scale fermentation by three selected fungus (*Absidia glauca* ATCC 22752, *Rhizopus arrhizus* ATCC 11145, and *Rhizopus stolonifer* ATCC 24795) resulted in extracts with improved functionality compared to the resulting metabolites (El Sayed et al. 1998) (Tables 4.1 and 4.2).

Table 4.1 List of Red Sea Alcyoniidae natural products, extracts and their reported bioactivities

N°	Molecules	Bioactivity	Biological target	Source	Reference
	Metabolites isolated from Sarcophyton sp				
1	N-((2S0,3*R*,4*E*,6*E*)-1,3-Dihydroxyhenicosa-4,6-dien-2-yl)tridecanamide	Anticonvulsant, anxiolytic	–	*S. auritum*	Eltahawy et al. (2015)
2	(1*R*,2*E*,4*S*,6*E*,8*R*,11*R*,12*R*)-2,6-Cembradiene-4,8,11,12-tetrol	Cytotoxic	MCF-7, HepG2	*S. auritum*	Eltahawy et al. (2014)
3	2-Epi-sarcophine	Id.	MCF-7, HepG2	*S. auritum*	Eltahawy et al. (2014)
4	(+)-Emblide	Antiproliferative	KB cell	*S. ehrenbergi*	Shaker et al. (2010)
5	(+)-7,8-Epoxy-7,8- dihydrocembrene	Id	HUVEC, K562 cell	*S. ehrenbergi*	Shaker et al. (2010)
6	7α,8β-Dihydroxy-deepoxysarcophine	Id.	HepG2, MCF-7, B16F10	*S. ehrenbergi*	Abou El-Ezz et al. (2013), Eltahawy et al. (2014), Hegazy et al. (2017)
7	7-Keto-8α-hydroxy-deepoxysarcophine	Cytotoxic	MCF-7	*S. ehrenbergi*	Elkhateeb et al. (2014)
8	(*E*)-Methyl-3-(5-butyl-1-hydroxy-2,3-dimethyl-4-oxocyclopent-2-enyl)acrylate	Id.	MCF-7	*S. ehrenbergi*	Elkhateeb et al. (2014)
9	7β-Chloro-8α-hydroxy-12-acetoxy deepoxysarcophine	Id.	MCF-7	*S. ehrenbergi*	Elkhateeb et al. (2014)
10	Sarcophine	Antifungal, antibacterial, antitumor, cytotoxic, antipredator, anti-acetylcholine	*Cryptococcus neoformans*, CV-1 cells, Erlish cell line, MCF-7, HepG2	*S. glaucum, S. auritum, S. ehrenbergi*	Abou El-Ezz et al. (2013), Al-Footy et al. (2015), Eltahawy et al. (2014), Hegazy et al. (2015), Ne'eman et al. (1974), Shaaban et al. (2015), Shaker et al. (2010)
11–13	Sarcoehrenbergilid A-C	Antiproliferative	HepG2	*S. ehrenbergi*	Hegazy et al. (2017)
14	Guajacophine	Moderate antiproliferative and cytotoxic	HUVEC, K-562, and HeLa cell lines	*S. ehrenbergi*	Shaker et al. (2010)

(continued)

Table 4.1 (continued)

N°	Molecules	Bioactivity	Biological target	Source	Reference
15	Sarcoglaucol-16-one	Antiproliferative, cytostatic	HM02, HepG2, MCF7	*S. ehrenbergi*	Shaker et al. (2010)
16	Sardisterol	Cytotoxic	A549	*S. ehrenbergi*	Hegazy et al. (2017)
17	Peridinin	Antiproliferative, cytotoxic, antitumor	HUVEC, K-562, Hela, DLD	*S. ehrenbergi*	Shaker et al. (2010)
18	Sarcotrocheliol	Antibacterial, cytotoxic	*S. aureus, Actinobacter spp and MRSA*, MCF-7 cell	*S. trocheliophorum, S. glaucum*	Abdel-Lateff et al. (2015), Al-Footy et al. (2015)
19	Sarcophytolide	Antimicrobial, antitumor, cytotoxic	*S. aureus, P. aeruginosa* and *S. cerevisiae*, B16F10, CV-1 cells	*S. glaucum*	Abou El-Ezz et al. (2013), Badria et al. (1997)
20–21	Sarcophytolide B, C	Antitumor, cytotoxic	MCF-7, HepG2	*S. glaucum*	Al-Lihaibi et al. (2014)
22	(1*S*,2*E*,4*R*,6*E*,8*R*,11*S*,12*R*)-8,12-Epoxy-2,6-cembradiene-4,11-diol	Antitumor	B16F10	*S. glaucum*	Abou El-Ezz et al. (2013)
23	(1*S*,4*R*,13*S*)-Cembra-2*E*,7*E*,11*E*-trien-4,13-diol	Cytotoxic, antitumor	CV-1 cells, B16F10 cells	*S. glaucum*	Abou El-Ezz et al. (2013)
24	(1*S*,2*E*,4R,6*E*,8*S*,11*R*,12*S*)-8,11-Epoxy-4,12-epoxy-2,6-cembradiene	Cytotoxic	CV-1 cells, B16F10 cells	*S. glaucum*	Abou El-Ezz et al. (2013)
25	7β-Acetoxy-8α-hydroxydeepoxysarcophine	Id.	HepG2, HCT-116, and HeLa cells	*S. glaucum*	Hegazy et al. (2011a)
26	Sarcophytolol	Id.	HepG2	*S. glaucum*	Al-Lihaibi et al. (2014)
27	Deoxosarcophine	Potent cytotoxic	MCF-7, HCT116, MCF-7	*S. glaucum*	Abdel-Lateff et al. (2015), Al-Lihaibi et al. (2014)
28	Sarcophytol A	Antitumor	C3H/HeNCrj mice3 and *N*-methyl-*N*-nitrosurea-induced large bowel cancer in rats	*S. glaucum*	El Sayed et al. (1998)
29	8-Epi-Sarcophinone	Antitumor	Cyp1A	*S. glaucum, S. trocheliophorum*	Hegazy et al. (2012, 2013)
30	Ent-sarcophine	Antitumor	Cyp1A	*S. glaucum, S. trocheliophorum*	Hegazy et al. (2012, 2013)
31	12(*S*)-Hydroperoxylsarcoph-10-ene	Antitumor	Cyp1A	*S. glaucum*	Hegazy et al. (2012)

(continued)

Table 4.1 (continued)

N°	Molecules	Bioactivity	Biological target	Source	Reference
32	11(*S*)-Hydroperoxylsarcoph-12(20)-ene	Cytotoxic	Cytochrome P450 1A	*S. glaucum*	Hegazy et al. (2012)
33	Sarcophinediol	Cytotoxic	HCT116, HepG2	*S. glaucum*	Abdel-Lateff et al. (2015)
34	10(14)-Aromadendrene	Cytotoxic, antitumor	PC-3, HepG2	*S. glaucum*	Al-Lihaibi et al. (2014)
35	(+)-alloaromadendrene	Antiproliferative	+ SA mammary epithelial cells at a dose of 20 μM	*S. glaucum*	Sawant et al. (2007)
36	6-Oxo-germacra-4(15),8,11-triene	Cytotoxic	HCT116	*S. glaucum*	Abdel-Lateff et al. (2015)
37	Sarcoglane	Cytotoxic	Fertilized sea urchin eggs	*S. glaucum*	Fridkovsky et al. (1996)
38	Crude extract	Antibacterial, antifouling paint formulation	*P. aeruginosa* ATCC6538, *S. aureus* ATCC6538, *E. coli*	*S. glaucum*	Soliman et al. (2017)
39	Sarcotrocheliol acetate	Antibacterial, cytotoxic	*S. aureus, Actinobacter* spp *and MRSA*	*S. trocheliophorum*	Al-Footy et al. (2015)
40	Cembrene C	Antifungal	*E. coli, P. aeruginosa, K. pneumonia, S. aureus, and S. epidermidis, E. coli*	*S. trocheliophorum*	Al-Footy et al. (2015)
41	(*Z*)-Cembrene C	Weak cytotoxic	Brine shrimp	*S. trocheliophorum*	Shaaban et al. (2015)
42–43	Zahramycins A-B	Antibacterial, antibiotic	Gram +ve bacteria *S. aureus* and *B. subtilis*	*S. trocheliophorum*	Shaaban et al. (2013a)
44	Palustrol	Antitumor, antiproliferative, antifungal, antibacterial	Lymphoma and Erlish Cell line	*S. trocheliophorum*	Al-Footy et al. (2015)
45	(5*S*)-3-[(3*E*,5*S*)-5-Hydroxy-3-hepten-6-yn-1-yl]-5-methyl-2(5H)-furanone	Antibacterial	Pathogenic bacterial strains, i.e., *B. cereus, S. typhi, E. coli, S. aureus and P. aeruginosa*	*S. trocheliophorum*	Gomaa et al. (2016)
46	Crude extract	Antifouling	–	*S. trocheliophorum*	Mohamed Ali and Soliman (2010)
47	Guaianediol	Cytotoxic	P-388, A549, HT-29, MEL-28	*S. trocheliophorum*	El Sayed and Hamann (1996)

(continued)

Table 4.1 (continued)

N°	Molecules	Bioactivity	Biological target	Source	Reference
	Metabolites from Sinularia sp.				
48	Singardin	Cytotoxic	P-388, A549, HT-29, MEL-28, *C. albicans* B311 and *C. neoformans*	*S. gardineri*	El Sayed and Hamann (1996)
49	24-Methylenecholestane-1α, 3β,5α,6β,11α-pentol	Cytotoxic	Brine shrimp	*S. polydactyla*	Shaaban et al. (2013b)
50	24-Methylenecholestane-3β,5α, 6β-triol	Cytotoxic	Brine shrimps (4–7% mortality at 10 μg/mL)	*S. polydactyla*	Shaaban et al. (2013b)
51	Hurgadacin	Cytotoxic	Brine shrimps (4–7% mortality at 10 μg/mL)	*S. polydactyla*	Shaaban et al. (2013b)
52	24-Methylcholestane-3β,5α,6β,25-tetrol 25-monoacetate	Cytotoxic, antitumor	Hep2 and HCT	*S. polydactyla*	Aboutabl et al. (2013), Hegazy et al. (2015)
53	24-Methylcholestane-5-en-3β,25-diol	Antimicrobial	Gram +ve: *Bacillus subtilis* and *Bacillus megaterium*	*S. polydactyla*	Aboutabl et al. (2013)
54	Peridinin	Cytotoxic	Brine shrimps (4–7% mortality at 10 μg/mL)	*S. polydactyla*	Shaaban et al. (2013b)
55	Lactiflorenol	Cytotoxic	Brine shrimps (4–7% mortality at 10 μg/mL)	*S. polydactyla*	Shaaban et al. (2013b)
56	Durumolide C	Selective toxicity	HepG2	*S. polydactyla*	Aboutabl et al. (2013)
57	Crude extract	Antibiotic, antifouling	Fungus *R. solani*	*S. polydactyla*	Mohamed Ali and Soliman (2010)
58	1S, 4S, 5S, 10R-4, 10-guaianediol	Antileishmanial, strong cytotoxic	*Leishmania donovani*, K562	*S. terspillii*	Mohammed et al. (2017)
59	5,7 Eduesm- 11(13) en-4-ol	Weak antileishmanial	*L. donovani*	*S. terspillii*	Mohammed et al. (2017)
60	Ergost-24(28)-ene-1, 3, 5, 6, 11-pentol (1α, 3β, 5α, 6β, 11α)	Weak antileishmanial, antifungal, cytotoxic	*L. donovani, C. neoformans*, HL60	*S. terspillii*	Mohammed et al. (2017)
61	Ergost-24(28)-ene-3, 5, 6-triol (3β, 5α, 6β)-triol	Antileishmanial, cytotoxic	*L. donovani*, K562, HL60	*S. terspillii*	Mohammed et al. (2017)

(continued)

Table 4.1 (continued)

N°	Molecules	Bioactivity	Biological target	Source	Reference
62	Alismol	Strong cytotoxic (>80%inhibition), High antibacterial	K562, Gram +ve (*S. aureus, S. S. epidermis and S. pneumonia*) and Gram −ve (*P. aeruginosa*)	*S. terspillii, Lobophytum sp*	Al-Footy et al. (2016), Mohammed et al. (2017)
63	Crude extract	Moderate cytotoxic	U937 and HeLa	*S. maxima*	Ellithey et al. (2014)
64	(*R*)-20-Hydroxy-N-[(2S,3S,4R)-1,3,4-trihydroxypentacosan-2-yl] nonadecanamide	Antiviral	Influenza H5N1, 100% inhibition at 1 μg/mL	*S. candidula*	Ahmed et al. (2013)
65	3b-25-Dihydroxy-4-methyl-5a,8a-epidioxy-2-ketoergost-9-ene	Antiviral	Id.	*S. candidula*	Ahmed et al. (2013)
66	*N*-[(2S,3R,E)-1,3-Dihydroxyhexacos-4-en-2-yl] icosanamide	Antiviral	Id.	*S. candidula*	Ahmed et al. (2013)
67	Crude extract	Antibacterial	–	*S. compressa*	Soliman et al. (2017)
68	Crude extract	Antifouling paint formulation	–	*S. cruciata*	Soliman et al. (2017)
69	Crude extract	Selective antifouling	–	*S. heterospiculata*	Mohamed Ali and Soliman (2010)
70	Crude extract	Antifouling activity	–	*S. variabilis*	Mohamed Ali and Soliman (2010)
	Metabolites from Cladiella sp				
71–75	Pachycladins A-E	Antimigration, antimetastatic	PC-3	*C. pachyclados*	Hassan et al. (2010)
76	(+)-Polyanthelin A	Antimigration, antimetastatic	PC-3	*C. pachyclados*	Hassan et al. (2010)
77	(6*Z*)-Cladiellin (cladiella-6*Z*,11(17)-dien-3-ol)	Antimigration	PC-3	*C. pachyclados*	Hassan et al. (2010)
78	3,6-Diacetyl cladiellisin	Antitumor, antimigration	PC-3	*C. pachyclados*	Hassan et al. (2010)
79	3-Acetylcladiellisin	Antimigration	PC-3	*C. pachyclados*	Hassan et al. (2010)
80–81	Klysimplexin E-F	Antimigration	PC-3	*C. pachyclados*	Hassan et al. (2010)
82	Patagonicol	Antitumor, anti-invasive	PC-3	*C. pachyclados*	Hassan et al. (2010)

(continued)

Table 4.1 (continued)

N°	Molecules	Bioactivity	Biological target	Source	Reference
83–84	Sclerophytin A-B	Antimigration, antimetastatic	PC-3	*C. pachyclados*	Hassan et al. (2010)
85	Sclerophytin F methyl ether	Antimigration	PC-3	*C. pachyclados*	Hassan et al. (2010)
86	Alismoxide	Antibacterial, selective cytotoxic	Gram +ve (*S. aureus, S. S. epidermis and S. pneumonia*) and Gram −ve (*P. aeruginosa*); HepG2	*C. pachyclados, L. crasssum*	Aboutabl et al. (2017), Hassan et al. (2010)
	Metabolites from Lobohyton sp. and Parerythropodium sp				
87	Aristol-9-ene	Antibacterial	Id	*Lobophytum sp*	Al-Footy et al. (2016)
88	Nardol	Antibacterial	Id	*Lobophytum sp*	Al-Footy et al. (2016)
89	Chalinasterol	Antibacterial	Id	*Lobophytum sp*	Al-Footy et al. (2016)
90	Nephalsterol C	Antibacterial	Id	*Lobophytum sp*	Al-Footy et al. (2016)
91	Cembrene A	cytotoxicity	*Artemia salina* and Erhlich carcinoma cells	*Lobophytum sp*	Al-Footy et al. (2016)
92	Epi-thunbergol	Mass Spawning	–	*L. crassum*	Coll et al. (1995)
93	5-Hydroxy-8-methoxy-calamenene	Specie recognition	–	*P. fulvum fulvum gray morph*	Kelman et al. (2000)
94	5-Hydroxy-8-methoxy-calamenene and 5- hydroxy-8-methoxy-calamenene-6-al	Specie recognition	–	*P. fulvum fulvum gray morph*	Kelman et al. (2000)
95	Crude extract	Feeding deterrence, antibacteria	–	*P. fulvum fulvum gray morph*	Kelman et al. (2000)
96	Fulfulvene	Antibacteria, chemical defense, allelopathy	–	*P. fulvum fulvum Yellow morph*	Kelman et al. (2000)
97	N-[(2*S*,3*S*,4*R*)-1,3,4-trihydroxyhexacosan-2-yl] icosanamide	Antivirus	Influenza H5N1, 100% inhibition at 1 μg/mL	*S. candidula*	Ahmed et al. (2013)
98	24-Methylenecholest-5-ene-1α,3β,11α-triol 1-acetate	Selective cytotoxic	HepG2	*L. crassum*	Aboutabl et al. (2017)
99	24-Methylenecholest-5-ene-3β-ol	Strong cytotoxic	HepG2, Hep-2 and HCT-116	*L. crassum*	Aboutabl et al. (2017)
100	24-Methylenecholestane-3β,5α,6β-triol	Selective cytotoxic	HepG2 IC_{50} 3.00 μM	*L. crassum*	Aboutabl et al. (2017)

4.3 Ecological Functions of the Red Sea *Alcyoniidae* Secondary Metabolites

Soft corals, deprived from a protective exoskeleton, constitute an important class of marine invertebrates which have sophisticated biochemical as well as physiological mechanisms, enabling them to produce elaborate bioactive compounds endowed with structural diversity either on their surface of secreted to their surrounding for purposes such as survival against various stresses, protection against predation, competition, or chemical communication in symbiotic relationships (Farag et al. 2017a). Soft corals belonging to the genus *Sarcophyton* are among the largest biodiversity contributors to many tropical coral habitats including the Red Sea and the Indo-Pacific region. Being a geologically "young" sea that is the Red Sea located in the warmest zone on Earth, unraveling the ecological importance or its benthic organism will be an asset to the global understanding of such marine environment. However, the status of the reef ecology research in the zone lag behind compared to the progress accomplished in other coastal regions, viz. GBR, Caribbean and the China Sea. Berumen et al. (2013) reported that almost half of the research related to the ecological aspect of the Red Sea is done in the Gulf of Eilatre presents only 2% of the Red Sea surface. Herein, we review the reported ecological function, viz. feeding deference, antibacterial, antifouling in Red Sea Alcyoniidae as well as their interaction among their benthic environment.

4.3.1 *Predator Defense*

Being sessile organisms located at the Seabed, soft corals cannot readily escape in time when attacked by their predators. Their chemical defense strategy relies therefore on secreting metabolites that display ichthyotoxic and antifeeding properties (Changyun et al. 2008). In (1974), Kashman and coworkers reported from *Sarcophyton glaucum* the isolation of furanocembranoid diterpenes **10** with remarkable yields of up to 3% dry weight, with (**10**) being suggested as a toxin that constitutes the major chemical defenses against coral natural predators (El Sayed et al. 1998; Hegazy et al. 2012; Kashman et al. 1974). To counteract cembrane diterpene, i.e., sarcophine toxicity inside corals themselves, cyclopropane-containing sterols found in corals (Farag et al. 2016) are suggested to be linked with coral adaptation to the membranolytic activities of their own toxins, that is, cembranoids. Such phenomenon, which involves the interdependent presence of two different types of secondary metabolites in an organism, that is, a "biochemical coordination" of the type "membranolytic toxins-unusual sterols," is evidenced for several marine sponges (Santalova et al. 2004).

With regard to antifeedant chemicals, the palatability of the Red Sea soft coral *Parerythropodium fulvum fulvum* on two generalist reef fish species *Thallasonia klunzingen* and *T. Lunare* was tested by Kelman et al. (1999). A comparative feeding assay of the coral organic extract and its sclerite were examined, with

Table 4.2 Red Sea Alcyoniidae metabolites for which bioactivity has yet to be reported, for structures refer to Fig. 4.6

N°	Identification	Coral source	Reference
101	(*Z*)-cembrenene C	*S. trocheliophorum*	Shaaban et al. (2015)
102	Flaccidoxide	*A. flaccidum*	Kashman et al. (1981)
103–105	Alcyonol A, B, C	*A. utinomi*	Hegazy et al. (2015), Kinamoni et al. (1983)
106	Cladiellisin	*C. pachyclados*	Hassan et al. (2010)
107	(3*E*,7*E*,11*E*)-18-Hydroxy-3,7,11,15(17) cembratetraen-16-14-olide	*L. crassum*	Kinamoni et al. (1983)
108	(7*E*,11*E*)-13,18-Dihydroxy-3, 4-epoxy-7,11,15(17)-cembratrien-16,14-olide	*L. crassum*	Kinamoni et al. (1983)
109	3-Deoxy-20-acetylpresinularolide B	*L. crassum*	Hegazy et al. (2015)
110	3-Deoxypresinularolide B	*L. crassum*	Hegazy et al. (2015)
111	Labolide	*L. crassum*	Hegazy et al. (2015)
112	Sinularolide C	*L. crassum*	Hegazy et al. (2015)
113	Sinularolide C diacetate	*L. crassum*	Hegazy et al. (2015)
114	(22*R*,24*E*)-24-Methylcholest-5-en-3β,22,25,28-tetraol	*L. depressum*	Hegazy et al. (2015)
115	(22*R*,24*E*,28*E*)-5β,6β-Epoxy-22,28-oxido-24 methyl-5α-cholestan-3β,25,28-triol	*L. depressum*	Hegazy et al. (2015)
116	5β,6β - Epoxy - 24b – methylchloestan -3β, 22(R),25-triol	*L. depressum*	Hegazy et al. (2015)
117	Depresosterol	*L. depressum*	Hegazy et al. (2015)
118	Lobophytosterol	*L. depressum*	Hegazy et al. (2015)
119	(+)-Alismol	*L. Lobophyton*	Hegazy et al. (2016b)
120	Gorgostan-5,25-dien-3b-ol	*L. Lobophyton*	Hegazy et al. (2016b)
121	Gorgosterol	*L. Lobophyton*	Hegazy et al. (2016b)
122–123	Pauciflorol A, B	*L. pauciflorum*	Hegazy et al. (2015), Kinamoni et al. (1983)
124	Tumbergol	*L. pauciflorum*	Hegazy et al. (2015)
125	(7*E*,11*E*)-18-Acetoxy-3, 4-epoxy-13-hydroxy-7,11,15 (17)-cembratrien-16,14-olide	*L. crassum*	Kashman et al. (1981)
126	(7*E*,11*E*)-18-Acetoxy-3, 4-epoxy-13-epihydroxy-7,11,15-(17)-cembratrien-16,14-olide	*L. crassum*	Kashman et al. (1981)
127	Lobolide	*L. crassum*	Kashman et al. (1981)
128–133	Lobophylins A, B, C, F, G, H	*L. crassum*	Mohammed et al. (2017)
134	1,4-Peroxymuurol-5-ene	*S. ehrenbergi*	Shaker et al. (2010)

(continued)

Table 4.2 (continued)

N°	Identification	Coral source	Reference
135–136	Sinulolide A, B	*S. ehrenbergi*	Hegazy et al. (2017)
137	2*R*,7*R*,8*R*-Dihydroxydeep oxysarcophine	*S. glaucum*	Hegazy et al. (2011a, b)
138	7α,8β-Dihydroxydeep oxysarcophine	*S. glaucum*	Hegazy et al. (2011a, b)
139	Sarcophytol B	*S. glaucum, A. flaccidum*	El Sayed et al. (1998), Kashman et al. (1981)
140	Dioxosarcoguaiacol	*S. glaucum*	Sawant et al. (2007)
141	(+)-Sarcophytoxide	*S. ehrenbergi, S. trocheliophorum*	Shaaban et al. (2015), Shaker et al. (2010)
142	16-Oxosarcophytonin E	*S. trocheliophorum*	Hegazy et al. (2013)
143	Bisabolene	*S. trocheliophorum*	Shaaban et al. (2015)
144	Alloaromadendrene	*S. trocheliophorum*	Shaaban et al. (2015)
145	Caryophyllene	*S. trocheliophorum*	Shaaban et al. (2015)
146	ß-Elemene	*S. trocheliophorum*	Shaaban et al. (2015)
147–148	Trochelioid A, B	*S. trocheliophorum*	Hegazy et al. (2013)
149	Cholesterol	*S. candidula*	Abou El Ezz et al. (2015)
150	24-methylene cholesterol	*S. candidula, S. terspilli*	Abou El Ezz et al. (2015), Mohammed et al. (2017)
151	Chimyl alcohol	*S. candidula*	Abou El Ezz et al. (2015)
152	5-Epi-sinuleptolide	*S. gardineri*	El Sayed and Hamann (1996)
153	Sinuleptolide	*S. gardineri*	El Sayed and Hamann (1996)
154	Ineleganolide	*S. polydactyla*	Hegazy et al. (2016a)
155	Scabrolide F	*S. polydactyla*	Hegazy et al. (2016a)
156–158	Sinularcasbane M, N, O	*S. polydactyla*	Hegazy et al. (2016a)
159	24-Methylenecholesterol	*S. terspilli*	Mohammed et al. (2017)
160	24α-Methyl cholesterol	*S. terspilli*	Mohammed et al. (2017)
161	Gorgosten-5(*E*)-3β-ol	*S. terspilli*	Mohammed et al. (2017)
162	24-Methylenecholest-5-ene-1α,3β,11α-triol	*L. crassum*	Aboutabl et al. (2017)
163	24-Methylenecholestane-1α,3β,5α,6β,11α-pentol	*L. crassum*	Aboutabl et al. (2017)

feeding deterrence found to be more associated with the organic extract. Further experimentation on the embryo expressed a higher response of deterrence mainly attributed to the mucus covering it that is likely to encompass more chemically active substance with antipredatory properties, with chemical structure or active substances yet to be elucidated. Compound [**96**] isolated from the yellow morph *Parerythropodium fulvum fulvum* was found active protectant of the coral species against two fish species (Kelman et al. 1999).

4.3.2 Interspecific Competition for Space

The competition for living space is frequent among benthic marine organisms in which a taxon can outcompete another one through the secretion of specific allelochemicals to a non-recognized species (Changyun et al. 2008). Allelochemicals are also found in terrestrial plants and to function in weed management system (Asaduzzaman et al. 2015). Nevertheless, much less is known regarding allelochemicals functioning in corals. This interspecific interaction is suggested to provide a margin and enough space to soft coral to grow with enough food (Sammarco et al. 1983).

A self/non self-mechanism was demonstrated by Frank et al. (1996) on the soft coral *P. fulvum fulvum*. This allogeneic recognition study, the first in Alcyoniidae family, consists of a tissue to tissue contact between 13 large colonies. While the isogenic repaired and fused themselves perfectly, the allogeneic species had two different responses. Allogeneic encounters were experimentally arranged in Eilat, Red Sea for the first time in the Alcyonacea (Frank et al. 1996). Two allopathic responses were observed in which the first exhibited a retreat growth ending with a separated two growing organisms. In contrast, the other reaction consisted of a unilateral or reciprocal tissue overgrowth.

4.3.3 Antifouling Activity

Antifouling property is the ability to hinder the growth or settlement of a biofilm on a given surface. The current method, of interest in the Naval industry to coat ships, requires the use of toxic chemicals mixed with the external paints which is not only expensive but ecologically unfriendly (Soliman et al. 2017). Following the observation that no organisms attach to soft corals bodies due to their chemical secretion (Changyun et al. 2008), they have been regarded as a potential source of environmentally friendly antifouling chemicals.

A MeOH:DCM extract from five soft corals species *Sinularia variabilis, S. polydactyla, S. heterospiculata, L arboretum, and S trocheliophorum* were studied as a paint formulation for their antifouling properties (Mohamed Ali and Soliman 2010) against the main fouling organism barnacle *Balanus amphitrite* and the tube worm *Hydroides elegans*. The responses were taken at 7, 17, 31, and 62 days post-exposure where the mass of the remaining fouling organism was measured. While the *S. heterospiculata* and *S. variabilis* exhibited the highest response to both fouling organisms, *S. polydactyla* was responsive to the barnacle only. Further analysis to identify the chemical nature of the antifouling agent should now follow in that crude extract (Mohamed Ali and Soliman 2010). A similar investigation was performed assessing seven Red Sea soft coral extracts by Soliman et al. (2017) where an Alcyoniidae *S. compressa* and a xenidae *H. fuscescens* were the most responsive to the marine biofouling barnacle and tubeworms.

4.3.4 Alcyoniidae Interaction and Biodiversity in Its Benthic Environment

Soft coral communication within its surrounding environment is of chemical nature, and besides their protective function, those chemicals have another purposes for coral life. In (2000), a study by Kelman et al. of the major secondary metabolite of *P. fulvum fulvum* led to the discovery of their intraspecific variation toward different geographical location. Fulfulvene **96** and calamenene (**93–94**) were, respectively, dominant of the yellow and gray morph of the soft coral found in the Red Sea, at Eilat. Following an ESI-MS analysis of the each colored species, a qualitative difference on their metabolic profile was noticed. Compound [**104**] occurred at 10.1 ± 4.1 and 2.9 ± 3.8% of crude organic extract in the shallow and deep colonies, respectively. This result corroborates with the compositional differences observed among *Sarcophyton* sp collected from different sites along the Egyptian Red Sea coast (Farag et al. 2016). The specific metabolites that contributed to discriminate between soft corals of *S. ehrenbergi* from the three different growing habitats belonged to cembrane-type diterpenes. Furthermore, compared to wild corals, aquarium grown species were found being less enriched in cembranoids and more enriched in oxylipids (Farag et al. 2016). Cembranoids therefore play an ecological role in coral life as discussed above and are more likely to be produced at higher levels where corals are under more stressful marine conditions compared to corals grown in an aquarium tank. This study is also the first to derive *Sarcophyton* species relatedness based on metabolite data not only from Red Sea area, as molecular phylogenetic analyses alone so far have been insufficient to clearly identify *Sarcophyton* species. The lack of understanding in both intraspecific variations of diagnostic morphological characters within that genus in addition to a lack of solid taxonomic and ecological work on *Sarcophyton* poses problems to derive a clear phylogenetic-based analysis.

Biopharmaceutical production of soft corals marine products or chemical, in order to give an efficient product, requires an in vitro culture followed by manipulation of the cultural parameters such as pH, temperature, nutrition, and chemical elicitors to enhance the production of the targeted molecule (Farag et al. 2017b). This method is already applied in the biomedical field and the extraction of natural compounds from endophyte, microorganisms or plants, whereas a promising research has been applied by Farag et al. (2016). The comparison between elicitations of several Red Sea corals produced varied responses where the oxylipins, viz. methyl jasmonate (MeJa) and its animal analogue prostaglandin (PG) showed higher product levels than other tested chemical elicitors. As a continuity of the report by Farag et al. (2017a), more analysis was done on two soft corals from the Red Sea. The effect of oxylipins on soft corals metabolism resulted in an upregulation of campestene-triol and a cembranoid in *Sarcophyton glaucum* compared to the use of geranylgeranyl phosphate (GGP) and arachidonic acid (AA) or wounding.

4.4 Conclusion and Future Perspectives

This review reports on the potential ecological and biological role of many coral metabolites influencing stress responses, virulence, and their further effects in humans as drug leads for treatment of various ailments. Improvement of coral-producing organisms and active constituents yield in natural extracts is ongoing challenge facing the nutraceutical industry. The production of metabolites in marine animals is also often low and depends greatly on ecological conditions where it survives. The response of sessile marine animals to stress conditions compared to plants is indeed in contrast much less explored. The biotechnological production of valuable secondary metabolites in coral using in vitro cultures is an attractive alternative, which has yet to be examined at a commercial scale. In addition, a detailed dissection of coral secondary metabolome is required to understand how coral metabolic responses are elicited. This issue is complicated considering that the exact origin of coral bioactive chemicals is not identified especially with coral acting as a holobiont encompassing several living organism algae, fungi, and bacteria, a challenge to decipher each organism role in producing these assortments of chemicals. Despite increasing reports, many aspects of corals metabolic pathways, regulation and perception, are still poorly characterized. The combined analysis of metabolic and gene expression profiles will likely be an increasingly powerful approach to identify candidate genes involved in the production and regulation of biologically active coral compounds. Classifying the Red Sea corals based on the characteristic metabolite signature of each species will bridge the gap between the complex relationship between the coral host and their symbiont as well as the chemical signature of the Alcyoniidae family. Moreover, systematic use of ^{13}C-mass isotopomers and ^{13}C-labeling associated with MS analysis should lead to progress in characterization of biochemical pathways involved in coral secondary metabolites production. This could also be facilitated by simultaneous monitoring, directly in crude extracts without the need for fractionation, of a large number of metabolites by spectroscopic techniques. An extended investigation on the chemical ecology of the Red Sea coral should also gain more attention as its unique environment and to impact its genetic makeup and adaptation differently from that of the other coastal regions. Finally, the Alcyoniidae family encompasses a wide array octocoral organism identified along the Red Sea that have yet to be reported for in terms of its chemical composition compared to results from other generic species. Therefore, identification and profiling of less investigated species in the Red Sea will provide a rich database of marine metabolites in that region.

The biological and chemical research of the coral reefs has made a remarkable progress as reviewed herein yet the support information of the biodiversity, functions profile and ecological landscapes still to be acquired. Flora, fauna, and fisheries are facing future threats by human activities such as heavy harvesting and environmental impact including global warming, habitats destruction and pollution. The decrease in water quality, salinity, and substrate changes are some of the factors that influenced negatively the oceans and could be easily translated to the reef corals, and thus, loss would be able to predict accompanied with shifting of the biodiversity and biological activity. Protecting the marine and particularly the corals natural properties is a multi-dimensional task where molecular biology and genomics could play a central role to report the configuration and accommodate the unique features. The first step is to identify the underlying biological and ecological processors/indices and create an evidence-based foundation for the coral population.

We anticipate that the future management and policy tools should be directed to restore the coral ecosystem. Future studies should attribute to the consistent complexity of the coral different habitats, and this is something we attempt to address in our ongoing projects in order to identify habitat indices and quantitative measures for the habitats structure. Future classification, mapping, and management could be planned to incorporate the importance of the coral community but also to assure the conservation and stability of the ecosystem. Identifying and monitoring the coral stock status at both national and international scales would aid the global goal to preserve the marine ecosystem and the parallel human activities in particular fisheries. The food web structure including fisheries is known as the ocean-based stocks where dietary nutrients, food security, and drug developments are all common features. Thus, the complementary expansion into the coral habitual complexity, sustainability in parallel to fisheries conservation would be highly recommended.

Acknowledgements M. A. Farag acknowledges the funding received from the Alexander von Humboldt foundation, Germany. H. R. El-Seedi is very grateful to the Swedish Research links grant 2016-05885 (VR for the years 2017–2019) for generous financial support and to Jiangsu University, China for Adjunct Professor fellowship.

References

Abdel-Lateff A, Alarif WM, Ayyad S-EN, Al-Lihaibi SS, Basaif SA (2015) New cytotoxic isoprenoid derivatives from the Red Sea soft coral *Sarcophyton glaucum*. Nat Prod Res 29 (1):24–30. https://doi.org/10.1080/14786419.2014.952637

Abou El Ezz R, Afifi M, Kamal A, Sannad E, Radwan M, Hassanin H, Ahmed S (2015) Isolation and structural elucidation of secondary metabolites obtained from the soft coral *Sinularia candidula* present in the Egyptian Red Sea. Planta Med 81(16):PM_38. https://doi.org/10.1055/s-0035-1565415

Abou El-Ezz RF, Ahmed SA, Radwan MM, Ayoub NA, Afifi MS, Ross SA, Szymanski PT, Fahmy H, Khalifa SI (2013) Bioactive cembranoids from the Red Sea soft coral *Sarcophyton glaucum*. Tetrahedron Lett 54(8):989–992. https://doi.org/10.1016/j.tetlet.2012.12.037

Aboutabl E-SA, Azzam SM, Michel CG, Selim NM, Hegazy MF, Ali AHA, Hussein AA (2013) Bioactive terpenoids from the Red Sea soft coral *Sinularia polydactyla*. Nat Prod Res 27 (23):2224–2226. https://doi.org/10.1080/14786419.2013.805333

Aboutabl EA, Selim NM, Azzam SM, Michel CG, Hegazy MF, Ali AM, Hussein AA (2017) Polyhydroxy sterols isolated from the Red Sea soft coral *Lobophytum crassum* and their cytotoxic activity. Nat Prod Commun 12(2):233–235

Ahmed S, Ibrahim A, Arafa AS (2013) Anti-H5N1 virus metabolites from the Red Sea soft coral, *Sinularia candidula*. Tetrahedron Lett 54(19):2377–2381. https://doi.org/10.1016/j.tetlet.2013.02.088

Al-Footy KO, Alarif WM, Asiri F, Aly MM, Ayyad S-EN (2015) Rare pyrane-based cembranoids from the Red Sea soft coral *Sarcophyton trocheliophorum* as potential antimicrobial-antitumor agents. Med Chem Res 24(2):505–512. https://doi.org/10.1007/s00044-014-1147-1

Al-Footy KO, Alarif WM, Zubair MS, Ghandourah MA, Aly MM (2016) Antibacterial and cytotoxic properties of isoprenoids from the Red Sea soft coral, *Lobophytum* sp. Trop J Pharm Res 15(7):1431–1438. https://doi.org/10.4314/tjpr.v15i7.11

Al-Lihaibi SS, Alarif WM, Abdel-Lateff A, Ayyad S-EN, Abdel-Naim AB, El-Senduny FF, Badria FA (2014) Three new cembranoid-type diterpenes from Red Sea soft coral *Sarcophyton glaucum*: isolation and antiproliferative activity against HepG2 cells. Eur J Med Chem 81:314–322. https://doi.org/10.1016/j.ejmech.2014.05.016

Asaduzzaman M, Pratley JE, An M, Luckett DJ, Lemerle D (2015) Metabolomics differentiation of canola genotypes: toward an understanding of canola allelochemicals. Front Plant Sci 5:765. https://doi.org/10.3389/fpls.2014.00765

Badria FA, Guirguis AN, El-Naggar WA (1997) Antibacterial and antifungal agents from Egyptian marine organisms. Pharma Biol 35(4):284–287. https://doi.org/10.1076/phbi.35.4.284.13307

Berumen ML, Hoey AS, Bass WH, Bouwmeester J, Catania D, Cochran JEM, Khalil MT, Miyake S, Mughal MR, Spaet JL, Saenz-Agudelo P (2013) The status of coral reef ecology research in the Red Sea. Coral Reefs 32(3):737–748. https://doi.org/10.1007/s00338-013-1055-8

Changyun W, Haiyan L, Changlun S, Yanan W, Liang L, Huashi G (2008) Chemical defensive substances of soft corals and gorgonians. Acta Ecol Sin 28(5):2320–2328. https://doi.org/10.1016/S1872-2032(08)60048-7

Cheng S-Y, Wen Z-H, Chiou S-F, Tsai C-W, Wang S-K, Hsu C-H, Dai C-F, Chiang MY, Wang W-H, Duh C-Y (2009) Ceramide and cerebrosides from the octocoral *Sarcophyton ehrenbergi*. J Nat Prod 72(3):465–468. https://doi.org/10.1021/np800362g

Cheng S-Y, Wang S-K, Duh C-Y (2014) Secocrassumol, a seco-cembranoid from the Dongsha Atoll soft coral *Lobophytum crassum*. Mar Drugs 12(12):6028–6037. https://doi.org/10.3390/md12126028

Coll JC, Leone PA, Bowden BF, Carroll AR, König GM, Heaton A, De Nys R, Maida M, Alino PM, Willis RH, Babcock RC, Alderslade PN (1995) Chemical aspects of mass spawning in corals. II. (-)-Epi-thunbergol, the sperm attractant in the eggs of the soft coral *Lobophytum crassum* (Cnidaria: Octocorallia). Mar Biol 123(1):137–143. https://doi.org/10.1007/BF00350332

DiBattista JD, Choat JH, Gaither MR, Hobbs J-PA, Lozano-Cortés DF, Myers RF, Paulay G, Rocha LA, Toonen RJ, Westneat MW, Berumen ML (2015) On the origin of endemic species in the Red Sea. J Biogeogr 43(1):13–30. https://doi.org/10.1111/jbi.12631

El Sayed KA, Hamann MT (1996) A new norcembranoid dimer from the Red Sea soft coral *Sinularia gardineri*. J Nat Prod 59(7):687–689. https://doi.org/10.1021/np960207z

El Sayed KA, Hamann MT, Waddling CA, Jensen C, Lee SK, Dunstan CA, Pezzuto JM (1998) Structurally novel bioconversion products of the marine natural product sarcophine effectively inhibit JB6 cell transformation. J Org Chem 63(21):7449–7455. https://doi.org/10.1021/jo9813134

El-Seedi HR, Gomaa M, Salem MM, Benchoula K, Keshk HM, Yosri N, Ayesh A, Asker AM, Soliman K, Hamza Z, Mansour HM (2016) Cytotoxic effects of the red sea soft coral sarcophyton trocheliophorum. Acta Pol Pharm 73(6):1587–1592

Elkhateeb A, El-Beih AA, Gamal-Eldeen AM, Alhammady MA, Ohta S, Pare PW, Hegazy M-EF (2014) New terpenes from the Egyptian soft coral *Sarcophyton ehrenbergi*. Mar Drugs 12 (4):1977–1986. https://doi.org/10.3390/md12041977

Ellithey MS, Lall N, Hussein AA, Meyer D (2014) Cytotoxic and HIV-1 enzyme inhibitory activities of Red Sea marine organisms. BMC Complement Altern Med 14:77. https://doi.org/10.1186/1472-6882-14-77

Eltahawy NA, Ibrahim AK, Radwan MM, ElSohly MA, Hassanean HA, Ahmed SA (2014) Cytotoxic cembranoids from the Red Sea soft coral, *Sarcophyton auritum*. Tetrahedron Lett 55 (29):3984–3988. https://doi.org/10.1016/j.tetlet.2014.05.013

Eltahawy NA, Ibrahim AK, Radwan MM, Zaitone SA, Gomaa M, ElSohly MA, Hassanean HA, Ahmed SA (2015) Mechanism of action of antiepileptic ceramide from Red Sea soft coral *Sarcophyton auritum*. Bioorg Med Chem Lett 25(24):5819–5824. https://doi.org/10.1016/j.bmcl.2015.08.039

Farag MA, Porzel A, Al-Hanimady MA, Hegazy M-EF, Meyer A, Mohamed TA, Westphal H, Wessjohann LA (2016) Soft corals biodiversity in the Egyptian Red Sea: a comparative MS and NMR metabolomics approach of wild and aquarium grown species. J Proteome Res 15 (4):1274–1287. https://doi.org/10.1021/acs.jproteome.6b00002

Farag MA, Al-Mahdy DA, Meyer A, Westphal H, Wessjohann LA (2017a) Metabolomics reveals biotic and abiotic elicitor effects on the soft coral *Sarcophyton ehrenbergi* terpenoid content. Sci Rep 7(1):648. https://doi.org/10.1038/s41598-017-00527-8

Farag MA, Fekry MI, Al-Hammady MA, Khalil MN, El-Seedi HR, Meyer A, Porzel A, Westphal H, Wessjohann LA (2017b) Cytotoxic effects of *Sarcophyton* sp. soft corals—is there a correlation to their NMR fingerprints? Mar Drugs 15(7):211. https://doi.org/10.3390/md15070211

Farag MA, Meyer A, Ali SE, Salem MA, Giavalisco P, Westphal H, Wessjohann LA (2018) Comparative metabolomics approach detects stress-specific responses during coral bleaching in soft corals. J Proteome Res 17(6):2060–2071. https://doi.org/10.1021/acs.jproteome.7b00929

Frank U, Bak RPM, Rinkevich B (1996) Allorecognition responses in the soft coral *Parerythropodium fulvum fulvum* from the Red Sea. J Exp Mar Biol Ecol 197(2):191–201. https://doi.org/10.1016/0022-0981(95)00153-0

Fridkovsky E, Rudi A, Benayahu Y, Kashman Y, Schleyer M (1996) Sarcoglane, a new cytotoxic diterpene from *Sarcophyton glaucum*. Tetrahedron Lett 37(38):6909–6910. https://doi.org/10.1016/0040-4039(96)01513-4

Georgopapadakou N (1995) The fungal cell wall as a drug target. Trends Microbiol 3(3):98–104. https://doi.org/10.1016/S0966-842X(00)88890-3

Gomaa MN, Soliman K, Ayesh A, Abd El-Wahed A, Hamza Z, Mansour HM, Khalifa SA, Mohd Ali HB, El-Seedi HR (2016) Antibacterial effect of the Red Sea soft coral Sarcophyton trocheliophorum. Nat Prod Res 30(6):729–734. https://doi.org/10.1080/14786419.2015.1040991

Hassan HM, Khanfar MA, Elnagar AY, Mohammed R, Shaala LA, Youssef DTA, Hifnawy MS, El Sayed KA (2010) Pachycladins A–E, prostate cancer invasion and migration inhibitory eunicellin-based diterpenoids from the Red Sea soft coral *Cladiella pachyclados*. J Nat Prod 73 (5):848–853. https://doi.org/10.1021/np900787p

Hegazy MEF, Su J-H, Sung P-J, Sheu J-H (2011a) Cembranoids with 3,14-ether linkage and a secocembrane with bistetrahydrofuran from the Dongsha Atoll soft coral *Lobophytum* sp. Mar Drugs 9(7):1243–1253. https://doi.org/10.3390/md9071243

Hegazy M-EF, El-Beih AA, Moustafa AY, Hamdy AA, Alhammady MA, Selim RM, Abdel-Rehim M, Pare PW (2011b) Cytotoxic cembranoids from the Red Sea soft coral *Sarcophyton glaucum*. Nat Prod Commun 6(12):1809–1812

Hegazy M-EF, Eldeen AMG, Shahat AA, Abdel-Latif FF, Mohamed TA, Whittlesey BR, Pare PW (2012) Bioactive hydroperoxyl cembranoids from the Red Sea soft coral *Sarcophyton glaucum*. Mar Drugs 10(1):209–222. https://doi.org/10.3390/md10010209

Hegazy M-EF, Mohamed TA, Abdel-Latif FF, Alsaid MS, Shahat AA, Pare PW (2013) Trochelioid A and B, new cembranoid diterpenes from the Red Sea soft coral *Sarcophyton trocheliophorum*. Phytochem Lett 6(3):383–386. https://doi.org/10.1016/j.phytol.2013.05.005

Hegazy MEF, Mohamed TA, Alhammady MA, Shaheen AM, Reda EH, Elshamy AI, Aziz M, Paré PW (2015) Molecular architecture and biomedical leads of terpenes from Red Sea marine invertebrates. Mar Drugs 13(5):3154–3181. https://doi.org/10.3390/md13053154

Hegazy M-E, Mohamed T, Elshamy A, Al-Hammady M, Ohta S, Paré P (2016a) Casbane diterpenes from Red Sea coral *Sinularia polydactyla*. Molecules 21(3):308. https://doi.org/10.3390/molecules21030308

Hegazy M-EF, Mohamed TA, Elshamy AI, Hassanien AA, Abdel-Azim NS, Shreadah MA, Abdelgawad II, Elkady EM, Pare PW (2016b) A new steroid from the Red Sea soft coral *Lobophytum lobophytum*. Nat Prod Res 30(3):340–344. https://doi.org/10.1080/14786419.2015.1046871

Hegazy M-EF, Elshamy AI, Mohamed TA, Hamed AR, Ibrahim MAA, Ohta S, Pare PW (2017) Cembrene diterpenoids with ether linkages from *Sarcophyton ehrenbergi*: an anti-proliferation and molecular-docking assessment. Mar Drugs 15(6):192. https://doi.org/10.3390/md15060192

Kashman Y, Zadock E, Néeman L (1974) Some new cembrane derivatives of marine origin. Tetrahedron 30(19):3615–3620. https://doi.org/10.1016/S0040-4020(01)97044-9

Kashman Y, Carmely S, Groweiss A (1981) Further cembranoid derivatives from the Red Sea soft corals *Alcyonium flaccidum* and *Lobophytum crassum*. J Org Chem 46(18):3592–3596. https://doi.org/10.1021/jo00331a003

Kelman D, Kushmaro A, Loya Y, Kashman Y, Benayahu Y (1998) Antimicrobial activity of a Red Sea soft coral, *Parerythropodium fulvum fulvum*: reproductive and developmental considerations. Mar Ecol Prog Ser 169:87–95. https://doi.org/10.3354/meps169087

Kelman D, Benayahu Y, Kashman Y (1999) Chemical defence of the soft coral *Parerythropodium fulvum fulvum* (Forskal) in the Red Sea against generalist reef fish. J Exp Mar Biol Ecol 238 (1):127–137. https://doi.org/10.1016/S0022-0981(99)00016-7

Kelman D, Benayahu Y, Kashman Y (2000) Variation in secondary metabolite concentrations in yellow and grey morphs of the Red Sea soft coral *Parerythropodium fulvum fulvum*: possible ecological implications. J Chem Ecol 26(5):1123–1133. https://doi.org/10.1023/A:1005423708904

Kelman D, Kashman Y, Rosenberg E, Kushmaro A, Loya Y (2006) Antimicrobial activity of Red Sea corals. Mar Biol 149(2):357–363. https://doi.org/10.1007/s00227-005-0218-8

Kinamoni Z, Groweiss A, Carmely S, Kashman Y, Loya Y (1983) Several new cembranoid diterpenes from three soft corals of the Red Sea. Tetrahedron 39(9):1643–1648. https://doi.org/10.1016/S0040-4020(01)88575-6

Laylin T (2011) The Red Sea is getting hotter faster than the rest. Retrieved 20 July 2018, from https://greenprophet.com/2011/09/red-sea-is-hotter/

Liang L-F, Guo Y-W (2013) Terpenes from the soft corals of the genus *Sarcophyton*: chemistry and biological activities. Chem Biodivers 10(12):2161–2196. https://doi.org/10.1002/cbdv.201200122

Mariottini GL, Grice ID (2016) Antimicrobials from Cnidarians. A new perspective for anti-infective therapy? Mar Drugs 14(3):48

Mohamed Ali A-HA-R, Soliman YA-A (2010) Antifouling activity of crude extracts from some Red Sea soft corals. Thalassia Salentina 32:73–89. https://doi.org/10.1285/i15910725v32p73

Mohammed TA-AA (2012) Coral reefs quantitatively assessment along the Egyptian Red Sea coast. Thalassas 28(2):17–26

Mohammed R, Radwan MM, Ma G, Mohamed TA, Seliem MA, Thabet M, ElSohly MA (2017) Bioactive sterols and sesquiterpenes from the Red Sea soft coral *Sinularia terspilli*. Med Chem Res 26(8):1647–1652. https://doi.org/10.1007/s00044-017-1876-z

Mustafa GA, Abd-Elgawad A, Ouf A, Siam R (2016) The Egyptian Red Sea coastal microbiome: a study revealing differential microbial responses to diverse anthropogenic pollutants. Environ Pollut 214:892–902. https://doi.org/10.1016/j.envpol.2016.04.009

Ne'eman I, Fishelson L, Kashman Y (1974) Sarcophine—a new toxin from the soft coral *Sarcophyton glaucum* (Alcyonaria). Toxicon 12(6):593-IN6. https://doi.org/10.1016/0041-0101(74)90192-5

Peng B-R, Lu M-C, El-Shazly M, Wu S-L, Lai K-H, Su J-H (2018) Aquaculture soft coral *Lobophytum crassum* as a producer of anti-proliferative cembranoids. Mar Drugs 16(1):15. https://doi.org/10.3390/md16010015

Řezanka T, Dembitsky VM (2001) γ-Lactones from the soft corals *Sarcophyton trocheliophorum* and *Lithophyton arboreum*. Tetrahedron 57(41):8743–8749. https://doi.org/10.1016/S0040-4020(01)00853-5

Riegl B, Bruckner A, Coles SL, Renaud P, Dodge RE (2009) Coral reefs threats and conservation in an era of global change. In: Ostfeld RS, Schlesinger WH (eds) Year in ecology and conservation biology 2009, vol 1162. Wiley-Blackwell, Malden, pp 136–186

Rocha LG, Almeida JRGS, Macêdo RO, Barbosa-Filho JM (2005) A review of natural products with antileishmanial activity. Phytomedicine 12(6):514–535. https://doi.org/10.1016/j.phymed.2003.10.006

Sammarco P, Coll J (1992) Chemical adaptations in the Octocorallia: evolutionary considerations. Mar Ecol Prog Ser 88:93–104. https://doi.org/10.3354/meps088093

Sammarco PW, Strychar KB (2013) Responses to high seawater temperatures in zooxanthellate octocorals. PLoS ONE 8(2):e54989. https://doi.org/10.1371/journal.pone.0054989

Sammarco PW, Coll JC, Barre SL, Willis B (1983) Competitive strategies of soft corals (Coelenterata: Octocorallia): Allelopathic effects on selected scleractinian corals. Coral Reefs 1 (3):173–178. https://doi.org/10.1007/BF00571194

Santalova EA, Makarieva TN, Gorshkova IA, Dmitrenok AS, Krasokhin VB, Stonik VA (2004) Sterols from six marine sponges. Biochem Syst Ecol 32(2):153–167. https://doi.org/10.1016/S0305-1978(03)00143-1

Sarmiento-Vizcaíno A, González V, Braña AF, Palacios JJ (2017) Pharmacological potential of phylogenetically diverse Actinobacteria isolated from deep-sea coral ecosystems of the submarine Avilés Canyon in the Cantabrian Sea. Microb Ecol 73(2):338–352. https://doi.org/10.1007/s00248-016-0845-2

Sawant S, Youssef D, Mayer A, Sylvester P, Wali V, Arant M, El Sayed K (2006) Anticancer and anti-inflammatory sulfur-containing semisynthetic derivatives of sarcophine. Chem Pharm Bull 54(8):1119–1123. https://doi.org/10.1248/cpb.54.1119

Sawant SS, Youssef DTA, Sylvester PW, Wali V, El Sayed KA (2007) Antiproliferative Sesquiterpenes from the Red Sea soft coral *Sarcophyton glaucum*. Nat Prod Commun 2 (2):117–119

Shaaban M, Ghani MA, Shaaban KA (2013a) Zahramycins A-B, two new steroids from the coral *Sarcophyton trocheliophorum*. Z Naturforsch B 68(8):939–945. https://doi.org/10.5560/znb.2013-3131

Shaaban M, Shaaban KA, Ghani MA (2013b) Hurgadacin: a new steroid from *Sinularia polydactyla*. Steroids 78(9):866–873. https://doi.org/10.1016/j.steroids.2013.05.006

Shaaban K, Ghani M, Shaaban M (2015) New cembranoid diterpenes from *Sarcophyton trocheliophorum*. Br J Pharm Res 5(3):192–201. https://doi.org/10.9734/BJPR/2015/14757

Shaaban M, Ghani MA, Shaaban KA (2016) Unusual pyranosyl cembranoid diterpene from Sarcophyton trocheliophorum. Zeitschrift für Naturforschung B 71(12):1211–1217

Shaker KH, Muller M, Ghani MA, Dahse H-M, Seifert K (2010) Terpenes from the soft corals *Litophyton arboreum* and *Sarcophyton ehrenbergi*. Chem Biodivers 7(8):2007–2015. https://doi.org/10.1002/cbdv.201000016

Soliman YAA, Brahim AM, Moustafa AH, Hamed MAF (2017) Antifouling evaluation of extracts from Red Sea soft corals against primary biofilm and biofouling. Asian Pac J Trop Biomed 7 (11):991–997. https://doi.org/10.1016/j.apjtb.2017.09.016

Tchernov D, Gorbunov MY, de Vargas C, Narayan Yadav S, Milligan AJ, Häggblom M, Falkowski PG (2004) Membrane lipids of symbiotic algae are diagnostic of sensitivity to thermal bleaching in corals. Proc Natl Acad Sci USA 101(37):13531–13535. https://doi.org/10.1073/pnas.0402907101

Wooldridge SA (2010) Is the coral-algae symbiosis really "mutually beneficial" for the partners? BioEssays 32(7):615–625. https://doi.org/10.1002/bies.200900182

Zubair MS, Al-Footy KO, Ayyad S-EN, Al-Lihaibi SS, Alarif WM (2016) A review of steroids from *Sarcophyton* species. Nat Prod Res 30(8):869–879. https://doi.org/10.1080/14786419.2015.1079187

Chapter 5
Woody Perennial Diversity at Various Land forms of the Five Agro-Climatic Zones of Rajasthan, India

Manish Mathur and S. Sundaramoorthy

Abstract Across the globe, the plant biodiversity have been gauged with many gradients like altitude/slope, grazing, nutrients and land uses. However, relationships between land forms and phyto-diversity have rarely been addressed. We compared the diversity of woody perennials at six different land forms of five agro-climatic zones of arid and semi-arid areas of Rajasthan, India. Characteristics of land forms (Hills and Piedmonts, Older Alluvial Plains [OAP], Sandy Undulating Hummocky Plains [SUHP], Younger Alluvial Plains [YAP], Sand Dune and Sandy Plains) along with generalized plant community succession trends were described. Five agro-climatic zones included arid Western, irrigated North Western plains, hyper arid partial irrigated, internal drainage dry zone and transitional plain of Luni basin. Thus, phyto-diversity was studied with land form specific as well as zonal specific approaches. Diversity patters were analyzed with help of species richness, diversity index (Shannon-Weaver index, H) and with evenness (E). SHE analysis was performed to quantify the patterns of these diversity patterns for a specific land form among the agro-climatic zones. Further, plant community behaviors under zonal and land form specific approaches were visualized through Principal Component Analysis (PCA). Our study identifies threshold limits (upper and lower) for different diversity parameters among agro climatic zones and at different land forms. Among the agro-climate zone, log normal (increase in dominance (S) and H but a decline in E) patterns were identified at SUHP, sandy plain, YAP and sand dune. While broken stick models (both S and H are expected to increase and E to stay constant) were identified at Hills and Piedmonts and at OAP. With this study, we identified zonal specific indicator species at different land forms. We linked the magnitude of land forms heterogeneity and their impacts on phyto-diversity. The present work can be extended with scientific inventories pertaining to effects of land forms on ecological roles of dominant/indicator species

M. Mathur
ICAR-Central Arid Zone Research Institute, Jodhpur 342003, India
e-mail: eco5320@gmail.com

M. Mathur · S. Sundaramoorthy (✉)
Department of Botany, Center of Advanced Studies, JNV University, Jodhpur 342001, India
e-mail: jnvusundar@rediffmail.com

K. G. Ramawat (ed.), *Biodiversity and Chemotaxonomy*, Sustainable Development and Biodiversity 24, https://doi.org/10.1007/978-3-030-30746-2_5

and effects of different pedological factors of such land forms on community composition and plant bio-diversity.

Keywords Phyto-diversity · Land forms · Agro-climatic zone · SHE analysis · Species richness · Principal Component Analysis

5.1 Climatic Properties of Arid Zone

In general, arid regions are often classified as those dry regions where rainfall is scarce and meet less than one-third of the annual water need (PE) of that area (Mathur 2005). In such areas, precipitation magnitude shows distinct seasonal patterns which are termed as pulse (rainy season), inter-pulse (winter) and non-pulse (summer or dry period). Such episodes significantly influence the structure and function of arid and semi-arid ecosystems (Mathur 2018). The Great Indian Desert spreads (24° and 29° N latitude and 70° and 76° E longitude) across the state of Rajasthan and parts of Gujarat in western India covers about 200,000 km^2. About 61% of the Indian Desert is in Rajasthan covering 12 districts (Singh et al. 2012). This western part of the country experiences varying climatic features of arid and humid. The western desert receiving average 100 mm rain annually while in southeast, some areas may receive 500 mm rain. Hot winds and dust storms occur in the summer, especially in the desert tract.

5.2 Agro-Climatic Zones of Rajasthan

On the basis of rainfall and soil properties, the Thar Desert can be subdivided into four sectors (Sharma and Mehra 2009).

- the Luni basin, comprising Pali, Jalore, the south-eastern part of Barmer, the eastern part of Jodhpur, the western part of Ajmer, Sirohi, and the southern part of Nagaur;
- Sikar, Jhunjhunu and northern Nagaur represented the northern drainage zone;
- the agriculturally rich district of Sri Ganganagar and Hanumangarh adjoining Punjab and Haryana; and
- the true desert or *marusthali*, consisting of Jaisalmer in its entirety, northern Barmer, and the western parts of the Jodhpur, Bikaner and Churu districts.

However, based on the rainfall, temperature, altitude, latitude, natural vegetation, soils, crops and livestock parameters, Rajasthan divided into ten agro-climatic zones (Table 5.1).

Table 5.1 Agro-climatic zones of Rajasthan

Agro-climatic zones	Rainfall (mm)	Districts
IA-Arid Western	200–370	Barmer and Jodhpur
IB-Irrigated North Western Plains	100–350	Ganganagar (IB_G) and Hanumangarh (IB_H)
IC-Hyper Arid Partial Irrigated Zone	100–350	Bikaner, Jaisalmer and Churu
IIA-Internal Drainage Dry Zone	300–500	Nagaur, Sikar (IIA_{Sik}) and Jhunjhunu (IIA_{Jhun})
IIB-Transitional Plain of Luni Basin	300–500	Jalor (IIB_{Jal}), Pali and Sirohi (IIB_{Sir})
IIIA-Semi-Arid Eastern Plain	500–700	Jaipur, Ajmer, Dausa and Tonk
IIIB-Flood Prone Eastern Plain	500–700	Alwar, Dholpur, Bharatpur, Savi Madhopur and Karauli
IVA-Sub Humid Southern Eastern Plain and Aravalli Hills	500–900	Bhilwara, Rajsamand and Chittoregarh
IVB-Humid Southern	500–1100	Dungarpur, Udaipur, Banswara and Pratapgarh
V-Humid Southern Eastern Plain	650–1000	Kota, Jhalawar, Bundi and Baran

Source Directorate of Agriculture, Government of Rajasthan
Underlying zones are included in the present study

5.3 Plant Community Attributes of Indian Desert

The natural vegetation of Indian hot arid and semi-arid areas is classed as Northern Tropical Thorn Forest (Champion and Seth 1968) occurring in small clumps scattered more or less openly and composed of tree, shrubs, and herbs. Dominated by shrubs like *Leptadenia pyrotechnica*, *Calligonum polygonoides*, *Calotropis procera*, *Acacia jacquemontii*, *Ziziphus nummularia*, etc. (Nawal et al. 2006). In spite of the adverse environmental conditions, the Western Rajasthan supports diversified flora and fauna, which is well adapted to xeric conditions. The plant life of the region belongs with 81 different families out of which 71 belong to dicot, while, nine and one belongs to monocot and gymnosperm groups, respectively. Poaceae being the largest family with 89 species followed by Fabaceae (61) and Asteraceae (38). These families consisting 597 different species and out of them 549 are indigenous to the region while 48 species are introduced. 292 species are being reported for their medicinal properties (Mathur and Sundaramoorthy 2013). Mathur (2015) have utilized various multi-variant approaches like Residual Value Analysis (RVA), Binomial, Bayesian and Imprecise Dirichlet Model for detecting the over and underuse plant taxa belongs to hot arid region of India and marked the Aizoaceae, Capparaceae, Chenopodiaceae, Combretaceae, and Polygonaceae as significant underused families. However, Menispermaceae, Verbenaceae, Burseraceae, Moringaceae, Salvadoraceae, Liliaceae designated as significant highly overused families.

Despite the other ecosystem services, *Acacia senegal*, *Prosopis juliflora*, *Albizzia lebbeck*, *Cordia rothii*, *Dalbergia sissoo*, *Zizyphus jujube*, *Calligonum polygonoides*, *Cassia auriculata*, *Ricinus communis*, *Zizyphus nummularia*, *Lasiurus sindicus*, *Panicum turgidum*, and *Erianthus munia* have proved successful in sand dune stabilization.

5.4 Land forms Characteristics and Their Attributes at Rajasthan: Succession Trends

Land forms are characterized by physical structures such as slope, elevation, stratification (Slaymaker et al. 2011; Pelffini and Bollati 2014). These features affect micro site condition development, ecosystem flow path and distribution of resources consequently affecting growth and diversity of biota. The relationships between land form attributes and ecological patterns are well documented (Swanson et al. 1988; Larkin et al. 2006; Solon et al. 2007; Ott and van Aarde 2014). The complex interdependent linkages between geomorphic (i.e. land forms) and biotic (i.e. organisms) requires clear understanding of geomorphic attributes and associated ecological linkages and organisms (Corenblit et al. 2008).

Within India hot arid and semi-arid region, Bawa et al. (1988) and Kolarkar et al. (1992) have identified 14 different types of land forms that includes: Hills, Piedmonts plains, Rocky or gravelly pediments, Flat buried pediments, Sandy undulating buried pediments, Flat aggraded older alluvial plains, Saline flat aggraded older alluvial plains, Sandy undulating aggraded older alluvial plains, Sand dunes, Flat inter-dunal, Sandy undulating inter-dunal plains, Shallow saline depressions, Graded river beds, and Younger alluvial plains. On the basis of physiognomy vegetation of the Indian arid zone can be classified into six characteristics types, i.e., mixed xenomorphic thorn forest, mixed xenomorphic wood lands on piedmonts and alluvial plains, mixed xenomorphic riverine thorn forest on younger alluvial plains around desertic rivers and water bodies, lithophytes scrub on eroded rocky gravelly plains, psammophytic scrub on sand dunes, hummocks, and sandy plains and halophytic scrub on low lying saline flats or ranns. Characteristics community composition at different land forms in the Indian Hot Desert is presented in Table 5.2.

Basic characteristics of some land forms of the Rajasthan and related succession trends in vegetation components are as follows:

5.4.1 *Hills and Piedmonts*

The hilly tract of the Aravalli mountain range comprising quartzite, phyllite, schist and micaschist rocks of the Aravalli and the Delhi age occur n the east of Sikar and northeast of Ajmer. These hill ranges generally have narrow ridges, conical shapes and height relative relief. The hill slopes are generally rectilinear in the form with 25–30° angles and are covered with colluvial deposits. In the west of the Aravalli range, rocks are found only in a scattered fashion consisting of rhyolite and granite rocks of the pre-Cambrian age. Sandstone and calcareous limestone occur in the Jodhpur, Pali, Jalore, Nagaur and Barmer districts. West parts of the Aravalli are usually of low height and easily accessible to the grazing animals. On denuded rock *Tragus biflorus*, *Enneapogon brachystachys*, *Melanocnechris jacquemontii*,

Table 5.2 Characteristics community composition at different land forms in the Indian Hot Desert

Land form	Trees	Shrubs	Under scrubs	Grasses	Forbs and Sedges
Hills rocks out crops	*Anogeissus pendula* *Acacia senegal* *Wrightia tinctoria*	*Euphorbia caducifolia* *Grewia tenax* *Maytenus emarginatas*	*Barleria acanthoides* *Tephrosia purpurea* *Sida cordifolia*	*Sehima nervosum* *Cymbopogon jwarancusa* *Aristida funiculate*	*Indigoera cardifoilia* *Tribulus terrestris* *Achyranthus aspera* *Cyperus rotundus*
Piedmonts plains					
1. Upper piedmont	*Acacia senegal* *Moringa concanensis* *Salvadora oleoides*	*Maytenus emarginatuas* *Commiphora wightii* *Grewia tenax*	*Tephrosia pentrosa* *Blepharis sindica*	*Aristida funiculate* *Heteropogon contortus* *Oropetium thomeaum*	*Indigofera cordifolia* *Boheravia diffusa* *Digera muricata* *Cyprus rotundus*
2. Lower piedmont	*Acacia lecophloea* *Acacia senegal* *Prosopis cineraria*	*Acacia jacquemontii* *Zizyphus nummularia* *Capparis decidua*	*Tephrosia purpurea* *Aerva persica*	*Cenchrus spp.* *Heteropogon contortus* *Eremopogon favelolotus*	*Indigofera cordifolia* *Clemove viscose* *Digera muricata* *Cyperus rotundus*
Pediment plains					
1. Flat exposed pediment plain	*Acacia senegal*	*Zizyphus nummularia* *Lycium barbarum* *Cassia aurivulata*	*Tephorisa purpurea* *Tephrosa pentrosa* *Clemome viscose*	*Enneapogon brachystachys* *Dactyloctenium sindicum* *Aristada funiculate*	*Indigofera cordifolia* *Mollugo cerviana* *Cleome papillosa*

(continued)

Table 5.2 (continued)

Land form	Trees	Shrubs	Under scrubs	Grasses	Forbs and Sedges
2. Buried pediment plain	*Acacia leucophloea* *Salvadora oleoides* *Prospois cineraria*	*Zizyphus nummularia* *Capparis decidua* *Maytenus mearginatas*	*Corchorus depressus* *Farsetia hamiltonii*	*Cenchrus biflorus* *Eleusine compressa* *Lasiurus sindicus*	*Indigofera cordifolia* *Boerhavia diffusa* *Gisekia pharnacoides* *Cyperus laevigatus*
Gravelly plain	–	*Zizyphus nummularia* *Capparis decidua* *Lycium barbarum*	*Corchorus depressus* *Fagonia cretica* *Blepharis sindica*	*Aristida mutabilis* *Dactyloctenium indicum*	*Indigofera cordifolia* *Tribulus terrestris*
Alluvial plains					
1. Younger alluvium	*Prosopis cineraria* *Acacia nilotica* *Salvadora oleoides*	*Acacia jacquemontii* *Calotropis procera* *Leptadenia pyrotechnica*	*Xanthium strumarium* *Tephrosia purpurea*	*Desmostachya bipinnata* *Digitaria adscendens* *Elusine compressa*	*Indigofera cordifolia* *Digera muricata* *Amaranthus viridis* *Cyperus rotundus*
2. Older alluvium with flat topography	*Prosopis cineraria* *Tecomella undulate* *Acacia nilotica*	*Balanites aegyptiaca* *Mimosa hamata* *Capparis decidua*	*Aerva persica* *Crotalaria burhia* *Tephrosia purpurea*	*Aristida funiculate* *Dactyloctenium aegyptium*	*Indigofera linifolia* *Cyperus laevigatus*

(continued)

Table 5.2 (continued)

Land form	Trees	Shrubs	Under scrubs	Grasses	Forbs and Sedges
3. Older alluvium with hummocky and undulating topography	*Prosopis cineraria* *Tecomella undulate* *Salvadora oleoides*	*Colligonum polygonoides* *Acacia jacquemontii* *Calotropis procera*	*Crotalaria burhia* *Aerva pseudotomentosa* *Aerva persica*	*Lasiurus sindicus* *Cenchrus biflorus* *Eleusing compressa*	*Indigofera linifolia* *Boerhavia diffusa*
Sand dunes	*Acacia senegal* *Prosopis cineraria* *Salvadora oleoides*	*Colligonum polygonoides* *Maytenus emarginatas* *Acacia jacquemontii*	*Aerva pseudotomentoda* *Crotalaria burhia* *Aerva persica*	*Panicum turgidum* *Cenchrus biflorus* *Aristida funiculate*	*Indigofera cordifolia* *Citrullus colocynthis* *Cyperus bulbosus*
Graded river bed	*Tamarix dioica* *Tamarix troupii*	*Tamarix ericoides* *Sesbania aegyptiaca*	*Xanathium strumarium*	*Cynodon dactylon* *Desmostachya bipinnata*	*Trianthema pentandra* *Amaranthus viridis* *Cyperus rotundus* *Fimbristylis sp.*
Saline flat and depression	*Prosopis juliflora* *Salvadora oleoides* *Zizyphus nummularia*	*Halozylon salicornicum* *Suaeda fruticosa* *Indigofera oblongifolia*	–	*Sporobolus marginatus* *Chloris virgata* *Aeluropus lagopoides*	*Trianthema portulacastrum* *Zygophyllum simplex* *Cyperus rotundus*

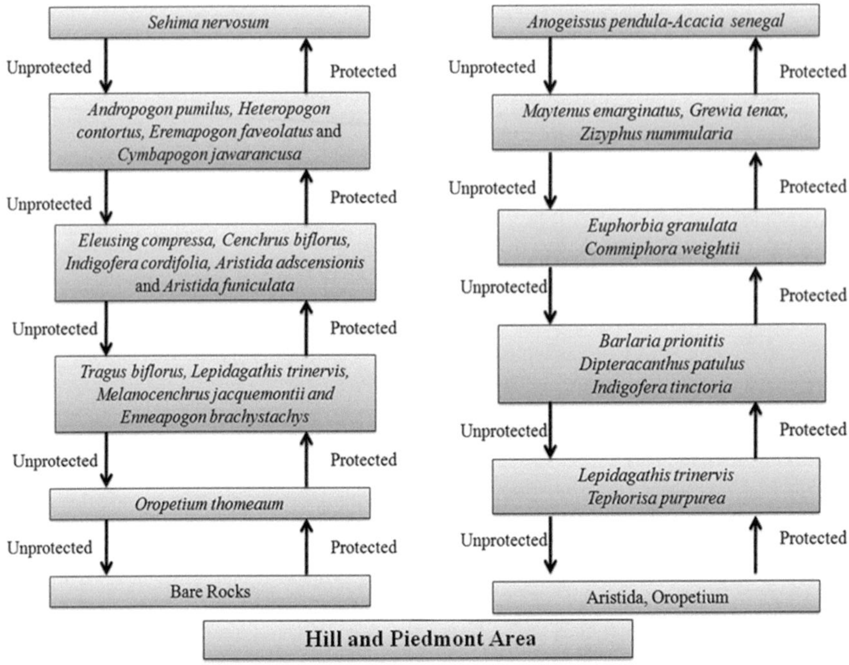

Fig. 5.1 Basic succession trend of woody perennials on Hills and Piedmonts

Lepdiagathis trinervis are the pioneers followed with shrubs species like *Zizyphus nummularia*, *Euphorbia caducifolia*, *Grewia tenax*, *Maytenus emarginatus*. Under protected conditions these rocky areas are dominated by *Acacia senegal-Anogeissus pendula* (>350 mm rainfall). Removal of such species facilitates the series of disclimax stages and *Euphorbia caducifolia* represents degradation stage while the *Tephrosia purpurea-Oropetium thomeaum* indicates the severe degradation (Fig. 5.1).

The piedmont plains of granite or rhyolite are generally covered with varying depths of soil and support mixtures of rocky and plant elements. *Acacia senegal-Salvadora oleodes* or *S. oloides-Prosopis cineraria* dominant under protection. Grazing and cutting are invariably practiced here which show a series of disclimax stage all over western Rajasthan. Moderate utilization of lands is represented by the presence of *Euphorbia caducifolia-Z. Nummularia/Capparis decidua* while dominance of *Tephrosia purpurea* and *Oropetium thomeaum* indicates severe degradation.

The pediments area are also related with Hills and Piedmonts and in the Rajasthan varied types that include rocky/gravelly pediment, flat buried pediments (high rainfall area) and sandy undulating buried pediments (low rainfall area).

Rocky/gravelly pediments: This type occurs along the base of hills of different formations. The slope varies from 3° to 8° in the upper part, 1° to 3° in the middle part and 0.8° to 1° in the lower part. The districts of Jodhpur, Jaisalmer and Nagaur are substantially occupied by this type while in other districts, the area under gravelly pediment is negligible. These areas are primarily designated as wastelands, however, at some area these are also served as grazing lands. Protected stands support *Acacia senegal-Prosopis cineraria*. While the degraded condition is represented by scattered shrubs of *Capparis decidua* and *Z. nummularia.*

Flat buried pediments: This unit is covered with 10–30 cm deep alluvial and colluvial sediments with less than 1° slope. These are mainly transported by stream channels from the adjoining hills and pediment surface. Here *Salvadora oleoides, S. persica, Prosopis cineraria, Z. nummularia* and *Capparis dedicua* are the major tree shrub cover and *Dichanthium annulatum* as grass cover (Shankar 1986).

Sandy undulating buried pediments: Basically the intense Aeolian activities by concealing the rocky/gravelly surface have created these pediments. *Haloxylon salicornicum, Z. nummularia* and *Leptadenia pyrotechnica* are the major shrub cover with *Lasiurus sindicus,* and *Panicum antidotale* grass cover.

5.4.2 Older Alluvial Plains (OAP)

These plains are covered with the alluvium of different thickness. Within the alluvial deposits, a thick layer of $CaCO_3$ has developed in the form of nodules which is locally known as *Kankar* pan. The nature of the sediments varies from loamy sand to and sandy loam and also loam but at certain places in pockets of silty clay loam. On the basis of soil type, this land form is of two types: older alluvial plains with light to medium soils and older alluvial plants with medium to heavy soils The first one supports the potential plant community of *Prospois cineraria-Zizyphus nummularia Capparis decidua.* Upon degradation the *P. cineraria* is replaced by *Caparris decidua. Tephrosia purpurea* and *Crotolaria burhia* are the major non-palatable species. While *Salvadora oleoides-P. cineraria* are the potential plant community on older alluvial plains with medium to heavy soil. On such type of land from the comparative stages of different succession stages with different soil types are presented in Figs. 5.2 (tree and shrubs) and 5.3 (grasses).

5.4.3 Sandy Undulating Hummocky Plains (SUHP)

The undulations due to sand in the form of dunes and hummocks that are formed by intense Aeolian activities have resulted in formation of this unit. The slope in this unit is irregular and varied from 0.6 to 0.59 mm. Larger part of western district including Barmer, Jaisalmer, Bikaner, and part of Jodhpur and Nagaur districts possesses sandy undulating hummocky plains (25%). Soils of such area are quite deep loamy sand which becomes quite loose on drying and helps in the formation of ridges and hummocks. The potential plant communities in these are represented

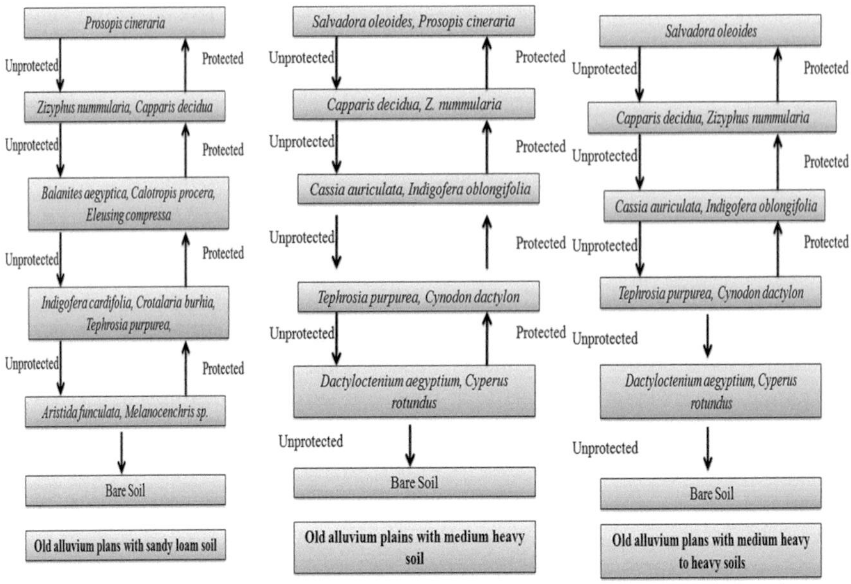

Fig. 5.2 Basic succession trend of woody perennials at OAP

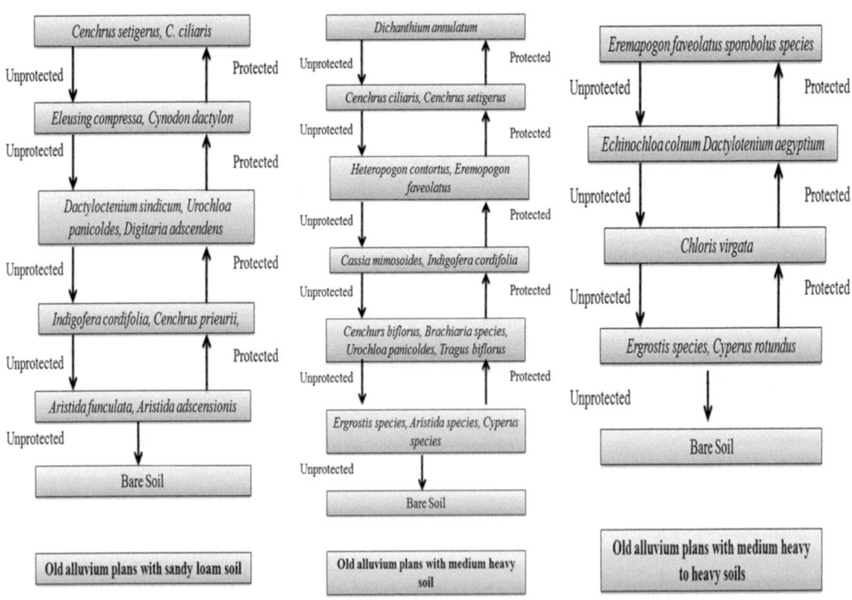

Fig. 5.3 Basic succession trend of grasses at OAP

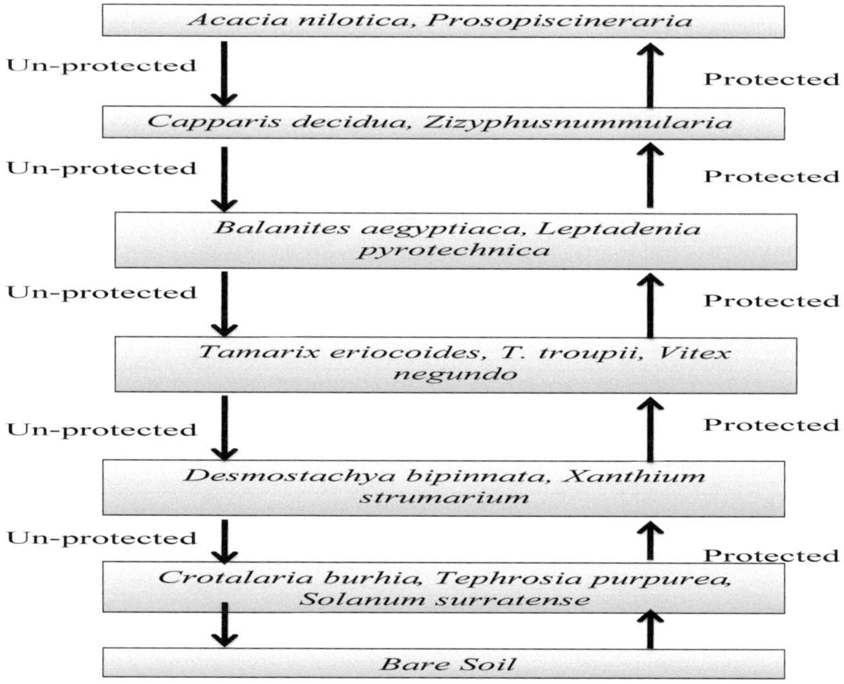

Fig. 5.4 Basic succession trend of woody perennials at SUHP

by *Prospois cineraria-Tecomella undulate* (Fig. 5.4). Common village grazing lands have virtually been maintained at moderate level of degradation where *Zizyphus nummularia* or *Acacia jacquemonti* or *Calligonum polygonoides. Aerva persica*, *A. pseyditinebtisa*, *Calotropis procera*, *Crotolaria burhia*, and *Tephrosia purpurea* represents the highly degraded conditions.

5.4.4 *Younger Alluvia Plains (YAP)*

The younger alluvial plains in the form of narrow strips are associated with the bank of Luni, Ghaggar and the west Banas rivers and their tributaries. At such land forms the soils are very deep sandy loam with good water potential. The areas are largely put to double cropping year after due to which the natural regeneration of the vegetation is always discouraged. Such habitats, being highly potential is never allowed to gain climatic climax. More often, the successionary process is held up at the pre-climax stage. The rate of succession is fairly rapid because of the availability of more moisture throughout the year. The climax community here represents by *Acacia nilotica-Prosopis cineraria* (Fig. 5.5).

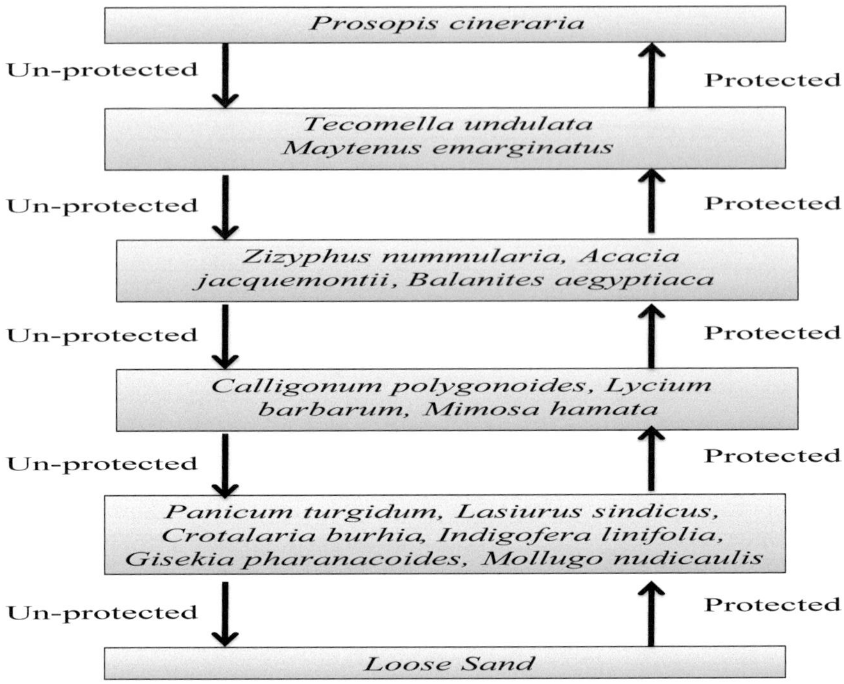

Fig. 5.5 Basic succession trend of woody perennials at YAP

5.4.5 *Sand Dune*

Based on the morphological and spatial orientation characteristics, six types of sand dunes viz., parabolic, shrub-coppice, longitudinal, obstacle, transverse, and barchans have been identified. The slopes on the leeward sides, flanks and windward sides of these dunes are 22–24°, 12–14° and 3–4°, respectively. The stabilized dunes are invariably dominated by the *Calligonum polygonoidum* community which forms an intermediate seral stage in succession. Some of the well-protected sand dunes support an open forest, dominated by the *Acacia senegal-Maytenus emarginatus*. Various stages of development or degradation of plant communities on the sand-dunes are encountered in western Rajasthan. Out of those stages, one can reconstruct its succession pattern. *Aristida funiculate*, *Cenchrus biflorus*, *Indigofera cordifolia*, *Faarsetia hamiltonii*, *Crotalaria burhia,* and *Aerva pseudotomentosa* are the pioneer species.

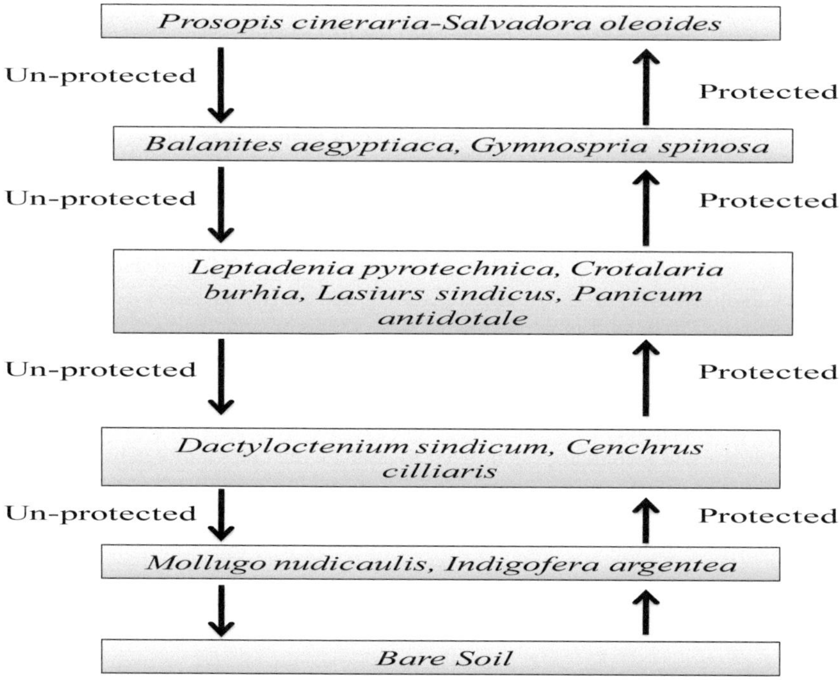

Fig. 5.6 Basic succession trend of woody perennials at sandy plain

5.4.6 Sandy Plains

On such type of land form the *Indigofera argentea* and *Mullugo nudicaulis* are the early colonizers which are succeeded by grasses such as *Dactyloctenium scindicum, Lasiurus sindicus,* and *Panicum antidotale. Balanites aegyptiaca and Gymnosporia spinosa* are late succession stages, which ultimately lead to the climax of *Prosopis cineraria* and *Salvadora oleoides* (Fig. 5.6).

Based on the above information's present endeavor was undertaken to evaluate the patterns of phyto-diversity of six land forms (Hills and Piedmonts, OAP, SUHP, YAP, sand dune and sandy plain) at five agro-climatic zones Jodhpur (IA); IB with two sub-zones i.e. Ganganagar IB_G, Hanumangarh IB_G; Hyper arid partial irrigated (IC); IIA-internal drainage dry zone with two sub-zones (Sikar IIA_{Sik} and Jhunjhunu IIA_{Jhun}); IIB transitional plain of Luni with two sub-zones (Jalore IIB_{Jal} and Sirohi IIB_{Sir}). At these places with land forms the present study was designed under three categories: Phyto-diversity: zonal and land form approach; community attributes: zonal specific approach and inter-zonal community attributes: land form approach. Information's on these phyto-sociological aspects were evaluated with help of previously related publications Satyanarayan and Gaur (1968), Prakash and Gupta (1976), Saxena (1977a, b), Kolarkar et al. (1992), Singh et al. (1995, 1996),

Khan and Ram (2003), Khan et al. (2003), Narain et al. (2005), Mathur (2005), Tiwari et al. (2007), Singh et al. (2012), Mathur and Sundaramoorthy (2016), Mathur (2018). The species richness is defined as the total number of species per sampling unit (Bhattarai et al. 2004). Shannon-Weaver is a diversity index and generally ranges from 1.5 to 3.5 and rarely up to 4.5. Its higher valued indicates the high diversity while the lower value represents the dominance of few species (Mathur 2005). Evenness or equitability represents the distribution of individuals among the species. When all species are equally abundant, an evenness index would be at a maximum and decrease to zero as the proportional abundance of the species diverges away from evenness (Ludwig and Reynolds 1999). SHE analysis was performed to quantify the patterns of these diversity patterns for a specific land form among the agro-climatic zones and this was conducted with Bio-Diversity Pro Software. Principal Component Analysis (PCA) is an Eigen value-based ordination method and this was performed with Pearson correlation coefficient (Xlstat 2007). This multi-variant technique was utilized to analyze the community behavior under zonal and land form specific approaches.

5.5 Phyto-Diversity: Zonal and Land Forms Approach

Phyto-diversity parameters pertaining to species richness, Shannon and Weaver diversity index (H^{I}) and evenness were quantified at different sites of different land forms of various zones. Results of these parameters were interpreted as average pertain to a land form of a specific zone.

5.5.1 *Hills and Piedmonts*

Diversity of such land forms particularly at IA-arid western zone (Jodhpur) having average richness of 10 with 2.3 Shannon and Weaver index (H') and evenness of 0.6 (Table 5.3). Within agro-climatic zone of IIB (transitional plain of Luni basin), great intra-zonal variations in richness, diversity and in evenness were recorded. For example, at the Sirohi sub-zone, the above three parameters were in average of 9, 2.13 and 1.07, respectively. While at Jalore sub-zone average richness increases up to 22 with diversity of 2.82 and evenness 0.84 (Table 5.3). Thus, within an agro-climatic zone of IIB, their sub-zones are differentiating with each other on the basis of their diversity patterns. At internal drainage zone (IIA_{Sik}) richness and diversity (H') were 13 and 2.56, respectively.

Table 5.3 Diversity parameters at Hills and Piedmonts pertains to different zones

Zones	Richness	*H*	Evenness
IIB_{Sir}	9	2.13	1.07
IIB_{Jal}	22	2.82	0.84
IA	10	2.3	0.6
IIA_{Sik}	13	2.56	0.14

Table 5.4 Diversity parameters at OAP pertains to different zones

Zones	Richness	*H*	Evenness
IIA_{Jhun}	5	1.38	0.92
IIA_{Sik}	16	2.77	0.72
IB_G	6	1.35	0.70
IB_H	5	1.38	0.92
IIB_{Jal}	20	2.75	0.84
IIB_{Sir}	9	1.94	0.86

5.5.2 *OAP*

Such land form at IIA (Internal drainage dry zone, Jhunjhunu) having average woody perennial richness 5 with 1.38 H^{I} and 0.92 evenness. More or less similar intra-zonal diversity patterns presented at agro-climatic zone IB (IB_G and IB_H). While at IIB agro-climatic zone high intra-zonal variations observed for average richness and H^{I}. At Jalore sub-zone average richness was 20 which dropped to 9 at Sirohi intra-zone. H^{I} at these two sub-zones was 2.75 and 1.94, respectively (Table 5.4).

5.5.3 *SUHP*

At this type of land form, more or less similar average diversity patterns were recorded particularly at IIA, $IB_{(Gang)}$, and $IB_{(Hanu)}$ zones. The minimum and maximum were recorded at IB_G and IC (Hyper arid partial irrigated zone, Churu, Table 5.5), respectively. Such trends revealed that regional phyto-diversity largely affected with irrigation patterns and its magnitude.

5.5.4 *YAP*

This type of land form present at IA and $IIB_{(Sirohi\ and\ Jalore)}$. Highest richness was recorded at $IIB_{(Jalore)}$ followed at IA and least was at $IIB_{(Sirohi)}$ (Table 5.6). Thus, again, intra-zonal diversity variations occurred at IIB Jalore and Sirohi. Richness and evenness of internal drainage dry zone (IIA_{Sik}) were recorded similar to IIB_{Sir}.

Table 5.5 Diversity parameters at SUHP pertains to different zones

Zones	Richness	*H*	Evenness
IIA_{Jhun}	6	1.5	0.81
IIA_{Sik}	10	2.30	0.72
IC	15	2.11	0.8
IB_G	4	1.11	0.71
IB_H	6	1.36	0.81

Table 5.6 Diversity parameters at YAP pertains to different zones

Zones	Richness	*H*	Evenness
IIB_{Sir}	7	1.6	0.79
IIB_{Jal}	18	2.59	0.74
IA	14	2.69	0.55
IIA_{Sik}	7	1.95	0.72

Table 5.7 Diversity parameters at sand dune pertains to different zones

Zones	Richness	*H*	Evenness
IIA_{Jhun}	6	1.41	0.82
IIA_{Sik}	10	2.3	0.80
IC	11	2.09	0.92
IB_G	5	1.19	0.8
IB_H	6	1.41	0.82
IIB_{Jal}	16	2.53	0.88
IA	16	2.77	0.53

5.5.5 *Sand Dune*

This type of land form is presented in agro-climatic zones of IA, IIA, IB, IIB, and IC. The highest average richness (16), H^{I} (2.77–2.53) and evenness (0.88–0.53) are presented on IA and IIB followed by IC (Table 5.7). More or less similar diversity patterns observed at $IIA_{(Jhu)}$, $IB_{(Gang\ and\ Hanu)}$.

5.5.6 *Sandy Plain*

This land form type is present at $IIA_{(Jhu)}$, $IB_{(Gang\ and\ Hanu)}$, $IIB_{(Sirohi)}$ and at IC. The highest diversity occurred at IC followed at IIB_{Sir} and least at IB and IIA (Table 5.8).

At YAP and Hills and Piedmonts, higher phyto-diversity does not necessarily correlate with agro-climatic zone or rainfall intensity, as we have noticed great intra-zonal variations between the Sirohi and Jalore (IIB). Jalore sub-zone is more diversified compared to Sirohi. These land forms at IA (Jodhpur-arid western region) are more diversified compared to the Sirohi. Phyto-diversity of SUHP and

Table 5.8 Diversity parameters at sandy plain pertain to different zones

Zones	Richness	H	Evenness
IIA_{Jhun}	5	1.08	0.59
IC	19	2.69	0.97
IB_G	4	0.96	0.72
IB_H	5	1.08	0.7
IIB_{Sir}	8	2.7	0.54

sandy plain are more positively related with geomorphologic characteristics of zone, as we have noticed higher value of these community attributes at IC, which gradually decreased at IB and IIB. At OAP, the phyto-diversity correlated with rainfall intensity being higher at Jalore (IIB) and gradually decreased in IIA and IB. Great intra-zonal variations are also rerecorded for this land form. Sand dune at IA and IIB are more diversified compared to IIA, IB, and IC. Thus, diversity at this land form related with rainfall as well as with geomorphologic feature of the area.

Thus, with this study, the threshold limits (upper and lower) for different diversity parameters among agro-climatic zones and at different land forms were identified. Upper limits of richness (22) and Shannon and Weaver diversity index (H' 2.82) were recorded at transitional plain of Luni basin (IIB_{Jal})-Hill and piedmonts. While lower limit of richness (4) and H' index (0.9) was recorded at irrigated northwestern plain (IB_G) at Sandy plain. Contrary to these results, upper (1.07) and lower limit (0.14) for evenness were recorded at same land form i.e. hill and piedmonts but from different zones i.e. transitional plain of Luni Basi (IIB_{Sir}) and internal drainage dry zone (IIA_{Sik}), respectively.

Buzas and Hayek (1998) developed SHE analysis for Bio-zone Identification (SHEBI), in which ln$S$ (species richness), H (diversity) and ln$E$ (evenness) are recalculated as samples are accumulated and the number of specimens N increases. They developed this technique for diversity assessment that allows independent yet simultaneous evaluation of the relative contributions of richness and evenness to community diversity across sampling scales. The basis for SHE is the linear decomposition equation, $H = \ln S + \ln E$. This decomposition is derived from the following conditions: (a) maximum H' diversity occurred when all species are uniformly distributed (H' Max = $\ln(S)$), and (b) E is correlated to H' by the equation $[(E) = e^H/S]$. Thus, the SHE decomposition formula, $H = \ln S + \ln E$ indicate that H' diversity equals its maximum value, $\ln(S)$, less the among of unevenness, $\ln(E)$ (subtracted because evenness ≤ 1 and $\ln(E)$ will be ≤ 0) in the sample. It is, therefore, an approach to look at the contribution of species number and accounted for changes in diversity.

With SHE analysis three patterns may be expected, a broken stick, log normal and log series model (Wilson et al. 2012). In broken stick both S and H^I are expected to increase and E to stay constant. The lognormal is associated with an increase in S and H^I but a decline in E. With the log series, S will increase, H^I remains constant, and E will decrease (Magurran 2004). Among the agro-climate zone, log normal (increase in S and H^I but a decline in E) pattern were identified at

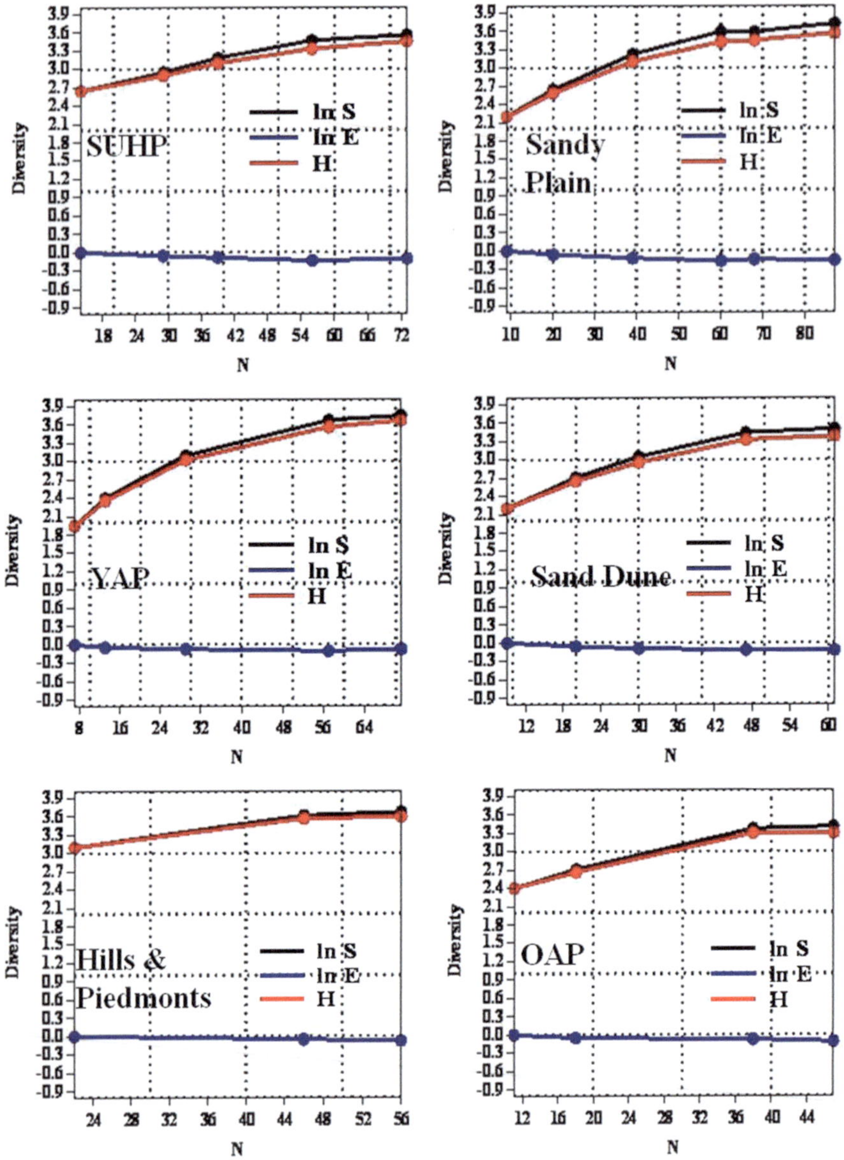

Fig. 5.7 SHE analysis at various land forms among different zones

SUHP, sand plain, YAP, and sand dune. While broken stick model (both S and H^{I} are expected to increase and E to stay constant) identified at Hills and Piedmonts and OAP (Fig. 5.7).

Baghani et al. (2009) utilized this tool for defining species diversity components of mountain rangelands (ZIARAT Basin, Gorgan) and they found that role of evenness was much more important than species richness in defining diversity at species and family levels. Javed (2016) used this tool to examine the relationships among diversity quantified from a single quadrat (micro-scale) to cumulative measure of community (macro-scale) across all vegetation units pertaining to alpine grassland at Bandipora, Kashmir. From conservation and management point of view, Salarian et al. (2015) suggested that this technique is very useful for planning of future trend of the rangeland ecosystem.

5.6 Community Attributes: Zonal Specific Approach

5.6.1 *IIA$_{Sik}$ (Internal Drainage Dry Zone)*

Based on community associates, OAP-YAP and Sand dune-SUHP of this agro-climatic zone are more positively related with each other (Fig. 5.8). While OAP and Hill are oppositely located to each other revealed different community-associated at these two land forms. Further, result of Bartlett's sphericity test (higher observed value of chi-square test compared to critical value) suggested significant relationships among different land forms located within this zone (Table 5.9). Hills and Piedmont vegetation comprises *H. integrifolia*, *G. tenax*, *B. monosperma*, *G. glavens*, *E. caducifolia*, *R. mysorensis*, *A. pendula*, *S. lecopyrus*, *W. tinctora,* and *A. vasica* species. SUHP and sand dune having

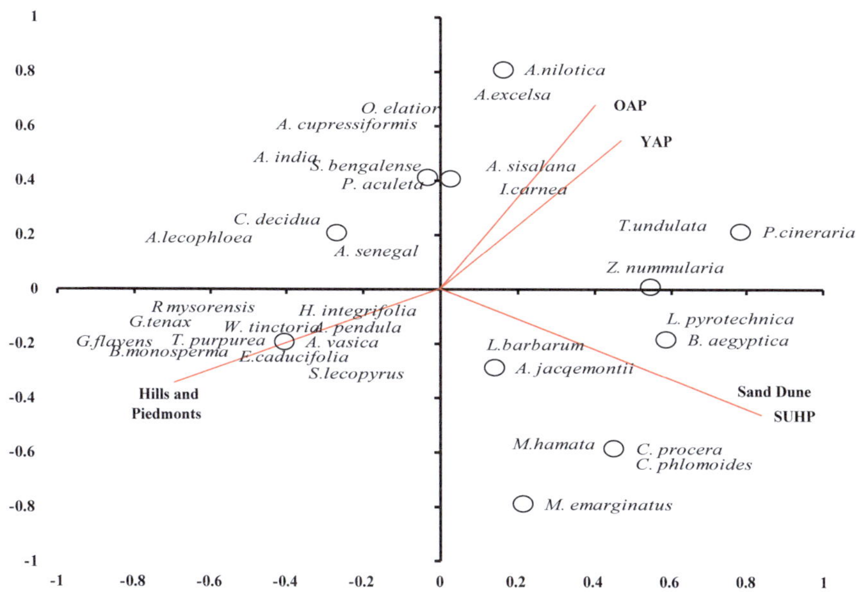

Fig. 5.8 Community attributes at IIA$_{Sik}$ agro-climatic zone

Table 5.9 Bartlett's sphericity test for different land form within a zone

Zones	Chi-square (observed value)	Chi-square (critical value)
IB_H	8.501	12.59
IB_G	18.32	12.59
IC	2.63	7.81
IIA_{Jhun}	6.2	7.8
IIA_{Sik}	54.14	18.30
IIB_{Jal}	41.4	18.3
IA	25.3	12.59
IIB_{Sir}	12.04	12.59

population of *C. procera*, *L. pyrotechnica*, *C. phlomoides*, *M. hamata*. *A. nilotica*, *C. decidua*, *A. cupressiformis*, *A. excelsa* are dominantly presented at OAP while *T. undulata-P. cineraria* are the indicator species of YAP.

5.6.2 *IIA_{Jhu} (Internal Drainage Dry Zone)*

At this agro-climatic zone, sandy plain and SUHP showed close proximity with each other (Fig. 5.9). YAP land form is characterized by species of *D. sisso*, *A. excela*, *P. cineraria*, *C. burhia*, *A. tortilis* and *A. pendula*. *C. procera*, *C. polygonides*, *A. jacquemontii*, *B. aegyptica*, *C. decidua*, *E. caducifolia*, *I. camera*, *A. nilotica* are

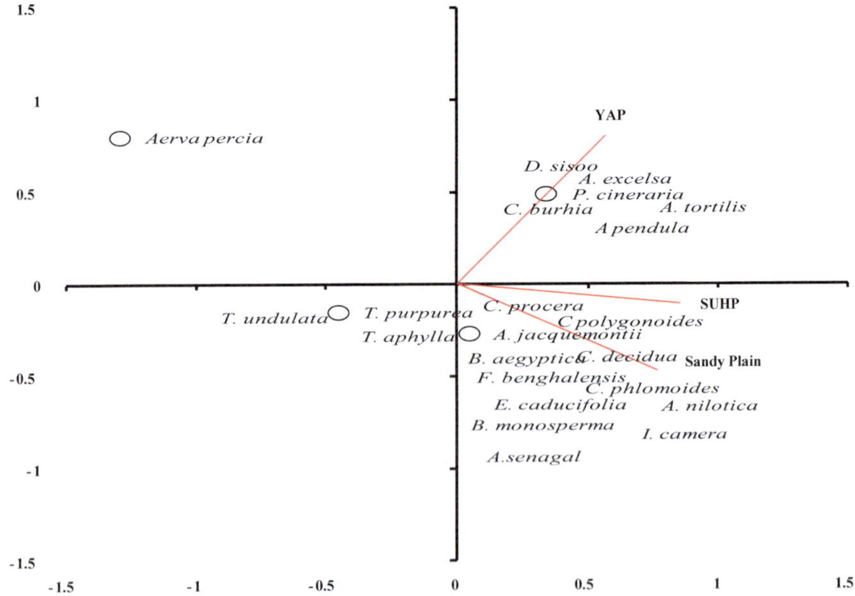

Fig. 5.9 Community attributes at IIA_{Jhun} agro-climatic zone

located at SUHP and Sandy plain. *Aerva persica* did not show any association with any land form. Contrary to IIA_{Sik} Bartlett's sphericity test for this zone showed non-significant relationships among different land forms of IIA_{Jhu}. Thus, by analyzing these two sub-zones of IIA, community attributes showed intra-zonal variability's.

5.6.3 IC (Hyper Arid Partial Irrigated Zone)

Here, SUHP and sand dune showed close proximity. Characteristics vegetation at these land forms are *P. cineraria*, *P. juliflora*, *L. pyrotechnica*, *A. tortilis*, *C. burhia*, *Z. nummularia*. However, *A. senegal*, *A. persica*, *C. decidua*, *L. barbarum*, *M. emarginatus*, *B. aegyptica* are indicator for sandy plain (Fig. 5.10) while *T. undulata*, *S. munja*, *E. caducifolia*, *G. tenax*, *A. pseudotomentosa* located on opposite direction. Bartlett's sphericity test was non-significant (Table 5.9). *A. nilotica*, *A. jacquemontii*, *M. hamata*, *S. fruticosa* were showed negative relationships with *E. caducifolia*, *G. tenax*, *A. pseudotomentosa*.

5.6.4 IB_G (Irrigated North Western Plain)

OAP land form of this zone characterized by species of *S. oliedis*, *H. salicornicum*, *L. pyrotechnica*, *P. juliflora*. *A. tortilia-A. nilotica,* and *C. polyalthea*, *A. pseudo-tomentosa* and *C. decidua* are the indictor for SUHP and Sandy plain, respectively.

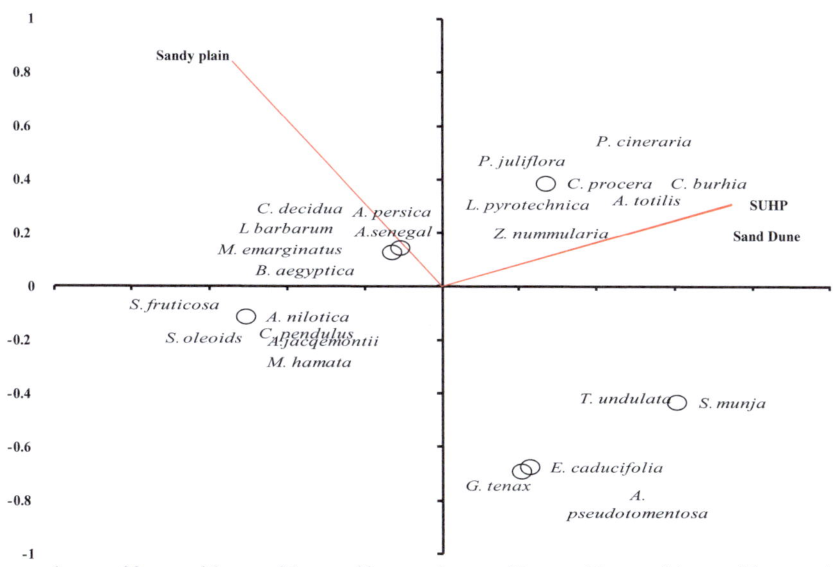

Fig. 5.10 Community attributes at IC agro-climatic zone

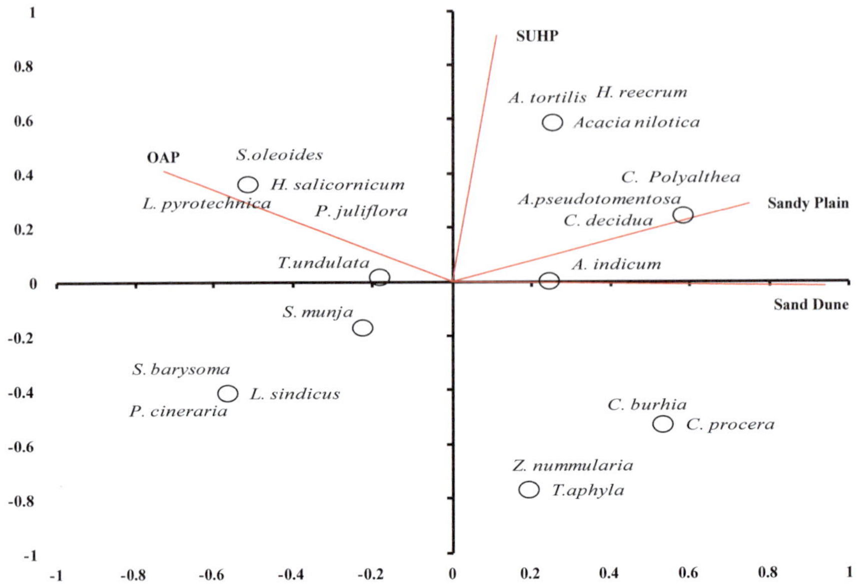

Fig. 5.11 Community attributes at IB_G agro-climatic zone

Negative relationships were recorded among *C. burhia*, *C. procera*, *Z. nummulara*, *T. aphyla* with *A. nilotica*, *A. tortilia* (Fig. 5.11). Bartlett's sphericity test also showed significant relationships among variables (Table 5.9).

5.6.5 *IB$_{Hanu}$* (Irrigated North Western Plain)

At this agro-climatic intra-zone, no significant relationships were observed among land forms (Fig. 5.12) which are also confirmed by Bartlett's sphericity test. *C. decidua* found as indicator species of sandy plain, while *A. persica-C. procera* and *L. barbarum-A. nilotica-T. amphyla-A. tortilis* for SUHP and OAP, respectively. *C. polygonides*, *Z. nummularia*, *L. sindicus*, *A. senegal*, *L. pyrotechnica* and *C. burhia* are oppositely located to *S. munja* and *S. baryosoma*. Similar opposite locations were recorded for the species *S. fruticosa* with *A. nilotica* and *A. tortilis*. Thus, intra-zonal variability's occurred at this agro-climatic zone (IB).

5.6.6 *IIB$_{Jal}$* (Transitional Plain of Luni Basi)

In this agri-climatic zone, YAP, OAP and pediments are closely located to each other (Fig. 5.13). While hills and sand dune as expected are oppositely located to

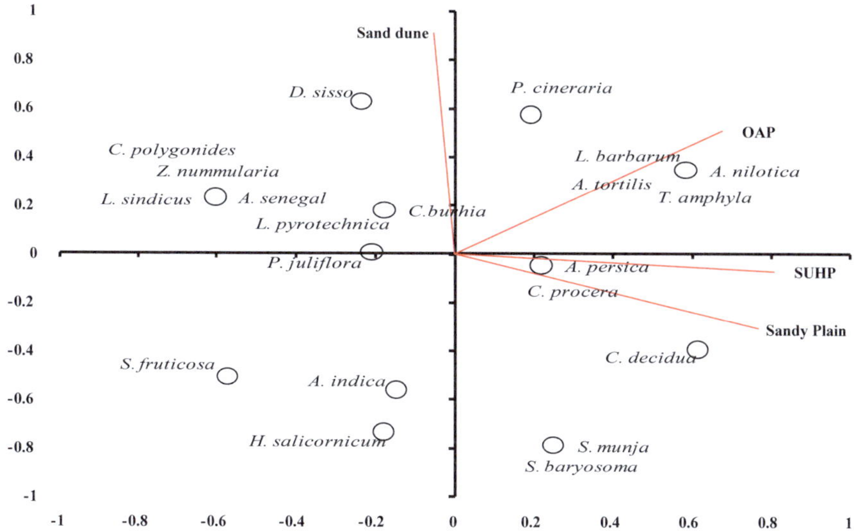

Fig. 5.12 Community attributes at IB_H agro-climatic zone

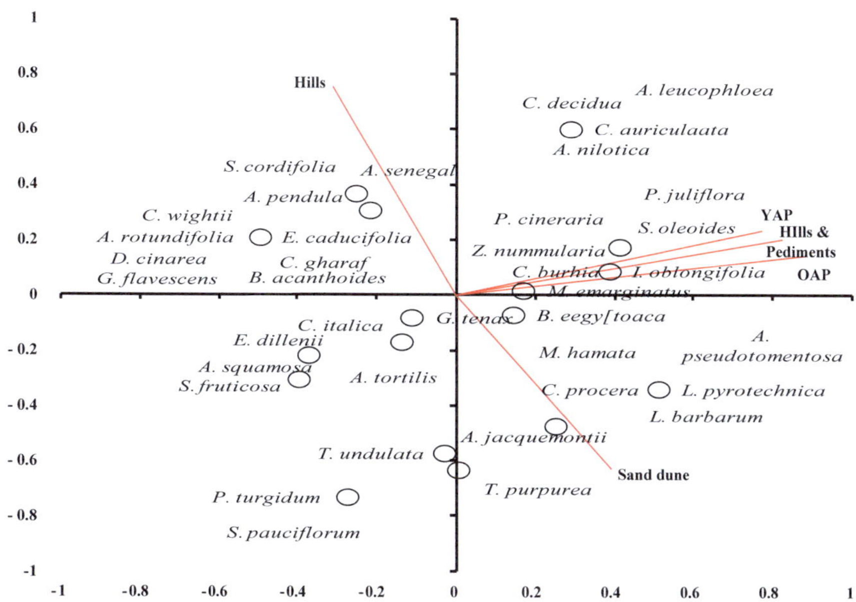

Fig. 5.13 Community attributes at IIB_{Jal} agro-climatic zone

each other. Result of Bartlett's sphericity test showed significant correlation among land form (Table 5.9). Hill land form having *S. cordifolia*, *A. senegal*, *A. pendula*, *C. weightii*, *A. rotunifolia*, *E. caducifolia*, *G. flavescens*, *B. acanthoide,s* and *C. gharaf* species, while sand dune is dominated with *A. jacquemontii*, *T. purpurea*, *L. barbarium*, *L. pyrotechnica*, and *C. procera*.

5.6.7 *IIB_{Sir} (Transitional Plain of Luni Basi)*

At this sub-zone of IIB, community of *G. teneax*, *A. tortilia*, *A. Senegal*, *M. emarginatus*, *A. nilotica*, *D. nutans* are presented at YAP ana Hills and Piedmonts. However, *Z. nummularia*, *P. juliflora*, *A. lecophloeea*, *P. cineraria*, *C. decidua*, *C. auriculata*, *S. oleoides,* and *B. aegyptica* are dominated at OAP and at sandy plain (Fig. 5.14). Four independent communities are *C. procera-B. monosperma-C. gharaf-R. mysorensis-A. pendula-M. hamata-A. indica*, *D. sissoo-T. undulata-P. dulca-I. oblongifolia-A. lebbeck*, *C. weightii-A. pseudotomentosa-T. purpurea-I caducifolia* and *L. pyrotechnica-C. burhia-A. jacqemontii* are also present. At this sub-zone non-significant Bartlett's sphericity test suggested lack of relationships among land forms (Table 5.9).

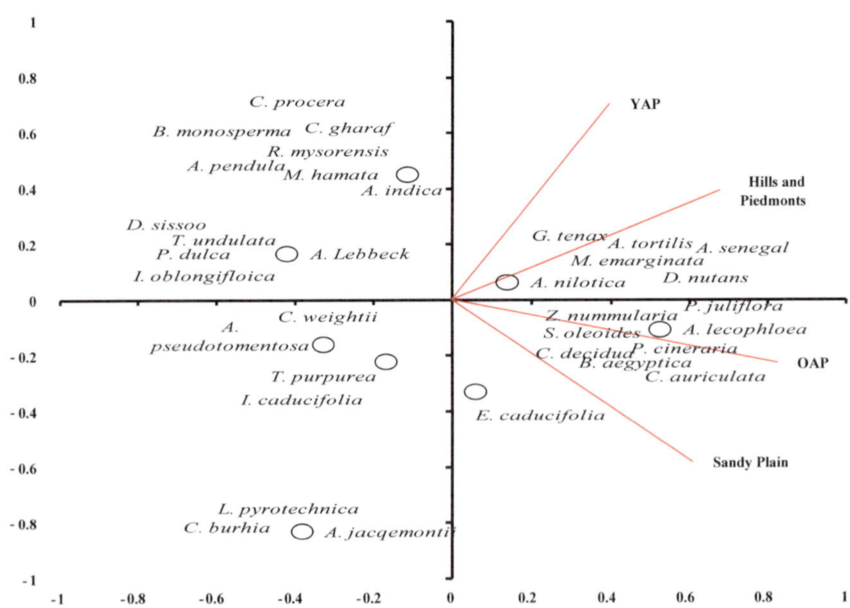

Fig. 5.14 Community attributes at IIB_{sir} agro-climatic zone

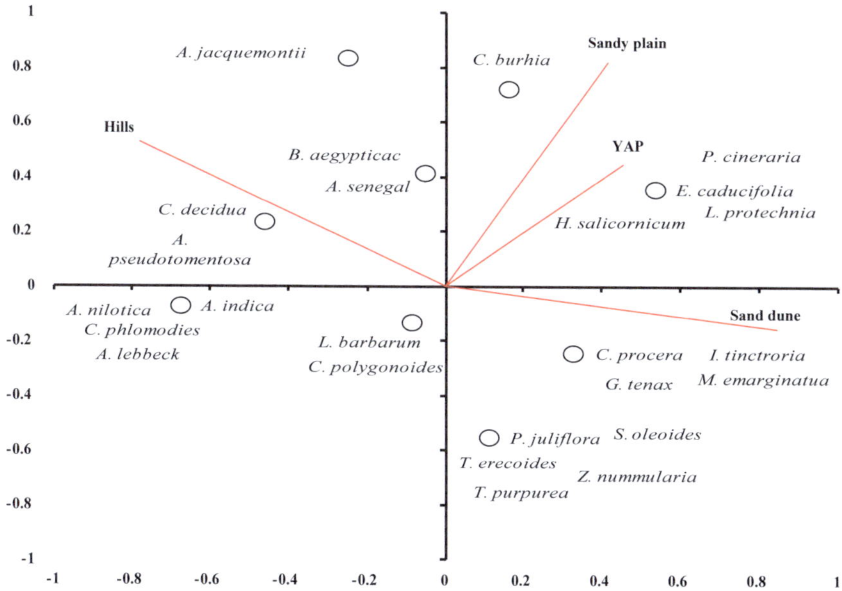

Fig. 5.15 Community attributes at IA agro-climatic zone

5.6.8 *IA (Arid Western)*

Here sandy plain and YAP showed close proximity with each other (Fig. 5.15). Sand dune and hills again as expected located on opposite direction to each other. Bartlett's sphericity test also showed the significant relationships among land forms (Table 5.9). *C. procera*, *M. emarginata*, *T. purpurea*, *P. juliflora*, *I. tinctoria* and *G. tenax* are the major sand dune species. While *E. caducifolia*, *L. pyrotechnica*, *H. salicornicum* are dominated at YAP. *C. burhia* are the indicator species for sandy plain.

5.7 Inter-zonal Community Attributes: Land Forms Approach

5.7.1 *SUHP*

Among the different zones, *C. polyalthea* and *C. decidua* are the indicator species at IIA_{Sik}, while *S. nudi*, *M. hamata*, *P. juliflora*, *S. baryosoma* and *M. emarginata* are the indicator species of IC (Fig. 5.16). *C. burhia*, *A. tortilis*, *A. jacquemontii*, *A. nilotica*, *C. polygonoides*, *D. sisso* are related with IB_G. On the other hand, *C. phlomoides*, *C. procera*, *B. aegyptica*, *A. excelsa* are related with IB_H. *G. tenax*, *H.*

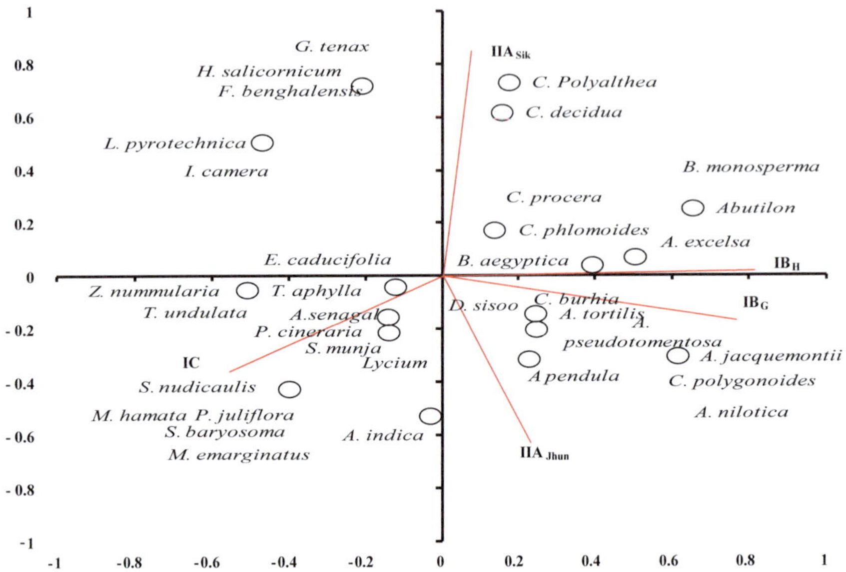

Fig. 5.16 Community attributes at SUHP of different agro-climatic zones

Table 5.10 Bartlett's sphericity test for different land forms among the different zones

Land forms	Chi-square (observed value)	Chi-square (critical value)
SUHP	17.31	18.3
Sandy plain	30.0	24.9
YAP	17.69	18.3
Sand dune	9.5	18.3
OAP	9.89	12.59
Hills Piedmont	10.66	7.81

salicornicum, *F. benghalensis*, *L. pyrotechnica* and *I. camera* are not linked with any zone. Result of Jaccard similarity index are presented in Table 5.11, revealed that IB_G is more similar with IB_H (0.52) followed with IIA_{Jhu} (0.34) and IC (0.24). IIA_{Sik} had lowest similarity was found with IIA_{Jhun} (012) and with IC (0.12). However, Bartlett's sphericity test revealed the non-significant correlations among variables (Table 5.10).

5.7.2 *Sandy Plain*

At this land form IA and IIA_{Jhun} zones are oppositely located to each other (Fig. 5.17). *C. burhia*, *H. salicornicum* and *L. pyrotechnica* are indicator species for

Table 5.11 Jaccard similarity index of SUHP land form among the various agro-climatic zones

Zones	IB_G	IB_H	IIA_{Sik}	IIA_{Jhun}
IB_H	0.52	–	–	–
IIA_{Sik}	0.2	0.19	–	–
IIA_{Jhun}	0.34	0.28	0.12	–
IC	0.24	0.18	0.12	0.25

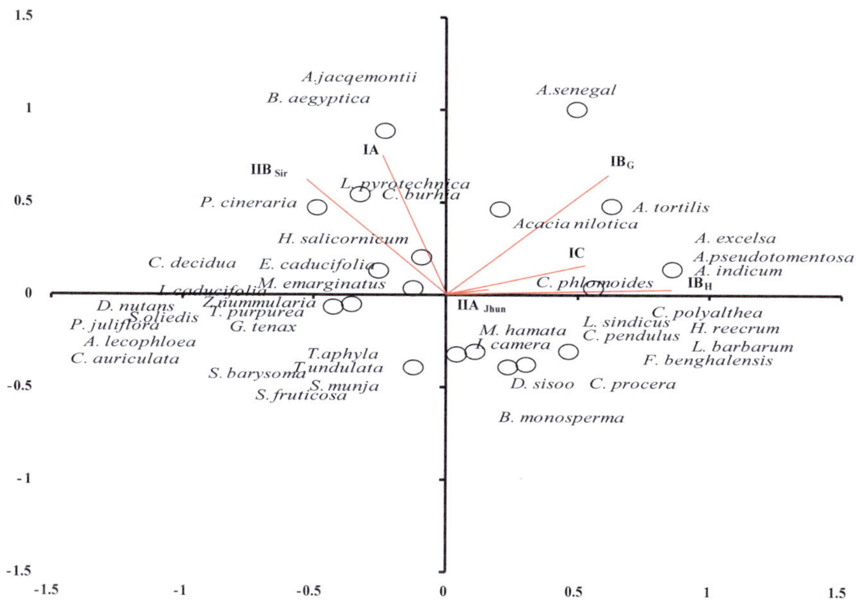

Fig. 5.17 Community attributes at sandy plains of different agro-climatic zones

IA and *D. sisso* for IIA_{Jhun}. Similarly, *A. indicum*, *A. psedotomentosa*, *A. excels* for IIB_H and *C. phlomoides*, *C. procera*, *B. monosperma*, *C. phlamoides* for II_G. *T. aphyla*, *T. purpurea*, *S. fruticosa*, *Z. nummularia*, *P. juliflora*, *S. munja*, *S. barysoma*, *T. undulata* are located away from IB_G and IB_H, i.e. gradual decrease of woody perennial as we approach sand plain toward IB (irrigated northwestern plain). Jaccard similarity index revealed (Table 5.12) high similarity of IB_G with

Table 5.12 Jaccard similarity index of sandy plain land form among the various agro-climatic zones

Zones	IB_G	IB_H	IC	IIA_{Jhun}	IA
IB_H	0.42	–	–	–	–
IC	0.27	0.30	–	–	–
IIA_{Jhun}	0.25	0.23	0.29	–	–
IA	0.21	0.05	0.17	0.11	–
IIB_{Siro}	0.21	0.07	0.22	0.33	0.28

IB_H (0.42). IB_H was least similar with IA (0.05) and with IIB_{Siro} (0.07). For other zones, similarity index among different zones are ranging from 0.11 to 0.33. Bartlett's sphericity test indicated as significant correlations (Table 5.10).

5.7.3 YAP

For this land form, different agro-climatic zones showed their own characteristic feature (Fig. 5.18). For example, IIB_{Siro} having indicator species like *C. felmentois*, *B. monosperma*, *D. nutans*, *C. procera*, *C. auriculata*. *A. senegal* is the indicator species for IIA_{Sik}, while, *A. leucophloa* and *A. pendula* for IIA_{Jhun}. IIB_{Jal} included species like *A. lebbeck* and *A. excelsa*, however, the second sub-zone of IIB i.e. IIB_{sir} had have species like *C. felmentois*, *B. monosperma*, *D. nutans*, *C. procera* and *C. auriculata*. *S. olides*, *A. cupressiformis*, *C. polyhonoides*, *C. burhia*, *Z. mauritiana*, *T. undulata*, *T. purpurea* are for IA. At this land form, major outlier species are *L. pyrotechnica*, *P. cineraria*, *R. mysorensis*, *G. tenax*, *M. hamata*, *S. cordifolia*, *S. oleoides*, *E. caducifolia*, *P. dulca*, *L. barbarium*. Results of Jaccard similarity index suggested very less similarity among the zones (Table 5.13) i.e. very less similarity and this also suggested by Bartlett's sphericity test (Table 5.10).

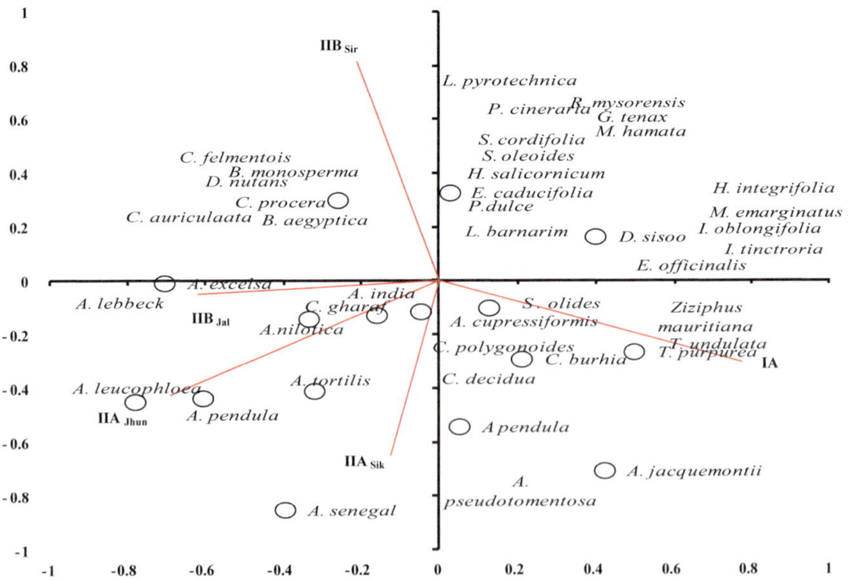

Fig. 5.18 Community attributes at YAP of different agro-climatic zones

Table 5.13 Jaccard similarity index of YAP land form among the various agro-climatic zones

Zones	IIA_{Sik}	IIA_{Jhun}	IIB_{Jal}	IIB_{Siro}
IIA_{Jhun}	0.18	–	–	–
IIB_{Jal}	0.09	0.22	–	–
IIB_{Siro}	0.09	0.09	0.29	–
IA	0.10	0	0.11	0.16

5.7.4 *Sand Dune*

PCA bi-plot for this land form at different zones revealed that IIA_{Sik}-IC and IIB_{Jhun}-IA are more closely associated with each other (Fig. 5.19). *G. tenax* and *L. sindicus* are indicator species for IIA_{Jhun}. Zone IIA_{Sik} is the leas similar with IIB_{Jal} (Table 5.14). Bartlett's sphericity test was recorded non-significant (Table 5.10).

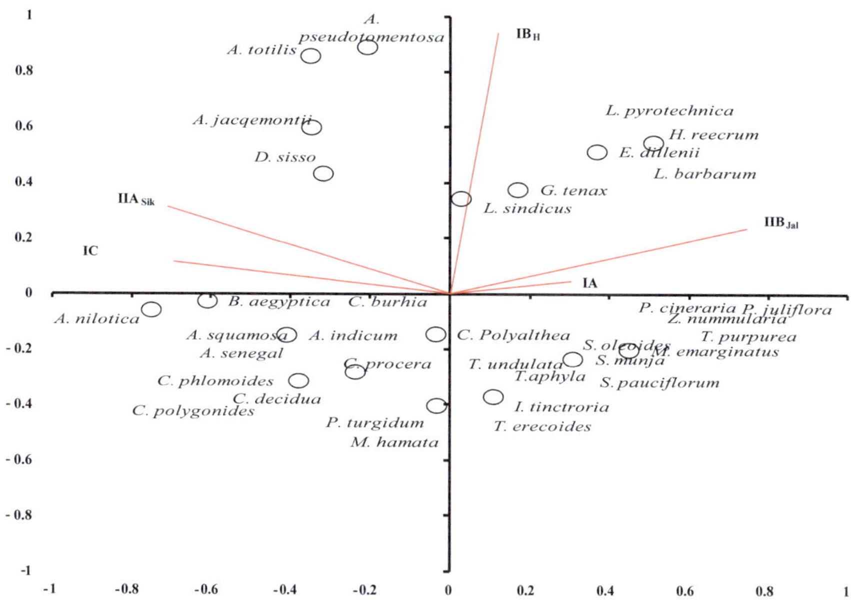

Fig. 5.19 Community attributes at sand dune of different agro-climatic zones

Table 5.14 Jaccard similarity index of sand dune land form among the various agro-climatic zones

Zones	IIA_{Sik}	IC	IB_H	IIB_{Jal}
IC	0.33	–	–	–
IB_H	0.18	0.16	–	–
IIB_{Jal}	0.08	0.12	0.28	–
IA	0.15	0.19	0.2	0.34

5.7.5 OAP

Here, *C. decidua* and *C. auriculata* are the indicator species of IIB_{Sir}, similarly *B. aegyptica* and *A. leucophloea* for IB_G and *C. procera* and *A. pseudotomentosa* for IIB_{Jal}. *D. sisso*, *E. caducifolia* and *H. salicornicum* for IB_{Hanu}. *S. barysoma*, *S. oliedis*, *P. juliflora*, *I. oblongifolia*, *L. barbarum*, *M. emarginata*, *T. amphyla*, *L. pyrotechnica* are the major outliersm (Fig. 5.20). Based on the Jaccard similarity test (Table 5.15) IB_{Hanu} is least similar with IIB_{Jal}. Bartlett's sphericity test was recorded non-significant.

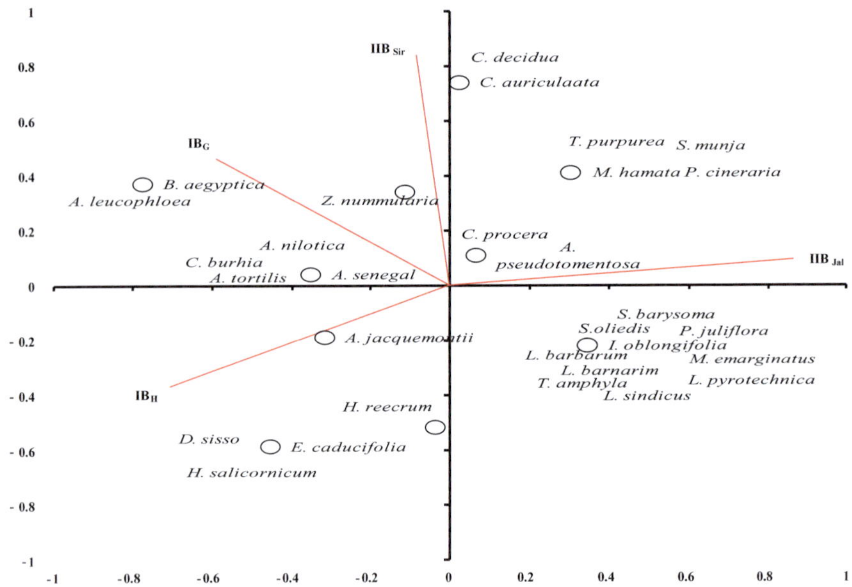

Fig. 5.20 Community attributes at OAP of different agro-climatic zones

Table 5.15 Jaccard similarity index of OAP land form among the various agro-climatic zones

Zones	IB_G	IB_H	IIB_{Jal}
IB_H	0.2	–	–
IIB_{Jal}	0.19	0.08	–
IIB_{Siro}	0.25	0.14	0.26

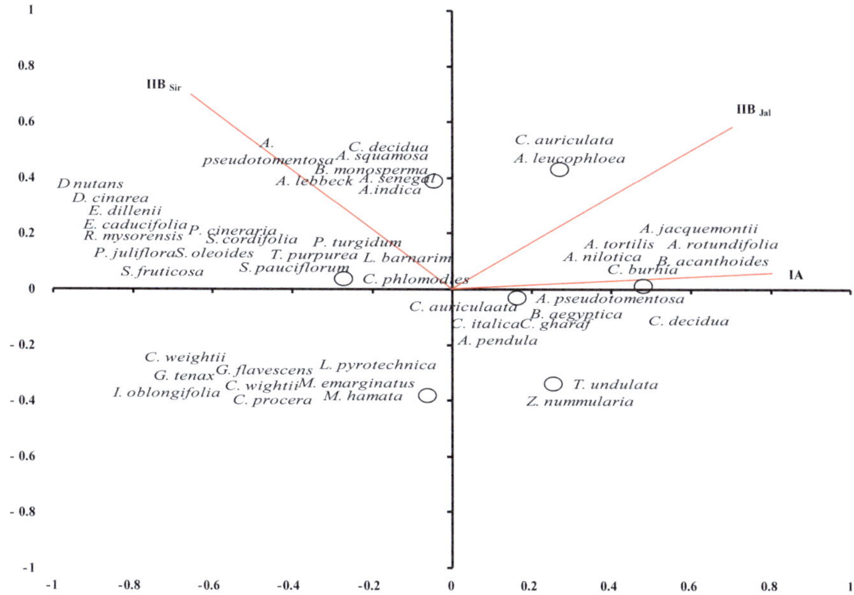

Fig. 5.21 Community attributes at Hills and Piedmonts of different agro-climatic zones

Table 5.16 Jaccard similarity index of Hills and Piedmonts land form among the various agro-climatic zones

Zones	IIB_{Jal}	IIB_{Siro}
IIB_{Siro}	0.24	–
IA	0.33	0.06

5.7.6 *Hills and Piedmont*

C. auriculata and *A. lecuophloea* are indicator species for IIB_{Jal}, Similarly, *A. pseudotomentosa*, *C. decidua*, *B. monosperma*, *A. senegal* and *A. indica* for IIB_{Sir} (Fig. 5.21). Maximum similarity was related between IIB_{Jal} and IA (0.33, Table 5.16) and least between IIB_{Sir} and IA. Barlett's sphericity test (Table 5.16) suggested significant correlation among variables.

5.8 Future Research Directions

In this holistic study, we found some newer information which may guide us for future endeavor.

- Zonal specific indicator species at different land forms.

- Need to study pertain to zone-specific succession trends at different land forms as we noticed some new species at different land forms related to different zones like:
  - *A. excels*, *B. aegyptica*, *L. pyrotechnica* at SUHP.
  - *A. pendula*, *T. undulate*, *A. cupressiformis*, *R. mysorensis* and *P. dulca* at YAP.
  - *C. auriculata*, *B. aegyptica*, *A. leucophloea*, *D. sissoo*, *M. emarginatus*, *S. barysoma* at OAP.
  - *B. monosperma* at Hills and Piedmonts.
- Linked the magnitude of land form heterogeneity and their impacts of phyto-diversity.
- Conversely, linked the impacts of phyto-diversity on land form heterogeneity.
- Studied the pedological behavior of different land from of different agro-climatic zones on community composition and diversity.
- Study the effect of land form on ecological roles of dominant species (like nutrient cycling, wind and soil erosion-resistant, etc.).
- Spatial patterns of dominant community associates at different land form at different agro-climatic zone.

Acknowledgement The authors gratefully acknowledge the Department for Facilities, UGC-CAS for financial support and CAZRI for granting study leave to the first author

References

Baghani M, Adel S, Barani H (2009) The role of SHE analysis in defining species diversity components of mountain rangelands (ZIARAT basin, Gorgan). J Agric Sci Nat Resour 6 (1):212–220

Bawa AK, Gupta IC, Sharma BD (1988) Natural resources and their management in rangelands of western Rajasthan. Curr Agric 12:33–56

Bhattarai KR, Vetaas OR, Grytnes JA (2004). Relationship between plant species richness and biomass in arid sub-alpine grassland of the central Himalayas Nepal. Folia Geobot 39:57–71

Buzas MA, Hayek LAC (1998) SHE analysis for biofacies identification. J Foramin Res 28: 233–239

Champion HG, Seth SK (1968) A revised survey of the forest types of India. The Manager of Publications, New Delhi, p 404

Corenblit D, Gurnell AM, Steiger J, Tabacchi E (2008) Reciprocal adjustments between landforms and living organisms: extended geomorphic evolutionary insights. Catena 73:261–273

Javed MD (2016) Distribution, species diversity and composition of plant communities in relation to various affecting factors in an alpine grassland at Bandipora, Kashmir. Pak J Bot 48(2): 551–560

Khan MA, Ram B (2003) Integrated natural resources and environmental impact assessment for sustainable development of Hanumangarh District, Rajasthan. Central Arid Zone Research Institute, Jodhpur, India, p 172

Khan MA, Moharana PC, Singh SK (2003) Integrated natural resources and environmental impact assessment for sustainable development of Ganganagar District, Rajasthan. Central Arid Zone Research Institute, Jodhpur, India, p 111

Kolarkar AS, Joshi DC, Sharma KD (1992) Rehabilitation of degraded arid ecosystem. Scientific Publishers, Jodhpur, p 203

Larkin D, Vivian-Smith G, Zedler JB (2006) Topographic heterogeneity theory and ecological restoration. Island Press

Ludwig JA, Reynolds JF (1999) Statistical ecology. A primary methods and computing. Wiley, New York, p 337

Magurran AE (2004) Measuring biological diversity. Blackwelll Publishing, USA, p 248

Mathur M (2005) Ecology and prospecting of some medicinal plants of aphrodisiac potential. Ph. D. thesis, Jai Narain Vyas University, Jodhpur, India, p 465

Mathur M (2015) Comportments of probability approaches in ethno-botanical inventories and the validations' of outcome through internal matrix exploration. Med Plants 7(2):79–94

Mathur M (2018) Magnitude of temporal fluctuations of composite environment and their impacts on *Cassia auriculata* (Linn.) performance. Trop Ecol 59(1):1–12

Mathur M, Sundaramoorthy S (2013) Economic assessment and conservation priorities of the Indian Thar desert medicinal plants. Indian J Nat Prod Resour 4(3):283–294

Mathur M, Sundaramoorthy S (2016) Pattern of herbaceous species richness and productivity along gradients of soil moisture and nutrients in the Indian Thar desert. J Arid Environ 125: 80–87

Narain P, Singh M, Khan MS, Kumar S (2005) Shrubs of Indian arid zone. Arid Agro-ecosystem Directorate, Central Arid Zone Research Institute, Jodhpur, p 178

Nawal K, Al-Amin N, Stigter CJ, El-Tayeb Mohammed A (2006) Establishment of tress for sand settlement in a completely decertified environment. Arid Land Res Manag 20:209–227

Ott T, Van Aarde RJ (2014) Coastal dune topography as a determinant of abiotic conditions and biological community restoration in northern KwaZulu-Natal, South Africa. Landscape Ecol Eng 10:17–28

Pelffini M, Bollati I (2014) Landforms and geomorphosites ongoing changes: concepts and implications for geoheritage promotion. Quaestiones Geogr 33(1):131–143

Prakash I, Gupta RK (1976) An ecological exploration of an extremely arid tract of Rajasthan. Indian J Ecol 3(1):11–21

Salarian T, Jouri MH, Askarizadeh D, Mahmoudi M (2015) The study of diversity indices of plant species using SHE method (case study: Javaherdeh Rangeland, Ramsar, Iran). J Range Sci 5 (1):27–37

Satyanarayan Y, Gaur YD (1968) Phytosociological changes of monsoon vegetation in semi-rocky habitats. J Indian Bot Soc 3–4:371–384

Saxena S (1977a) Desertification due to ecological changes in the vegetation of the Indian desert. Ann Arid Zone 16(3):367–373

Saxena S (1977b) Vegetation and its succession in the Indian desert. Desertification and its control. Indian Council of Agricultural Research, New Delhi, pp 176–192

Shankar V (1986) Vegetation recovery under protection. In: Shankarnarayan KA, Shankar V (eds) Desert environment conservation and management. Central Arid Zone Research Institute Monograph, p 134

Sharma KK, Mehra SP (2009) The thar of Rajasthan (India): ecology and conservation of a desert ecosystem. In: Sivaperuman C et al (eds) Functional ecology and conservation of the great Indian desert. Springer-Verlag, Berlin Heidelberg, pp 1–10

Singh S, Sharma KD, Joshi DC (1995) Integrated natural and human resources appraisal for sustainable development of Jalore district. Central Arid Zone Research Institute, Jodhpur, India, p 108

Singh S, Vangani NS, Singh N (1996) Integrated natural and human resources appraisal for sustainable development of Sikar district. Central Arid Zone Research Institute, Jodhpur, India, p 172

Singh G, Singh K, Mishra D, Shukla S (2012) Vegetation diversity and role of *Leptadenia pyrotechnica* in biomass contribution and carbon storage in arid zone of India. Arid Ecosyst 2 (4):264–272

Slaymaker O, Spencer T, Embleton-Hamann C (2011) Geomorphology and global environmental change. Cambridge University Press, p 211

Solon J, Degórski M, Roo-Zielińska E (2007) Vegetation response to a topographical-soil gradient. Catena 71:309–320

Stallins JA (2006) Geomorphology and ecology: unifying themes for complex systems in biogeomorphology. Geomorphology 77:207–216

Swanson FJ, Kratz TK, Caine N, Woodmansee RG (1988) Landform effects on ecosystem patterns and processes. Bioscience 38:92–98

Tiwari JC, Tripathi D, Sharma AK, Harsh LN, Khan MA (2007) Woody vegetation in forest of Kumbhalgarh central hills lying in transitional zone between arid and semi-arid tract. In: Trivedi PC (ed) Global biodiversity: status and conservation. Pointer Publishers, Jaipur, pp 1–29

Wilson B, Orchard K, Phillip J (2012) SHE analysis for biozone identification among foraminiferal sediment assemblages on reefs and in associated sediment around St. Kitts, Eastern Caribbean Sea, and its environmental significance. Mar Micropalentol 83:38–45

Xlstat (2007). Data analysis and statistical solution for microsoft excel. Addinsoft, Paris, France

Chapter 6 Chemotaxonomic Significance of Alkaloids in Plants

Ram Singh, Poonam and Geetanjali

Abstract The classification of plants on the basis of their chemical constituents is a powerful weapon for plant taxonomists. The chemotaxonomic classification, when applied with proper care and phytochemical inputs, is more useful than morphological- and anatomical-based classifications. However, chemotaxonomy also suffers from some noticeable disadvantages like the presence of common compounds in many plants. These compounds have small taxonomic values. Alkaloids are nitrogen-containing secondary metabolites present in plant kingdom. These alkaloids play significant role in chemotaxonomic classification of plants. The present chapter discusses the distribution of alkaloids in plants and their applications in plant taxonomy.

Keywords Plants · Taxonomy · Chemotaxonomy · Secondary metabolites · Alkaloids

6.1 Introduction

Secondary metabolites are the treasure of nature. The secondary metabolites synthesized by the plants constitute a major part of chemicals and they are useful for the survival of the plants (Jing et al. 2014; Croteau et al. 2000; Matsuura and Fett-Neto 2015). Alkaloids are plant secondary metabolites possessing basic nitrogen atom(s) mainly in the ring (Acamovic et al. 2004; Aniszewski 2007). The term 'alkaloids' was coined by the German chemist Carl F. W. Meissner in 1819 and the word is derived from the Arabic name al-qali that is related to the plant from which soda was first isolated (Aniszewski 2007). These biomolecules are obtained from amino acids or from the transamination process and serve as secondary

R. Singh (✉) · Poonam
Department of Applied Chemistry, Delhi Technological University, Delhi 110042, India
e-mail: ramsingh@dtu.ac.in

Geetanjali
Department of Chemistry, Kirori Mal College, University of Delhi, Delhi 110007, India

K. G. Ramawat (ed.), *Biodiversity and Chemotaxonomy*, Sustainable Development and Biodiversity 24, https://doi.org/10.1007/978-3-030-30746-2_6

metabolites. These compounds possess low molecular weight structures and provide defence against the herbivores and pathogens that are harmful to the plants. The isolated alkaloids and their synthetic derivatives are either directly or in combination also utilized as medicinal agents (Singh et al. 2011; Kaur and Arora 2015; Okwu and Okwu 2004). Starting from neuroactive molecules like nicotine (**1**) and caffeine (**2**) to the anti-malarial molecule like quinine (**3**), plant alkaloids have given its utility to human health. In this way, alkaloids have played its role to both plants and humans equally. The application of alkaloids for plant classification is another field waiting to get exploited.

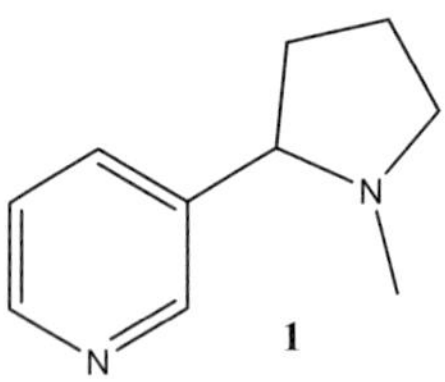

1

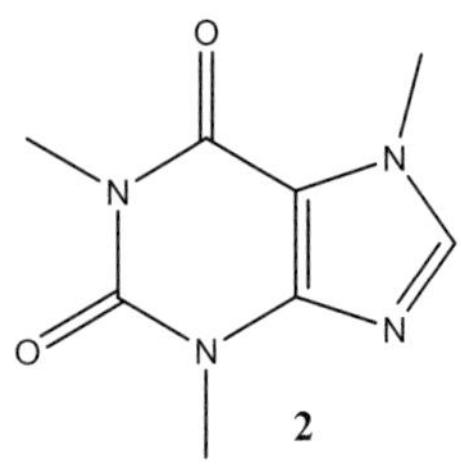

2

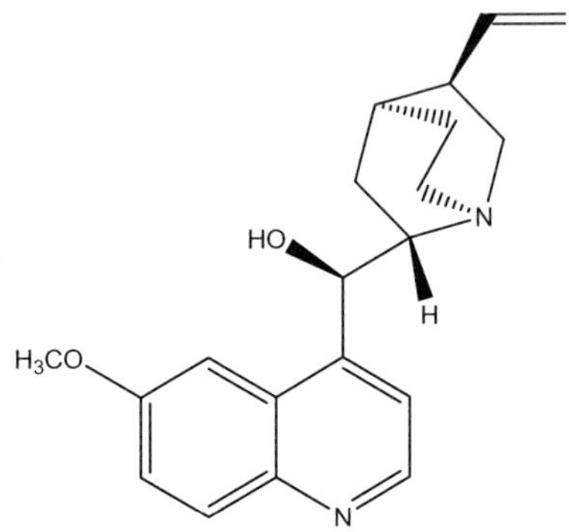

3

The classification of plants is one of the important aspects of botany people. To name and organize the plants is always a tedious work due to their huge number and varieties. From centuries, botanists attempted to formulate groupings based upon physiology, morphology, paleobotany and so on (Sivarajan 1991). The morphological and anatomical classifications are the traditional plant classifications. A comparatively newer approach towards plant classification is based on the chemical constituents or secondary metabolites synthesized by the plants. This classification is known as chemotaxonomy (Wink 2003; Misra and Srivastava 2016; Singh 2016; Singh and Geetanjali 2018). Chemotaxonomy includes the principles and procedures involved in the applications of chemical compounds produced by plants as evidence for classificatory purposes. The study of patterns of compounds isolated from various individual parts of plant have been considered for understanding the relationship of plants and categorize them into a particular group. The chemotaxonomic conclusion cannot be drawn by simply working on a few plants, isolating their secondary metabolites, rather adequate sampling is required from different developmental stages of a large number of species present in various ecological systems. This chapter will discuss, the secondary metabolite, alkaloids as chemical tool for chemotaxonomy by taking adequate evidences and a few examples from the literature.

6.2 Occurrence and Distribution of Alkaloids in Plants

Alkaloids have been defined by chemists as a heterocyclic nitrogenous compound that retain their basic chemical properties. However, the retention of basic properties is no longer a prerequisite for alkaloids. Since earlier time, the definitions of alkaloids have been under academic controversy. The definitions carry a few exceptions and omissions of the compounds belong to amino sugars, peptides, amino acids and nucleosides (Hampel and Hawley 1976; Hesse 1981; Jakubke et al. 1994).

The angiospermae, flowering plants are the major source of alkaloids in the past but with the discovery of better isolation methods and characterization techniques, alkaloids have also been obtained from other sources like animals, insects, lower plants, etc. (Aniszewski 2007; Roberts and Wink 1998). The alkaloids have been classified on the basis of various aspects. One important aspect is the biological pathway to synthesize the alkaloid molecule either from amino acids or without amino acids (Fig. 6.1) (Aniszewski 2007). True alkaloids and protoalkaloids are derived from amino acids whereas pseudo alkaloids are not derived from amino acids. In true alkaloids, nitrogen derived from amino acids is a part of heterocyclic ring (Pelletier 1983; Dewick 2002). These alkaloids are present as N-oxide or nitrogenous salt or in free state in plant species and families. Quinine (**3**), morphine (**4**) and cocaine (**5**) are a few examples of this alkaloid.

The protoalkaloids and nitrogen derived from amino acids like L-tyrosine and L-tryptophan are not the part of heterocyclic ring (Jakubke et al. 1994). Hordenine (**6**) and yohimbine (**7**) are the examples of protoalkaloids. In pseudoalkaloids, amino acids play indirect role and the nitrogen atom gets inserted at later stage (Jakubke et al. 1994). The precursors of amino acid are involved in their synthesis. In some plants, they are also derived from non-amino acid precursors. Caffeine (**2**), theobromine (**8**), ephedrine (**9**) and coniine (**10**) are a few examples of pseudoalkaloids.

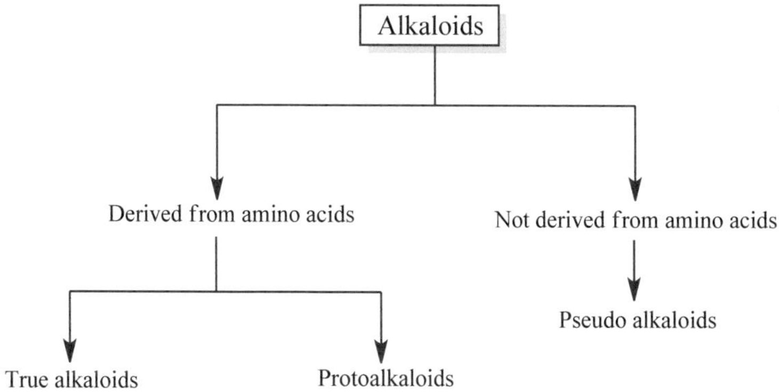

Fig. 6.1 Classification of alkaloids

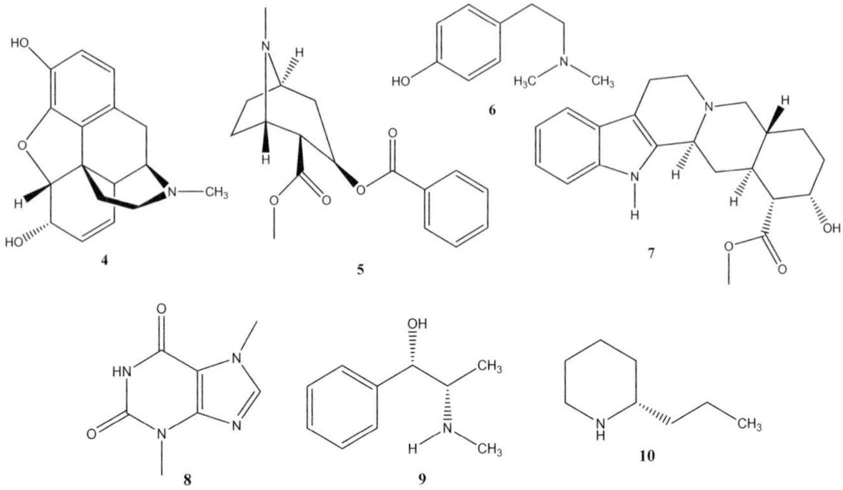

6.3 Alkaloids in Chemotaxonomy

According to Hegnauer, the alkaloid plants are those plants that contain alkaloids of more than 0.01% (Hegnauer 1988). More than 25% of the higher plants possess the alkaloid molecules. Examples of some alkaloid containing plant families are given in Fig. 6.2. With the development of advanced analytical techniques, the presence of alkaloids is easily detected even in very less amount. Alkaloids help better interaction of plants with the environment along with their fitness and survival (Kliebenstein 2012; Costa et al. 2012). The important groups of alkaloids based on

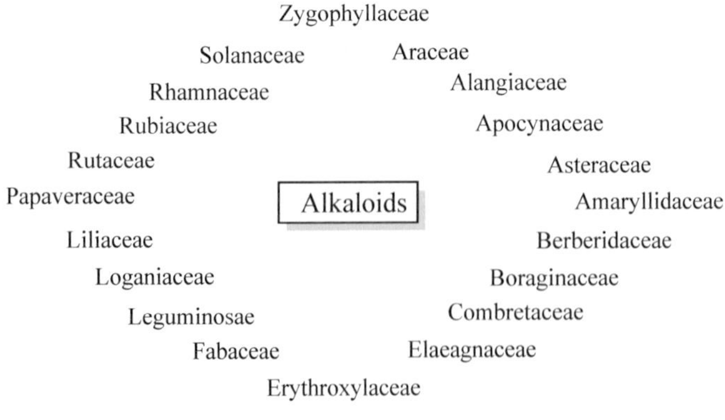

Fig. 6.2 Examples of some alkaloids containing plant families

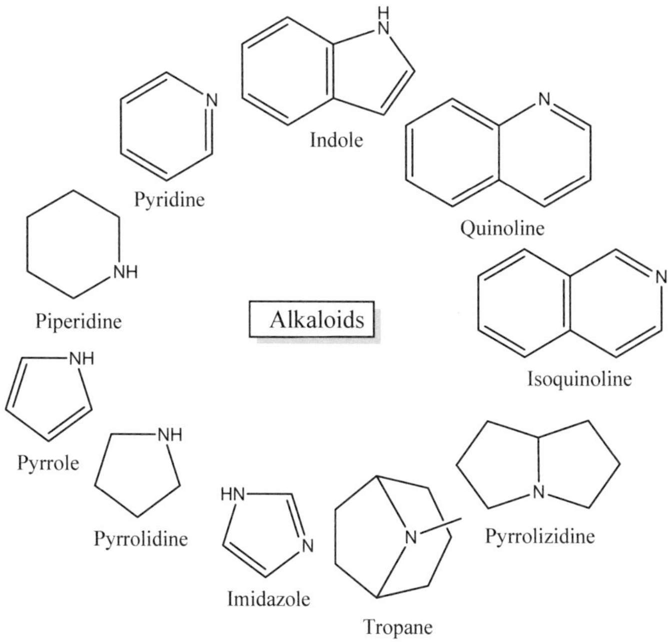

Fig. 6.3 Important of groups of alkaloids based on structure present in plant families

their structure present in plant families are given in Fig. 6.3. The following sections are going to discuss the utility of indole alkaloids, tropane alkaloids, pyrrolizidine alkaloids and piperidine alkaloids as model example in chemotaxonomy to understand the role of alkaloids in chemotaxonomic classifications.

6.3.1 Indole Alkaloids

Indole (Fig. 6.3) is an aromatic heterocyclic organic compound with molecular formula C_8H_7N, having a benzene ring fused with five-membered pyrrole ring. The precursor for the synthesis of various indole alkaloids is the amino acid L-tryptophan (**11**). This alkaloid is widely present in many plant species and helpful in their chemotaxonomy. A typical family for this alkaloid is *Apocynaceae*. The family *Apocynaceae* consists of flowering plants comprising some 5100 species. As per 2014, this family has 366 genera and 25 tribes that includes trees, shrubs and herbs (Endress et al. 2014; Endress and Bruyns 2000).

Rauvolfia serpentina member of *Apocynaceae* family is rich in monoterpenoid indole alkaloid and contains more than 50 different alkaloids. Many times, this genus is named as genus of the monoterpenoid indole alkaloid family. The major alkaloids isolated from this genus include yohimbine (**11**), reserpine (**12**), reserpiline (**13**), ajmaline (**14**), ajmalimine (**15**), ajmalicine (**16**), rescinnamine (**17**), serpentine (**18**), deserpidine (**19**), indobinine (**20**) and indobine (**21**) (Kumari et al. 2013). Apart from these alkaloids, another five alkaloids that included two ajmaline derivatives N_b-methylajmaline and N_b-methylisoajmaline and three others like 3-hydroxysarpagine, yohimbinic acid and isorauhimbinic acid were isolated from the dried roots of *R. serpentina* (Itoh et al. 2005). The roots of *Rauwolfia hookeri*, *R. micrantha*, *R. serpentina*, *R. verticillata*, *R. tetraphylla* and *R. vomitoria* also contain monoterpene indole alkaloids (Bindu et al. 2014; O'Connor and Maresh 2006).

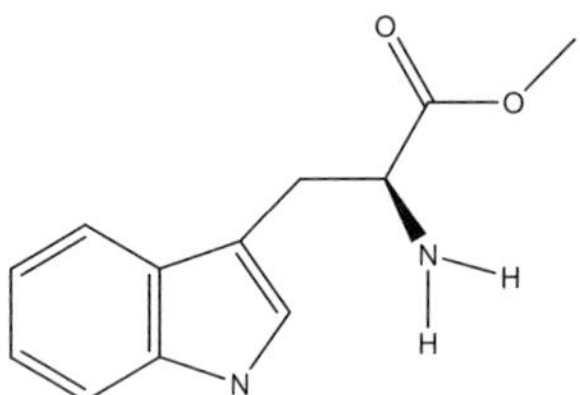
11

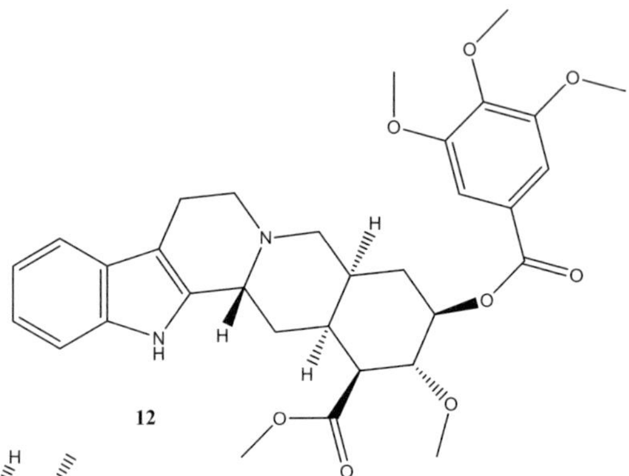
12

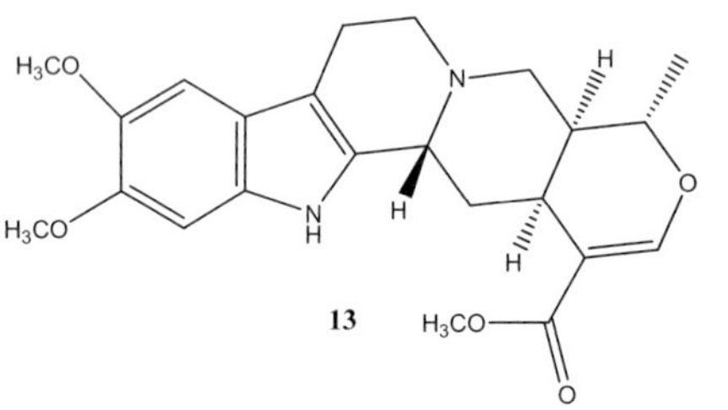
13

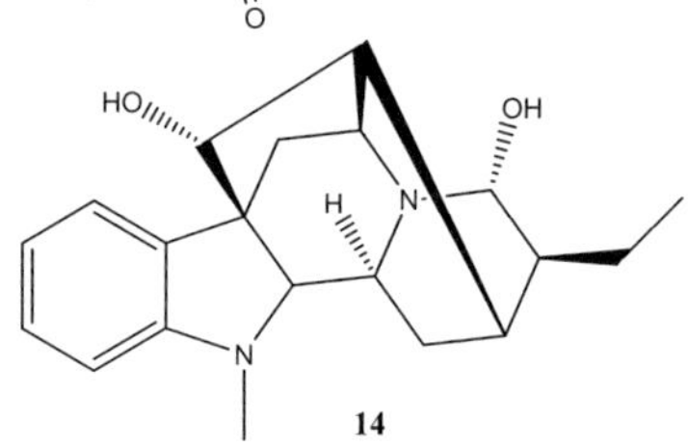
14

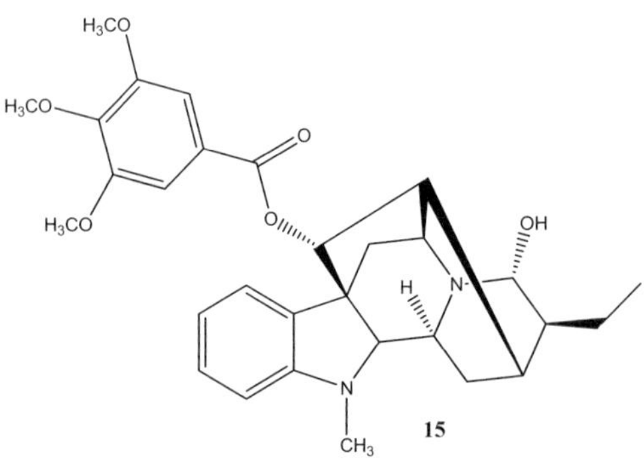
15

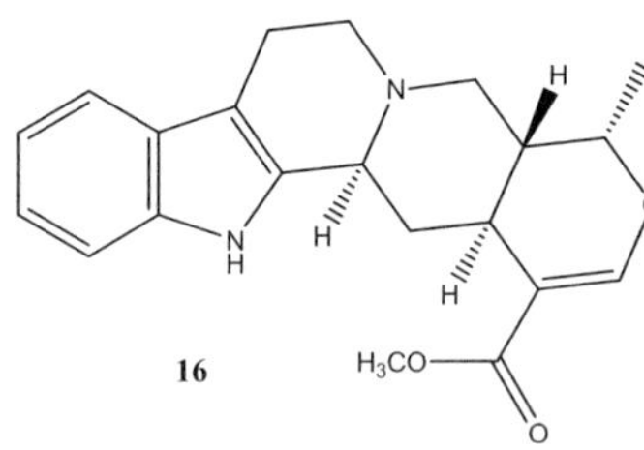
16

Another genus from the family *Apocynaceae*, *Catharanthus* produces more than 130 different terpenoid indole alkaloids including serpentine (**18**), alstonine (**22**), secodine (**23**) and perivine (**24**) (van der Heijden et al. 2004). Some of the important alkaloids isolated from *C. roseus*, which is a rich source of alkaloids among the genus *Catharanthus*, includes ajmalicine (**16**), vinblastine (**25**), vincristine (**26**) and anhydrovinblastine (**27**). The genus *Catharanthus* has close relation with the genera *Vinca* and *Amsonia*. The species of these genera are associated chemotaxonomically with Catharanthus indole alkaloids (Taylor and Farnsworth 1975; Zhu et al. 2015).

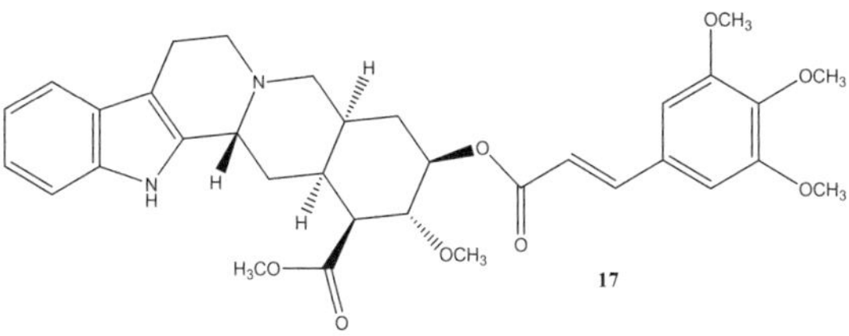
17

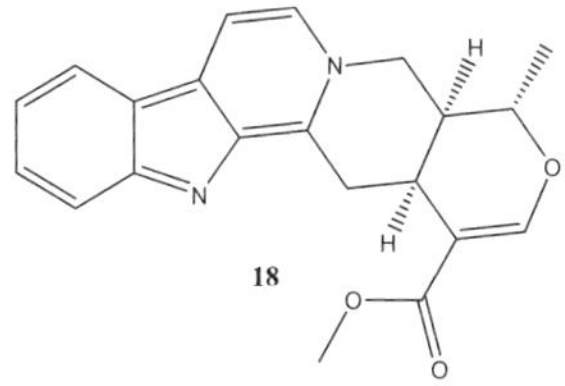
18

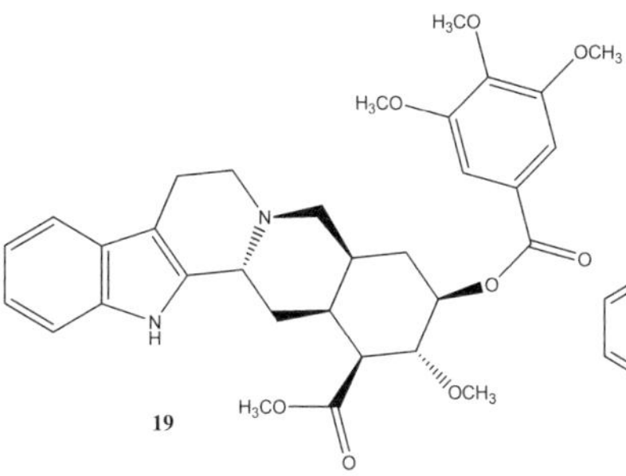
19

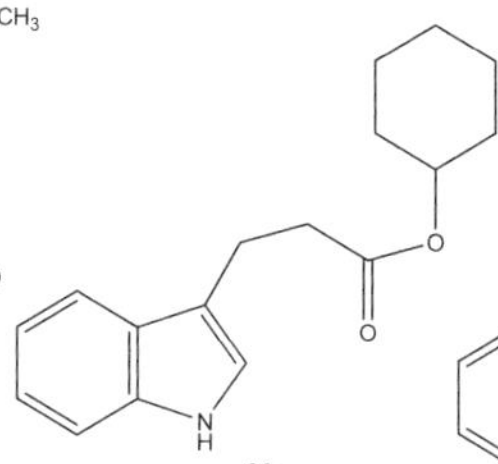
20

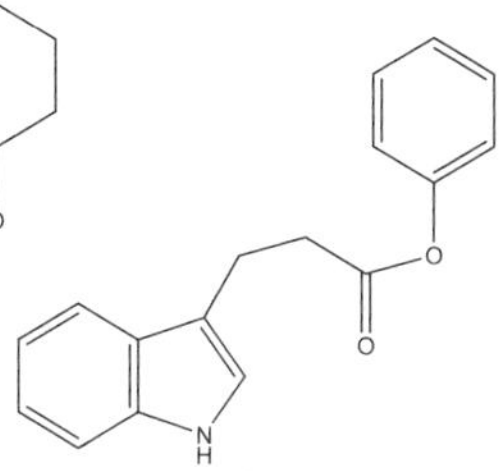
21

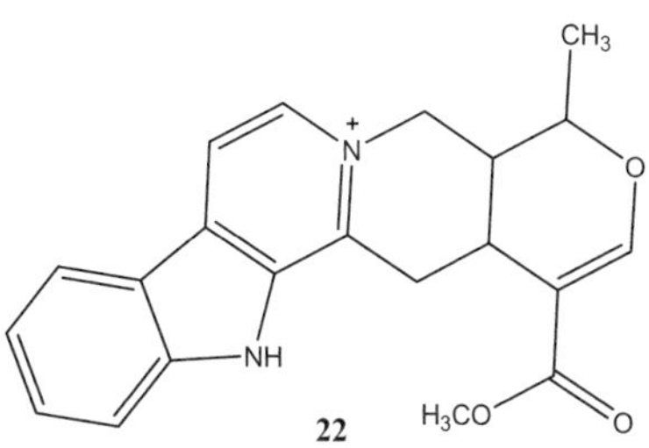
22

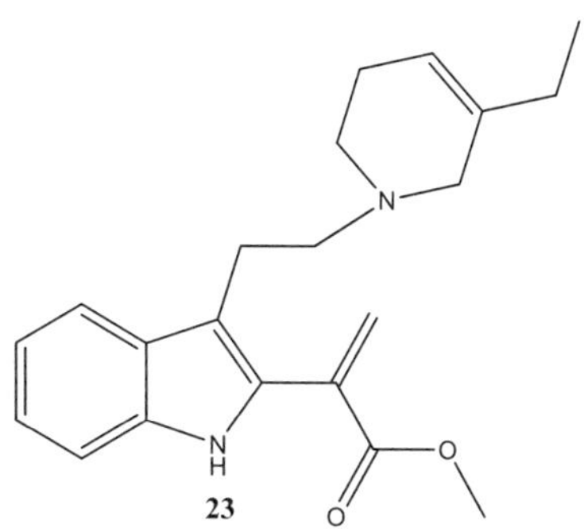
23

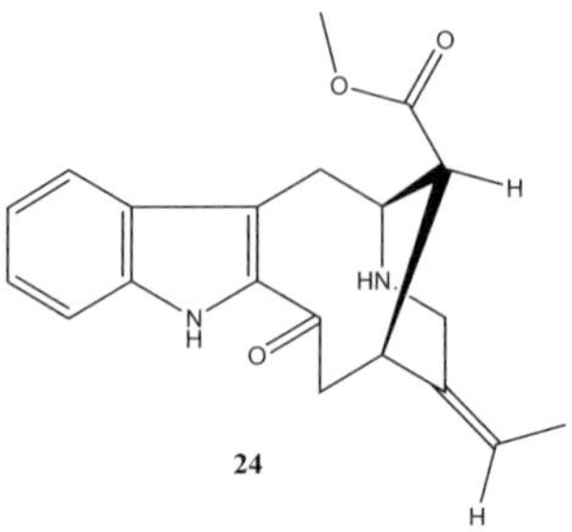

24

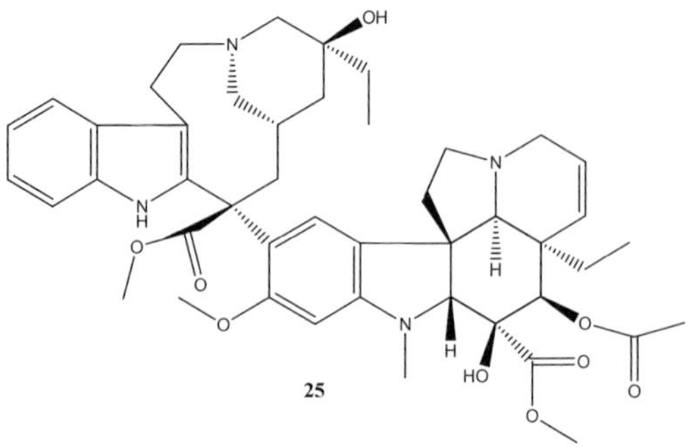

25

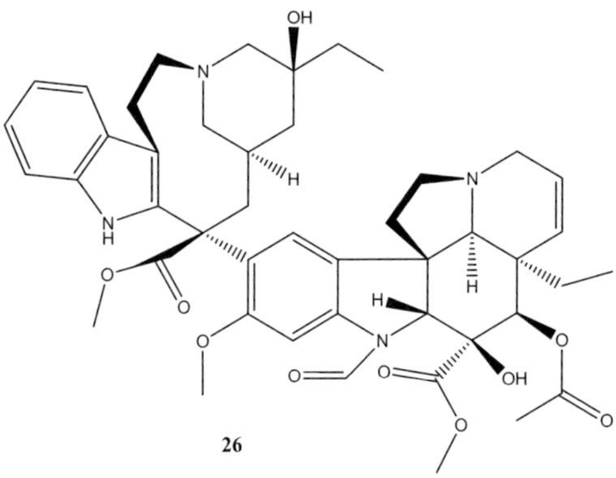

26

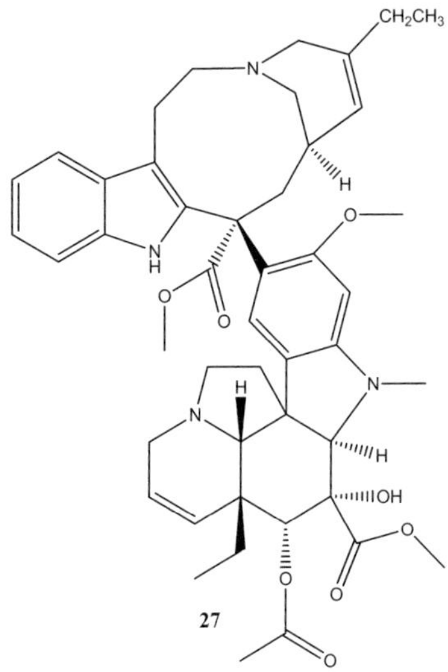

27

6.3.2 Tropane Alkaloids

Tropane (8-methyl-8-azabicyclo[3.2.1]octane, Fig. 6.3) is a bicyclic nitrogenous organic compound that comprises of a pyrrolidine and a piperidine ring sharing one nitrogen and two carbons atoms (Lounasmaa and Tamminen 1993). All the tropane alkaloid contains this moiety where this is present as its ester in most of the secondary metabolites isolated (Griffin and Lin 2000). The plant families like *Convolvulaceae*, *Erythroxylaceae*, *Euphorbiaceae*, *Proteaceae*, *Rhizophoraceae* and *Solanaceae* are the rich sources of tropane alkaloids (Griffin and Lin 2000; Pigatto et al. 2015). The most studied family with respect to this alkaloid is *Solanaceae,* which comprises about 100 genera and 3000 species and secondary metabolites isolated has its potential to be used for taxonomic classification (Pigatto et al. 2015).

From the family, *Solanaceae*, the genera like *Anthocercis* Labill., *Brugmansia* Pers., *Datura* L., *Hyoscyamus* L. and *Solandra* are rich in tropane alkaloids (Pigatto et al. 2015). The compounds like hyoscyamine (**28**) and atropine (**29**) have been isolated from about 70 species of 17 genera belonging to this family. Another compound scopolamine (**30**) has been isolated from about 18 genera, making this as a chemotaxonomic marker. The genus, *Datura* produced more than 30 tropane alkaloids from its different species (Berkov and Zayed 2004; Berkov et al. 2006). The alkaloids like hyoscyamine (**28**), scopolamine (**30**) and atropine (**29**) have been isolated from *Datura stramonium* (Soler-Rodríguez et al. 2006). Their concentrations were appreciable in *Datura ferox*, as well as in *Datura innoxia* also and mainly found in the leaves and flowers of the plants.

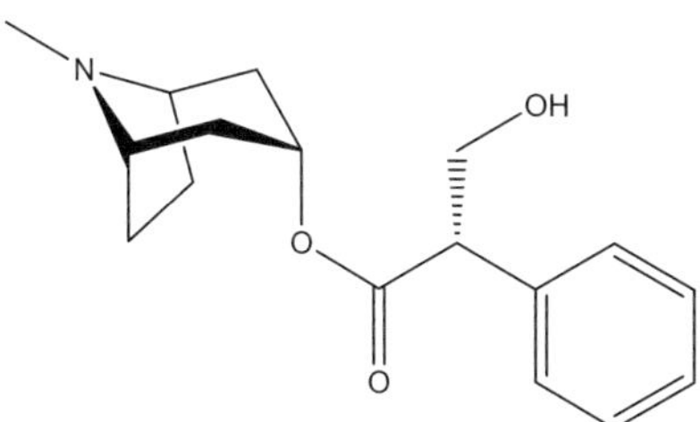

28

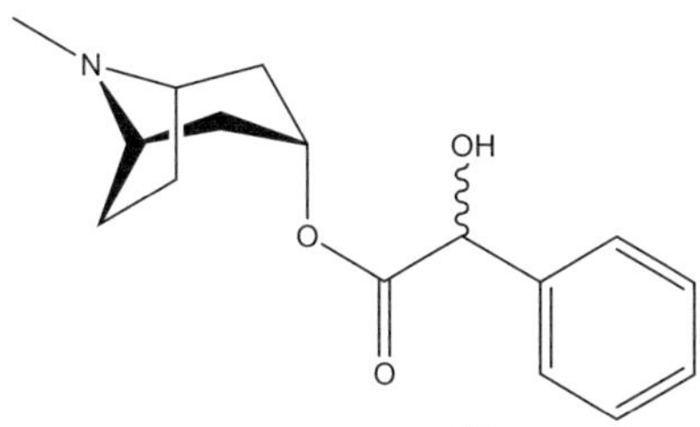

29

The family *Erythroxylaceae* has also been reported to be the source of tropane alkaloids (Jirschitzka et al. 2012). The genus *Erythroxylum,* which belongs to this family, has about 250 species. The phytochemical investigation of *E. monogynum* resulted in the isolation of several tropane alkaloids (Bringmann et al. 2000). Another species *Erythroxylum coca* produces tropane alkaloid cocaine (**5**) in appreciable amount (Plowman 1981).

6.3.3 Pyrrolizidine Alkaloids

The pyrrolizidine alkaloids (PAs) are the derivatives of pyrrolizidine (Fig. 6.3). These alkaloids are present in a large number of plant families which includes *Asteraceae*, *Apocynaceae*, *Boraginaceae*, *Euphorbiaceae*, *Fabaceae*, *Poaceae*, *Ranunculaceae* and *Scrofulariaceae* along with some other families where this is found in lower amounts (Smith and Culvenor 1981; Hartmann 1999). These alkaloids are usually formed by the esterification of necic acid (**31**) and a necine base and are present in about 6000 species (Langel et al. 2011).

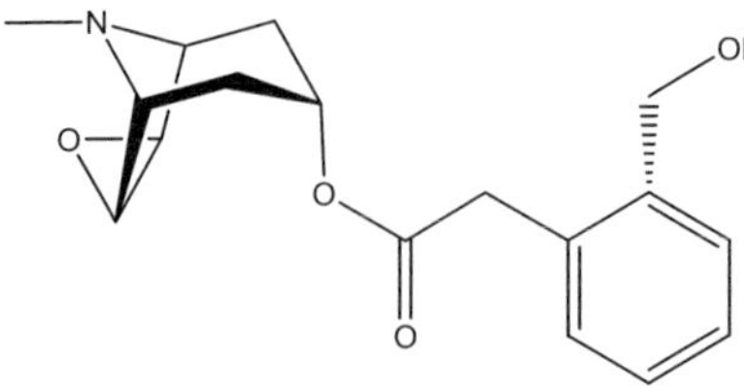

30

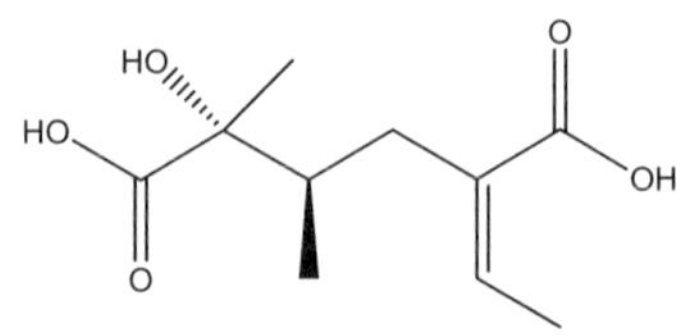

31

In the family of *Asteraceae*, the genera *Senecio* possess pyrrolizidine alkaloids in the macrocyclic diester form (Hartmann and Ober 2000; Reimann et al. 2004). The presence of PAs as secondary metabolites had already been reported in many species of *Senecio* such as *S. brasiliensis*, *S. conzyifolius*, *S. heterotrichius*, *S. oleosus*, *S. oxyphyllus*, *S. riograndensis*, *S. bonariensis*, *S. grossidens*, *S. icoglossus*, *S. juergensii*, *S. pulcher*, *S. ceratophylloides*, *S. crassiflorus* (Trigo et al. 2003); *S. alpines* (Pelser et al. 2005); *S. madagascariensis* (Gardner et al. 2006); *S. nemorensis*, *S. aquaticus*, *S. vernalis*, *S. jacobaea* (Kostova et al. 2006); *S. scandens*, *S. chrysanthemoides* (Roeder 2000; Tundis et al. 2009) and *S. pterophorus* (Castells et al. 2014).

Another family rich in pyrrolizidine alkaloids is *Boraginaceae*, flowering plant family. This family contains about 145 genera and above 2200 species. The alkaloids isolated from these species were either free necines or along with their N-oxides possessing either single esters at C-9 or C-7 or diesters linking and/or open chain with C-7 and C-9 (McLean 1970; Roeder 1999, 2000; Roeder and Wiedenfeld 2009, 2013). The alkaloids retronecine (**32**), supinidine (**33**), and trachelanthamidine (**34**) were isolated from about 20 species of genus *Amsinckia* (*Boraginaceae*) (El-Shazly and Wink 2014; Roitman 1988; Kelley and Seiber 1992). The retronecine type of alkaloids has also been isolated from genus *Echium* (*Boraginaceae*). The alkaloids supinidine (**33**) and trachelanthamidine (**34**) were isolated from the genus *Cynoglossum* (*Boraginaceae*) (Mroczek et al. 2004). Another genus of *Boraginaceae* family, genus *Heliotropium* also found to be rich is pyrrolizidine alkaloids like retronecine (**32**), supinidine (**33**), and trachelanthamidine (**34**) (Hartmann and Witte 1995). The alkaloids helindicine was isolated from *H. indicum* (Souza et al. 2005), incanine from *H. olgae* and trichodesmine from *H. arguzioides* (Hartmann and Witte 1995). The heliotridine and echinatine were isolated from the genus *Rindera* (Akramov et al. 1965; Mandić et al. 2013). The retronecine type of alkaloids were isolated from the genus *Symphytum* of *Boraginaceae* family, however, some species like *S. asperum*, *S. caucasium*, and *S. officinale* were found to produce both retronecine and heliotridine type alkaloids (Hartmann and Witte 1995; Kurucu et al. 2002; Liu et al. 2009; Huizing et al. 1982; Mel'kumova et al. 1974).

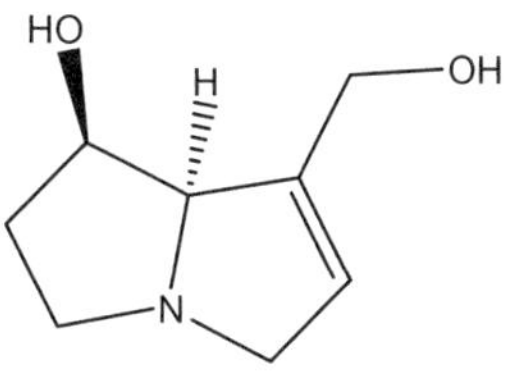

32

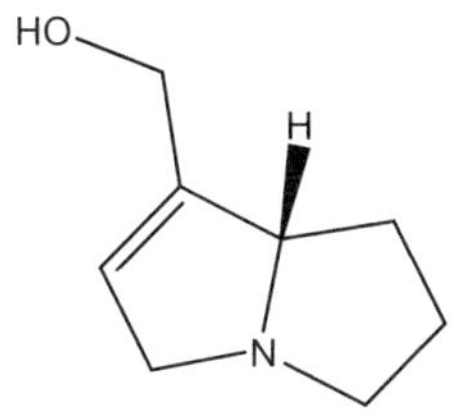

33

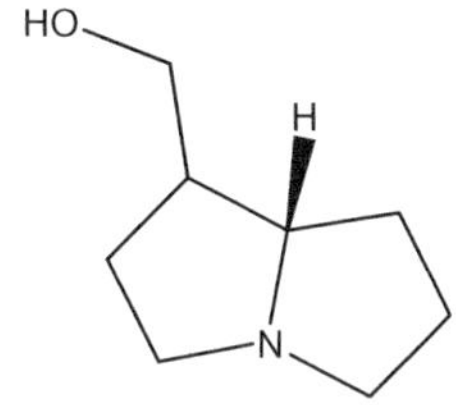

34

6.3.4 *Piperidine Alkaloids*

Piperidine is a heterocyclic molecule with the molecular formula $(CH_2)_5NH$ (Fig. 6.3). The alkaloids possessing this moiety are known as piperidine alkaloids. They are widely distributed in plant kingdom. One of the important family, bean family, which is largest in terms of a number of species, produces this type of alkaloids (Xu and Deng 2017). The piperidine alkaloids such as 2-piperidine carboxylic acid (**35**), 4-hydroxy-2-piperidine carboxylic acid (**36**), and ammodendrine (**37**) were produced from the three subfamilies Genistoid clade, Astragaleae, and Castanospermum of *Fabaceae* (Wink 1993, 2013). These alkaloids can be a chemotaxonomic marker for the family. The various genera rich in piperidine alkaloids are *Ammodendron*, *Carica*, *Cassia*, *Conium*, *Collidium*, *Dichroa*, *Duboisia*, *Genista*, *Withania,* and so on (Panter et al. 1988; Hotti and Rischer 1962).

Another important piperidine alkaloid, coniine (**10**) is also a potential chemotaxonomic marker due to its presence in a diversity of plants, including the genera *Aloe* (Nash et al. 1992), *Conium* (Cromwell 1956) and *Sarracenia* (Hotti et al. 2017). The naturally occurring coniine (**10**) is S-isomer with respect to C-2 position (Reynolds 2005). The genus *Aloe* (*Xanthorrhoeaceae*) has about 400 species but a few comparatively less alkaloids have been isolated. The piperidine alkaloids have been isolated from about twelve species (Nash et al. 1992). The alkaloid γ-coniceine has been isolated from about ten species and coniine (**10**) from five species. These alkaloids may act as chemotaxonomic marker for these species (Nash et al. 1992; Dring et al. 1984). About eight species of the genus *Sarracenia* from *Sarraceniaceae* family also possess coniine (**10**) (Mody et al. 1976; Hotti et al. 2017).

6.4 Limitations of Alkaloid Chemotaxonomy

The actual presence or absence of a particular alkaloids in a plant species can only be detected by using large quantity of plant materials. It is most likely that traces of alkaloids may not be detected by using a small amount of plant materials. The quantitative distribution of alkaloids in any plant depends upon various factors. This is also true for other secondary metabolites whose quantity varies with change in ecological aspects (Broun et al. 2006). The ecological aspects play important role in the quantitative accumulation and biosynthetic pathways of alkaloids (Verma and Shukla 2015). The abiotic and biotic factors influence the biosynthetic pathways of alkaloids and other secondary metabolites (Zhi-Lin et al. 2007; Radusiene et al. 2012).

The variation in the production of alkaloids causes chemotaxonomic limitations based on alkaloids. The abiotic factors like temperature, soil condition, humidity, altitudes, etc., sometimes lead to the environmental stress for plants and hence production of secondary metabolites varies from species to species. At the time of stressful condition, the secondary metabolites of the plants help them to survive like *Artemisia annua*, *Hypericum perforatum,* and *Catharanthus roseus* showed during difficult ecological condition increase in their secondary metabolites production (Verma and Shukla 2015).

The alkaloids production is also affected by the salinity of the water. For example, the concentration of reserpine (**12**) and vincristine alkaloids from the genera *Rauvolfia tetraphylla* and *Catharanthus roseus* increases due to the salt stress. The salt effect even affects the concentration of alkaloids in different parts of the plant also. For example, the amount of ricinine (**38**) alkaloids was found to increase in shoots in comparison with the roots with change in the salt concentration of the water (Said-Al Ahl and Omer 2011).

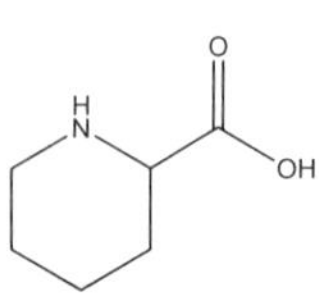

35

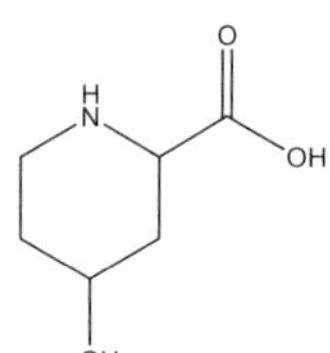

36

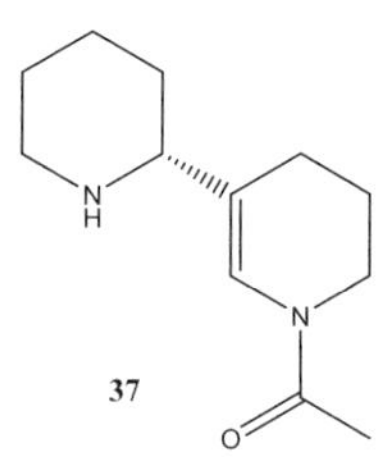

37

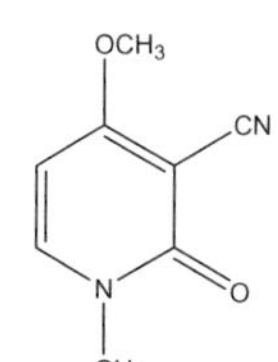

38

6.5 Conclusions

The chemotaxonomic classification is the classification of plants on the basis of the chemical constituents isolated from them. Due to the advancement in the isolation methods and analytical techniques, the isolation and characterization of secondary metabolites become less cumbersome in comparison to the traditional methods. This makes this classification more approachable. The chemotaxonomic classification based on alkaloids as secondary metabolites is more useful due to its wide presence in plant kingdom as this helps the plants to fight with pathogens. Detailed understanding of alkaloid biosynthesis is essential to understand the presence of a particular type of alkaloids in a particular plant(s) to get chemotaxonomic help. This will help in sustainably exploit them for plant classification. The merits of chemotaxonomic classification are always there if the systematic classification contradicts among themselves.

The chemotaxonomic classifications also suffer from some serious limitations as the presence or absence of secondary metabolites including alkaloids can only be detected by using large quantity of plant materials. Also, the quantity varies with change in ecological aspects. The ecology always influences the quantitative accumulation and biosynthetic pathways of alkaloids. Hence, to make sure about the presence, plant materials must be sampled from different places and analysed.

Acknowledgements The author Poonam is thankful to Delhi Technological University for financial support.

References

Acamovic T, Stewart CS, Pennycott TW (2004) Poisonous plants and related toxins. CABI, Wallingford, UK

Akramov ST, Kiyamitdinova F, Yunusov SY (1965) Study of *Rindera* and *Lindelofia*. Dokl Akad Nauk Uzb 22:35–38

Aniszewski T (2007) Alkaloids-secrets of life: alkaloids chemistry, biological significance, applications and ecological role. Elsevier, Amsterdam, The Netherlands

Berkov S, Zayed R, Doncheva T (2006) Alkaloid patterns in some varieties of *Datura stramonium*. Fitoterapia 77:179–182

Berkova S, Zayed R (2004) Comparison of tropane alkaloid spectra between *Datura innoxia* grown in Egypt and Bulgaria. Z Naturforsch C 59:184–186

Bindu S, Rameshkumar KB, Kumar B, Singh A, Anilkumar C (2014) Distribution of reserpine in *Rauvolfia* species from India—HPTLC and LC–MS studies. Ind Crops Prod 62:430–436

Bringmann G, Günther C, Mühlbacher J, Gunathilake MLP, Wickramasinghe A (2000) Tropane alkaloids from *Erythroxylum zeylanicum* OE Schulz (Erythroxylaceae). Phytochemistry 53:409–416

Broun P, Liu Y, Queen E, Schwarz Y, Abenes ML, Leibman M (2006) Importance of transcription factors in the regulation of plant secondary metabolism and their relevance to the control of terpenoid accumulation. Phytochem Rev 5:27–38

Castells E, Mulder PP, Pérez-Trujillo M (2014) Diversity of pyrrolizidine alkaloids in native and invasive *Senecio pterophorus* (Asteraceae): implications for toxicity. Phytochemistry 108: 137–146

Costa TDSA, Vieira RF, Bizzo HR, Silveira D, Gimenes MA (2012) Secondary metabolites. In: Dhanarasu S (ed) Chromatography and its applications. InTech Publishers, Croatia, pp 131–164

Cromwell BT (1956) The separation, micro-estimation and distribution of the alkaloids of hemlock (*Conium maculatum* L.). Biochem J 64:259–266

Croteau R, Kutchan TM, Lewis NG (2000) Natural products (secondary metabolites). Biochem Mol Biol Plants 24:1250–1319

Dewick PM (2002) Medicinal natural products. A biosynthetic approach, 2nd edn. Wiley, Chichester, New York

Dring JV, Nash RJ, Roberts MF, Reynolds T (1984) Hemlock alkaloids in *Aloes*. Occurrence and distribution of γ-coniceine. Planta Med 50:442–443

El-Shazly A, Wink M (2014) Diversity of pyrrolizidine alkaloids in the Boraginaceae structures, distribution, and biological properties. Diversity 6:188–282

Endress ME, Bruyns PV (2000) A revised classification of the Apocynaceae s.l. Bot Rev 66:1–56

Endress ME, Liede-Schumann S, Meve U (2014) An updated classification for Apocynaceae. Phytotaxa 159:175–194

Gardner DR, Thorne MS, Molyneux RJ, Pfister JA, Seawright AA (2006) Pyrrolizidine alkaloids in *Senecio madagascariensis* from Australia and Hawaii and assessment of possible livestock poisoning. Biochem Syst Ecol 34:736–744

Griffin WJ, Lin GD (2000) Chemotaxonomy and geographical distribution of tropane alkaloids. Phytochemistry 53(6):623–637

Hampel CA, Hawley GG (1976) Glossary of chemical terms. Nostrand Reinhold Company, New York

Hartmann T (1999) Chemical ecology of pyrrolizidine alkaloids. Planta 207:483–495

Hartmann T, Ober D (2000) Biosynthesis and metabolism of pyrrolizidine alkaloids in plants and specialized insect herbivores. Springer-Verlag, Berlin, pp 207–243

Hartmann T, Witte L (1995) Pyrrolizidine alkaloids: chemical, biological and chemoecological aspects. Alkaloids Chem Biol Perspect 9:155–233

Hegnauer R (1988) Biochemistry, distribution and taxonomic relevance of higher plant alkaloids. Phytochemistry 27:2423–2427

Hesse M (1981) Alkaloid chemistry. Wiley, New York

Hotti H, Rischer H (1962) The killer of Socrates: coniine and related alkaloids in the plant kingdom. Molecules 22. https://doi.org/10.3390/molecules22111962

Hotti H, Gopalacharyulu P, Seppänen-Laakso T, Rischer H (2017) Metabolite profiling of the carnivorous pitcher plants *Darlingtonia* and *Sarracenia*. PLoS ONE 12:e0171078

Huizing HJ, Gadella TW, Kliphuis E (1982) Chemotaxonomical investigations of the *Symphytum officinale* polyploid complex and *S. asperum* (Boraginaceae): the pyrrolizidine alkaloids. Plant Syst Evol 140:279–292

Itoh A, Kumashiro T, Yamaguchi M, Nagakura N, Mizushina Y, Nishi T, Tanahashi T (2005) Indole alkaloids and other constituents of *Rauwolfia serpentina*. J Nat Prod 68:848–852

Jakubke H-D, Jeschkeit H, Eagleson M (1994) Concise encyclopedia chemistry. Walter de Gruyter, Berlin, New York

Jing H, Liu J, Liu H, Xin H (2014) Histochemical investigation and kinds of alkaloids in leaves of different developmental stages in *Thymus quinquecostatus*. Sci World J 2014:1–6

Jirschitzka J, Schmidt GW, Reichelt M, Schneider B, Gershenzon J, D'Auria JC (2012) Plant tropane alkaloid biosynthesis evolved independently in the Solanaceae and Erythroxylaceae. Proc Nat Acad Sci 109:10304–10309

Kaur R, Arora S (2015) Alkaloids-important therapeutic secondary metabolites of plant origin. J Crit Rev 2:1–8

Kelley RB, Seiber JN (1992) Pyrrolizidine alkaloid chemosystematics in *Amsinckia*. Phytochemistry 31:2369–2387

Kliebenstein DJ (2012) Making new molecules—evolution of structures for novel metabolites in plants. Curr Opin Plant Biol 16:1–6

Kostova N, Christov V, Cholakova M, Nikolova E, Evstatieva L (2006) Pyrrolizidine alkaloids from Bulgarian species of the genus *Senecio*. J Serb Chem Soc 71:1275–1280

Kumari R, Rathi B, Rani A, Bhatnagar S (2013) *Rauvolfia serpentina* L. Benth. ex Kurz.: phytochemical, pharmacological and therapeutic aspects. Int J Pharm Sci Rev Res 23:348–355

Kurucu S, Kartal M, Choudhary MI, Topçu G (2002) Pyrrolizidine alkaloids from Symphytum sylvaticum Boiss. subsp. *sepulcrale*. (Boiss. & Bal.) Greuter & Burdetvar. *sepulcrale* and *Symphytum aintabicum* Hub. -Mor. & Wickens. Turk J Chem 26:195–200
Langel D, Ober D, Pelser PB (2011) The evolution of pyrrolizidine alkaloid biosynthesis and diversity in the Senecioneae. Phytochem Rev 10:3–74
Liu F, Wan SY, Jiang Z, Li SFY, Ong ES, Osorio JCC (2009) Determination of pyrrolizidine alkaloids in comfrey by liquid chromatography–electrospray ionization mass spectrometry. Talanta 80:916–923
Lounasmaa M, Tamminen T (1993) The tropane alkaloids: chemistry and Biology. In: Cordell GA (ed) The alkaloids, vol 44. Academic Press, New York, pp 1–114
Mandić BM, Simić MR, Vučković IM, Vujisić LV, Novaković MM, Trifunović SS, Milosavljević SM (2013) Pyrrolizidine alkaloids and fatty acids from the endemic plant species *Rindera umbellata* and the effect of lindelofine-N-oxide on tubulin polymerization. Molecules 18:10694–10706
Matsuura HN, Fett-Neto AG (2015) Plant alkaloids: main features, toxicity, and mechanisms of action. In: Gopalakrishnakone P, Carlini C, Ligabue-Braun R (eds) Plant toxins. Toxinology. Springer, Dordrecht
McLean EK (1970) The toxic actions of pyrrolizidine (Senecio) alkaloids. Pharmacol Rev 22:429–483
Mel'kumova ZV, Telzhenetskaya MV, Yunusov SY, Man'ko IV (1974) Refinement of the structure of asperumine. Khim Prir Soed 4:478–480
Misra A, Srivastava S (2016) Chemotaxonomy: an approach for conservation and exploration of industrially potential medicinal plants. J Pharmacogn Nat Prod 2:108
Mody NV, Henson R, Hedin PA, Kokpol U, Miles DH (1976) Isolation of insect paralysing agent coniine from *Sarracenia flava*. Experientia 32:829–830
Mroczek T, Baj S, Chrobok A, Glowniak K (2004) Screening for pyrrolizidine alkaloids in plant materials by electron ionization RP-HPLC-MS with thermabeam interface. Biomed Chromatogr 18:745–751
Nash RJ, Beaumont J, Veitch NC, Reynolds T, Benner J, Hughes CNG, Dring JV, Bennett RN, Dellar JE (1992) Phenylethylamine and piperidine alkaloids in *Aloe* species. Planta Med 58:84–87
O'Connor SE, Maresh JJ (2006) Chemistry and biology of monoterpene indole alkaloid biosynthesis. Nat Prod Rep 23:532–547
Okwu DE, Okwu ME (2004) Chemical composition of *Spondias mombin* Linn plant parts. J Sustain Agric Environ 6:140–147
Panter KE, Bunch TD, Keeler RF, Sisson DV (1988) Radio ultrasound observation of the fetotoxic effects in sheep from ingestion of *Conium maculatum* (poison-hemlock). J Toxicol Clin Toxicol 26:175–187
Pelletier SW (1983) The nature and definition of an alkaloid. In: Pelletier SW (ed) Alkaloids. Chemical and biological perspectives, vol 1. Wiley, New York, pp 1–31
Pelser PB, de Vos H, Theuring C, Beuerle T, Vrieling K, Hartmann T (2005) Frequent gain and loss of pyrrolizidine alkaloids in the evolution of *Senecio* section *Jacobaea* (Asteraceae). Phytochemistry 66:1285–1295
Pigatto AGS, Blanco CC, Mentz LA, Soares GLG (2015) Tropane alkaloids and calystegines as chemotaxonomic markers in the Solanaceae. An Acad Bras Ciênc 87:2139–2149
Plowman T (1981) Amazonian coca. J Ethnopharmacol 3:195–225
Radusiene J, Karpaviciene B, Stanius Z (2012) Effect of external and internal factors on secondary metabolites accumulation in St. John's worth. Bot Lith 18:101–108
Reimann A, Nurhayati N, Backenköhler A, Ober D (2004) Repeated evolution of the pyrrolizidine alkaloid–mediated defence system in separate angiosperm lineages. Plant Cell 16:2772–2784
Reynolds T (2005) Hemlock alkaloids from Socrates to poison aloes. Phytochemistry 66:1399–1406
Roberts MF, Wink M (1998) Alkaloids: biochemistry, ecology and medicinal applications. Planum Press, New York

Roeder E (1999) Analysis of pyrrolizidine alkaloids. Curr Org Chem 3:557–576
Roeder E (2000) Medicinal plants in China containing pyrrolizidine alkaloids. Pharmazie 55:711–726
Roeder E, Wiedenfeld H (2009) Pyrrolizidine alkaloids in medicinal plants of Mongolia, Nepal and Tibet. Pharmazie 64:699–716
Roeder E, Wiedenfeld H (2013) Plants containing pyrrolizidine alkaloids used in the Traditional Indian Medicine—including Ayurveda. Pharmazie 68:83–92
Roitman JN (1988) Longitubine and neolatifoline, new pyrrolizidine alkaloids from *Hackelia longituba*. Aust J Chem 41:1827–1833
Said-Al Ahl HAH, Omer EA (2011) Medicinal and aromatic plants production under salt stress. A review. Herba Pol 57:72–87
Singh R (2016) Chemotaxonomy: a tool for plant classification. J Med Plants Stud 4:90–93
Singh R, Geetanjali (2018) Chapter 6: Chemotaxonomy of medicinal plants: possibilities and limitations. In: Mandal SC, Mandal V, Konishi T (eds) Natural products and drug discovery—an integrated approach. Elsevier, pp 119–136
Singh R, Geetanjali, Singh V (2011) Exploring alkaloids as inhibitors of selected enzymes. Asian J Chem 23:483–490
Sivarajan VV (1991) Introduction to the principles of plant taxonomy. Cambridge University Press
Smith LW, Culvenor CCJ (1981) Plant sources of hepatotoxic pyrrolizidine alkaloids. J Nat Prod 44:129–152
Soler-Rodríguez F, Martín A, García-Cambero JP, Oropesa AL, Pérez-López M (2006) *Datura stramonium* poisoning in horses: a risk factor for colic. Vet Rec 158:132–133
Souza JSN, Machado LL, Pessoa OD, Braz-Filho R, Overk CR, Yao P, Lemos TL (2005) Pyrrolizidine alkaloids from *Heliotropium indicum*. J Braz Chem Soc 16:1410–1414
Taylor WI, Farnsworth NR (eds) (1975) The Catharanthus alkaloids. Botany, chemistry, pharmacology, and clinical use. In: Stearn WT. A synopsis of the genus Catharanthus (Apocynaceae). Marcel Dekker, Inc., New York, pp 9–44
Trigo JR, Leal IR, Matzenbacher NI, Lewinsohn TM (2003) Chemotaxonomic value of pyrrolizidine alkaloids in southern Brazil *Senecio* (Senecioneae: Asteraceae). Biochem Syst Ecol 31:1011–1022
Tundis R, Loizzo MR, Bonesi M, Menichini F, Dodaro D, Passalacqua NG, Menichini F (2009) In vitro cytotoxic effects of *Senecio stabianus* Lacaita (Asteraceae) on human cancer cell lines. Nat Prod Res 23:1707–1718
van der Heijden R, Jacobs DI, Snoeijer W, Hallard D, Verpoorte R (2004) The Catharanthus alkaloids: pharmacognosy and biotechnology. Curr Med Chem 11:607–628
Verma N, Shukla S (2015) Impact of various factors responsible for fluctuation in plant secondary metabolites. J Appl Res Med Aromat Plants 49:1–9
Wink M (1993) Quinolizidine alkaloids. In: Waterman PG (ed) Methods in plant biochemistry. Academic Press, London, pp 197–239
Wink M (2003) Evolution of secondary metabolites from an ecological and molecular phylogenetic perspective. Phytochemistry 64:3–19
Wink M (2013) Evolution of secondary metabolites in legumes (Fabaceae). S Afr J Bot 89:164–175
Xu Z, Deng M (2017) Identification and control of common weeds. Springer 2547-615
Zhi-Lin Y, Chuan-Chao D, Lian-Qing C (2007) Regulation and accumulation of secondary metabolites in plant-fungus symbiotic system. Afr J Biotechnol 6:1266–1271
Zhu J, Wang M, Wen W, Yu R (2015) Biosynthesis and regulation of terpenoid indole alkaloids in *Catharanthus roseus*. Pharmacog Rev 9:24–28

Chapter 7
Chemotaxonomic Implications of Methoxy Flavonoids in *Ageratina* and *Chromolaena*

Debora Cristina Baldoqui, Adriano Borges Meniqueti, Anderson Valdiney Gomes Ramos, Maria Helena Sarragiotto and Marta Regina Barrotto do Carmo

Abstract Flavonoids have been shown to be an important taxonomic marker for the identification of species, genera, tribes or even families. In this chapter, we evaluated the importance of flavonoids as chemotaxonomic marker of *Ageratina* and *Chromolaena*, which were previously classified as *Eupatorium*. The chemotaxonomic study of *Ageratina* genus showed that flavonoids occur in 28 out of 41 species investigated. Already in *Chromolaena*, this number is higher, of the 26 *Chromolaena* species studied, 20 showed the presence of flavonoid, which corroborates the importance of this class of secondary metabolites as chemotaxonomic marker for these genera. Flavonols are the predominant compounds, more than 50% of the flavonoids found in *Ageratina* and *Chromolaena* belongs to this class. Furthermore, about 80% of the flavonoids found in these two genera are methoxylated, and the methoxy group is usually attached at C-6, C-7 or/and C-4′. Highly methoxylated flavonoids were described from *Ageratina* and *Chromolaena*, however, pentamethoxy flavonoids were obtained exclusively from *Chromolaena* species. This review confirms that flavonoids are a useful marker for *Ageratina* and *Chromolaena* species, especially the mono- and di-methoxylated flavonoids.

D. C. Baldoqui (✉) · A. B. Meniqueti · A. V. G. Ramos · M. H. Sarragiotto
Departamento de Química, Universidade Estadual de Maringá, Av. Colombo 5790, Maringá, Paraná 87020-900, Brazil
e-mail: dcbaldoqui@uem.br

A. B. Meniqueti
e-mail: adrianomeniquetti@hotmail.com

A. V. G. Ramos
e-mail: anderson_ramos.19@hotmail.com

M. H. Sarragiotto
e-mail: mhsarragiotto@uem.br

M. R. B. do Carmo
Departamento de Biologia Geral, Universidade Estadual de Ponta Grossa, Av. Carlos Cavalcanti, 4748, Ponta Grossa, Paraná 84030-910, Brazil
e-mail: mrcarmo@uepg.br

K. G. Ramawat (ed.), *Biodiversity and Chemotaxonomy*, Sustainable Development and Biodiversity 24, https://doi.org/10.1007/978-3-030-30746-2_7

Keywords *Asteraceae* · *Eupatorium* · *Ageratina* · *Chromolaena* · Methoxy flavonoid

7.1 Introduction

The chemical structure of the secondary metabolites is often specific and restricted to taxonomically related organism, and the classification base on chemical constituents is known as chemotaxonomy. Phenolic compounds, alkaloids, and terpenoids are classes of secondary metabolites widely utilized in chemotaxonomic classification (Singh 2016).

Flavonoids are the most common and largely distributed group of plant phenolic compounds, which are found in several parts of the plant. They are synthesized by phenylpropanoid pathway and have fifteen-carbon skeleton consisting of two benzene rings linked by a heterocyclic pyrane ring (Fig. 7.1). Among the many types of flavonoids, those of particular interest are flavones, flavonols, flavanones, flavan-3-ols, isoflavones, chalcones, dihydrochalcones, dihydroflavonols, anthocyanidins, and aurones. Flavonoids occur as aglycones, glycosides, methylated, and prenylated derivatives, often hydroxylated in positions 3, 5, 7, 2, 3′, 4′, and 5′.

Flavonoids have been shown to be an important taxonomic marker for the identification of species, genera, tribes or even families (Viljoen et al. 1998; Kharazian 2014; Sareedenchai and Zidorn 2010; Martjnez and Swain 1985). Furthermore, flavonoids have been isolated in great scale from *Asteraceae*, and they are considered as a good chemical marker for this family (Emerenciano et al. 2007; Funk et al. 2009).

7.2 *Eupatorium Sensu Latu*

Eupatorium s.l. genus belongs to Eupatorieae tribe, and it is a taxonomically complex group. In traditional concept, *Eupatorium* has been a highly artificial, tending to include other members of the tribe of *Asteraceae* lacking rays and with

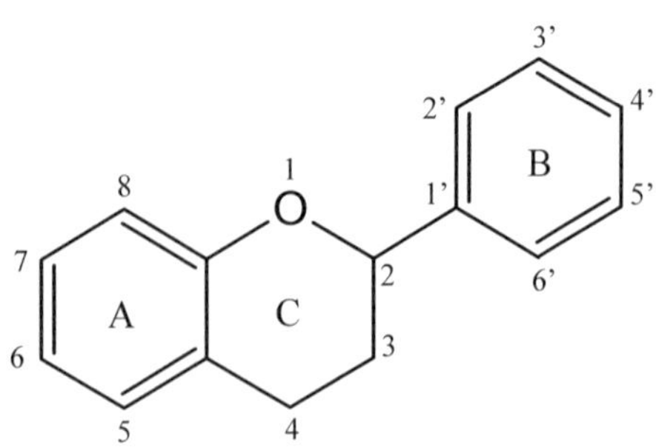

Fig. 7.1 Basic flavonoid structure

flowers not yellow (King and Robinson 1987). Thus, the concept of the genus has been allowed to degenerate in this century. *Eupatorium* was restructured and fragmented in approximately 80 genera, previously about 1200 species were classified as *Eupatorium*, and currently there are only 48 species (King and Robinson 1987).

Liu et al. (2015) showed that 321 substances were isolated from *Eupatorium* between 1904 and 2014. The terpenes were identified as the main chemical constituents of this genus, and flavonoid was appointed as the second major class of secondary metabolites found in *Eupatorium s.l.*

In the present chapter, information on *Eupatorium s.l.* genus was gathered via searching in SciFinder database by using the keyword "*Eupatorium*", which result in 1749 scientific papers published in Journals. The analysis of these scientific papers allowed to identify that flavonoids have been described mainly from species that now are classified as *Ageratina* Spach and *Chromolaena* DC., which were previously classified as member of *Eupatorium* genus.

Ageratina, the largest genus of the Oxylobinae subtribe, comprises about 290 species (King and Robinson 1987; Bremer 1994; Funk et al. 2009). *Ageratina* are perennial herbs or shrubs with florets white or lavender, usually serrate, glandular and opposite leaves (rarely alternate). They are distributed along the South and North America, West Indies, most species in Central America (Bremer 1994).

Chromolaena belongs to Praxelinae subtribe, and comprises ca. 165 species (King and Robinson 1987; Bremer 1994; Rodríguez-Cabeza et al. 2014), distributed North and South America, from Northeastern United States South to Argentina, and West Indies. They are perennial herbs, shrubs or climbers, with florets white, blue, lavender, or purple, and opposite, sometimes alternate or verticillate leaves (Bremer 1994).

7.3 Flavonoids Isolated from *Ageratina* and *Chromolaena*

Data on flavonoids on *Ageratina* and *Chromolaena* were obtained by searching in SciFinder database by using the keyword "*Ageratina*" and "*Chromolaena*", respectively. It was performed a search in www.tropicos.org, and only species with *Eupatorium* basionym was considerate in this review.

Chemical investigation on 41 *Ageratina* species showed the presence of flavonoid in 28 species (68%), and from *Chromolaena*, flavonoids have been described from 20 of the 26 species studied (77%), showing the importance of this class of secondary metabolite as a chemotaxonomic marker for these genera.

It was found that 190 flavonoids have been described from *Ageratina* and *Chromolaena*, more than 50% belongs to flavonol class (98 compounds). Flavone and flavanone classes have the same number of described compounds (35). The minor flavonoids were found to be chalcones (10), flavan-3-ols (9), and isoflavonoids (3). Table 7.1 shows the flavonoids found in species of *Ageratina* and *Chromolaena*, as well as their basionym.

Table 7.1 Flavonoids from *Ageratina* and *Chromolaena*

Flavonoids	Basionym	Accepted name	Reference
Non-methoxylated			
Apigenin (**1**)	*E. espinosarum*	*A. espinosarum*	Wollenweber et al. (1997)
	E. odoratum	*C. odorata*	Yuan et al. (2007)
	E. congestum	*C. congesta*	Oliveira et al. (2017)
Luteolin (**2**)	*E. espinosarum*	*A. espinosarum*	Wollenweber et al. (1997)
	E. altissimum	*A. altissima*	Wollenweber et al. (1996)
	E. hirsutum	*C. hirsuta*	Taleb-Contini et al. (2004, 2007)
	E. odoratum	*C. odorata*	Suksamrarn et al. (2004), Yuan et al. (2007), Fu et al. (2007)
	E. squalidum	*C. squalida*	Taleb-Contini et al. (2007)
	E. moritzianum	*C. moritziana*	Báez et al. (1998)
	E. congestum	*C. congesta*	Oliveira et al. (2017)
Galangin (**3**)	*E. chaseae*	*C. chaseae*	Bohlmann et al. (1982)
Kaempferol (**4**)	*E. espinosarum*	*A. espinosarum*	Wollenweber et al. (1997)
	E. havanense	*A. havanensis*	Yu et al. (1987)
	E. odoratum	*C. odorata*	Wollenweber et al. (1995), Yuan et al. (2007), Fu et al. (2007), Na et al. (2012), Heiss et al. (2014)
	E. congestum	*C. congesta*	Oliveira et al. (2017)
Quercetin (**5**)	*E. calophyllum*	*A. calophylla*	Fang et al. (1986)
	E. espinosarum	*A. espinosarum*	Bohlmann et al. (1977), Wollenweber et al. (1997)
	E. areolare	*A. areolaris*	Yu et al. (1986b)
	E. gracile	*A. gracilis*	Torrenegra et al. (1984)
	E. glandulosum[a]	*A. adenophora*	Mukherjee et al. (2001)
	E. hirsutum	*C. hirsuta*	Taleb-Contini et al. (2004, 2007)
	E. odoratum	*C. odorata*	Wollenweber et al. (1995), Yuan et al. (2007), Fu et al. (2007), Na et al. (2012)
	E. squalidum	*C. squalida*	Taleb-Contini et al. (2007)
Quercetagetin (**6**)	*E. gracile*	*A. gracilis*	Torrenegra et al. (1984)
Herbacetin (**7**)	*E. gracile*	*A. gracilis*	Torrenegra et al. (1984)
	E. chaseae	*C. chaseae*	Bohlmann et al. (1982)
Pinocembrin (**8**)	*E. dictyoneurum*	*A. dictyoneura*	Eiroa et al. (2018)
Naringenin (**9**)	*E. espinosarum*	*A. espinosarum*	Bohlmann et al. (1977)
	E. illitum	*A. illita*	Castillo et al. (2015)
	E. tyleri	*C. tyleri*	Perez and Espitia (1999)
	E. connivens	*C. connivens*	Tamayo-Castillo et al. (1989)
	E. odoratum	*C. odorata*	Heiss et al. (2014)

(continued)

Table 7.1 (continued)

Flavonoids	Basionym	Accepted name	Reference
Eriodictyol (**10**)	*E. espinosarum*	*A. espinosarum*	Wollenweber et al. (1997)
	E. arnottianum	*C. arnottiana*	Clavin et al. (2007)
	E. tyleri	*C. tyleri*	Perez and Espitia (1999)
Pinocembrin chalcone (**11**)	*E. chaseae*	*C. chaseae*	Bohlmann et al. (1982)
Phloretin (**12**)	*E. espinosarum*	*A. espinosarum*	Bohlmann et al. (1977)
Dihydrogalangin (**13**)	*E. chaseae*	*C. chaseae*	Bohlmann et al. (1982)
Apigenin 7-*O*-rhamnoside (**14**)	*E. hookerianum*	*C. hookeriana*	Ferraro et al. (1983)
Isoquercetin (**15**)	*E. havanense*	*A. havanensis*	Yu et al. (1987)
	E. saltillense	*A. saltillensis*	Yu et al. (1986a)
	E. areolare	*A. areolaris*	Yu et al. (1986b)
	E. moritzianum	*C. moritzianum*	Báez et al. (1998)
Quercitrin (**16**)	*E. hirsutum*	*C. hirsuta*	Taleb-Contini et al. (2004)
Astragalin (**17**)	*E. cylindricum*	*A. cylindrica*	Bustos-Brito et al. (2015)
	E. odoratum	*C. odorata*	Hung et al. (2011)
Hyperoside (**18**)	*E. arbutifolium*	*A. arbutifolia*	Perez and Molina (1995)
	E. saltillense	*A. saltillensis*	Yu et al. (1986a)
	E. areolare	*A. areolaris*	Yu et al. (1986b)
	E. havanense	*A. havanensis*	Yu et al. (1987)
	E. arnottianum	*C. arnottiana*	Wagner et al. (1972), Clavin et al. (2007)
3,5,7,2′,3′,4′-hexahydroxyflavonol 3-*O*-glucoside (**19**)	*E. sternbergianum*	*A. sternbergiana*	D'Agostino et al. (1994)
Nicotiflorin (**20**)	*E. odoratum*	*C. odorata*	Ling et al. (2007), Hung et al. (2011)
	E. moritzianum	*C. moritzianum*	Báez et al. (1998)
Rutin (**21**)	*E. odoratum*	*C. odorata*	Wollenweber et al. (1995), Suntornsuk and Anurukvorakun (2005), Ling et al. (2007), Hung et al. (2011)
	E. arnottianum	*C. arnottiana*	Clavin et al. (2007)
	E. laevigatum	*C. laevigata*	Bauer (1976)
	E. moritzianum	*C. moritzianum*	Báez et al. (1998)
Quercetin 3-*O*-robinobioside (**22**)	*E. hirsutum*	*C. hirsuta*	Taleb-Contini et al. (2004)
Kaempferol-3-*O*-glucoside-7-*O*-rhamnoside (**23**)	*E. hookerianum*	*C. hookeriana*	Ferraro et al. (1983)
Kaempferitrin (**24**)	*E. hookerianum*	*C. hookeriana*	Ferraro et al. (1983)
	E. subscandens	*C. subscandens*	Amaro-Luis and Delgado-Mendez (1993)
6-hydroxykaempferol-7-*O*-glucoside (**25**)	*E. adenophorum*	*A. adenophora*	Li et al. (1997)
Kaempferol 7-*O*-rhamnoside (**26**)	*E. subscandens*	*C. subscandens*	Amaro-Luis and Delgado-Mendez (1993)

(continued)

Table 7.1 (continued)

Flavonoids	Basionym	Accepted name	Reference
Quercetagitrin (**27**)	*E. adenophorum*	*A. adenophora*	Li et al. (1997)
	E. glandulosum[a]	*A. adenophora*	Nair et al. (1995)
Quercetagetin 7-*O*-(6″-O-acetyl)-glucoside (**28**)	*E. adenophorum*	*A. adenophora*	Wei et al. (2011)
	E. glandulosum[a]	*A. adenophora*	Nair et al. (1995)
Quercimetrin (**29**)	*E. glandulosum*[a]	*A. adenophora*	Nair et al. (1995)
Galactin 7-*O*-glucoside (**30**)	*E. pichinchense*	*A. pichinchensis*	Romero-Cerecero et al. (2013)
Gossypetin 7-*O*-glucoside (**31**)	*E. pichinchense*	*A. pichinchensis*	Romero-Cerecero et al. (2013)
6-hydroxykaempferol 7-*O*-(6″-E-caffeoyl)-glucoside (**32**)	*E. glandulosum*[a]	*A. adenophora*	Nair et al. (1993)
Herbacetin-5-*O*-(6″-(E)-caffeoyl)-glucoside (**33**)	*E. adenophorum*	*A. adenophora*	Wang et al. (2016)
Gossypetin 5-*O*-(6″-(E)-caffeoyl)-glucoside (**34**)	*E. adenophorum*	*A. adenophora*	Wang et al. (2016)
Quercetin 6-*C*-glucoside (**35**)	*E. calophyllum*	*A. calophylla*	Fang et al. (1986)
Mono-methoxylated			
3-methoxy			
Isokaempferide (**36**)	*E. congestum*	*C. congesta*	Oliveira et al. (2017)
	E. odoratum	*C. odorata*	Yuan et al. (2007)
3-methoxy quercetin (**37**)	*E. dictyoneurum*	*A. dictyoneura*	Eiroa et al. (2018)
	E. saltillense	*A. saltillensis*	Yu et al. (1986a)
	E. hirsutum	*C. hirsuta*	Taleb-Contini et al. (2004, 2007)
	E. squalidum	*C. squalida*	Taleb-Contini et al. (2007)
5,6,7,4′-tetrahydroxy-3-methoxyflavone (**38**)	*E. deltoideum*	*A. deltoidea*	Yang et al. (1991)
3-methoxy quercetagetin (**39**)	*E. espinosarum*	*A. espinosarum*	Wollenweber et al. (1997)
5-methoxy			
5-methoxy apigenin (**40**)	*E. congestum*	*C. congesta*	Oliveira et al. (2017)
5-methoxy kaempferol (**41**)	*E. erythropappum*[b]	*A. ligustrina*	Talapatra et al. (1985)
6-methoxy			
Hispidulin (**42**)	*E. angustifolium*	*A. angustifolia*	De Villarraga and De Perez (1994)
	E. calophyllum	*A. calophylla*	Fang et al. (1986)
	E. glyptophlebium	*A. glyptophlebia*	Perez and Pinilla (1990)
	E. tomentellum	*A. tomentella*	Fang and Mabry (1986)
	E. arnottianum	*C. arnottiana*	Elema et al. (1989), Stevens et al. (1995), Clavin et al. (2007)
Eupafolin (**43**)	*E. glyptophlebium*	*A. glyptophlebia*	Perez and Pinilla (1990)
	E. tomentellum	*A. tomentella*	Fang and Mabry (1986)
	E. areolare	*A. areolaris*	Yu et al. (1986b), Torrenegra et al. (1984)
	E. altissimum	*A. altissima*	Wollenweber et al. (1996)
	E. hirsutum	*C. hirsuta*	Taleb-Contini et al. (2004, 2007)
	E. squalidum	*C. squalida*	Taleb-Contini et al. (2007)
	E. arnottianum	*C. arnottiana*	Clavin et al. (2007)

(continued)

Table 7.1 (continued)

Flavonoids	Basionym	Accepted name	Reference
Alnusin (**44**)	*E. chaseae*	*C. chaseae*	Bohlmann et al. (1982)
	E. leivense	*C. leivensis*	Torrenegra and Rodriguez (2011), Torrenegra et al. (2016)
6-methoxy kaempferol (**45**)	*E. deltoideum*	*A. deltoidea*	Yang et al. (1991)
	E. saltillense	*A. saltillensis*	Yu et al. (1986a)
	E. areolare	*A. areolaris*	Yu et al. (1986b)
	E. altissimum	*A. altissima*	Wollenweber et al. (1996)
	E. espinosarum	*A. espinosarum*	Wollenweber et al. (1997)
Patuletin (**46**)	*E. calophyllum*	*A. calophylla*	Fang et al. (1986)
	E. havanense	*A. havanensis*	Yu et al. (1987)
	E. saltillense	*A. saltillensis*	Yu et al. (1986a)
Alnustinol (**47**)	*E. chaseae*	*C. chaseae*	Bohlmann et al. (1982)
6-methoxy kaempferol 3-*O*-glucoside (**48**)	*E. calophyllum*	*A. calophylla*	Fang et al. (1986)
	E. havanense	*A. havanensis*	Yu et al. (1987)
	E. saltillense	*A. saltillensis*	Yu et al. (1986a)
	E. areolare	*A. areolaris*	Yu et al. (1986b)
	E. adenophorum	*A. adenophora*	Zhang et al. (2015a)
6-methoxy kaempferol 3-*O*-galactoside (**49**)	*E. areolare*	*A. areolaris*	Yu et al. (1986b)
6-methoxy kaempferol 3-*O*-rhamnoside (**50**)	*E. saltillense*	*A. saltillensis*	Yu et al. (1986a)
Patuletin 3-*O*-glucoside (**51**)	*E. havanense*	*A. havanensis*	Yu et al. (1987)
	E. saltillense	*A. saltillensis*	Yu et al. (1986a)
	E. areolare	*A. areolaris*	Yu et al. (1986b)
Patuletin 3-*O*-galactoside (**52**)	*E. arbutifolium*	*A. arbutifolia*	Perez and Molina (1995)
	E. areolare	*A. areolaris*	Yu et al. (1986b)
6-methoxy kaempferol 7-*O*-glucoside (**53**)	*E. adenophorum*	*A. adenophora*	Zhang et al. (2015a)
	E. glandulosum[a]	*A. adenophora*	Nair et al. (1995)
Patuletin 7-*O*-glucoside (**54**)	*E. adenophorum*	*A. adenophora*	Zhang et al. (2015a)
7-methoxy			
Genkwanin (**55**)	*E. espinosarum*	*A. espinosarum*	Wollenweber et al. (1997)
	E. areolare	*A. areolare*	Yu et al. (1986b)
	E. angustifolium	*A. angustifolia*	De Perez and Cespedes (1984)
	E. congestum	*C. congesta*	Oliveira et al. (2017)
7-methoxy luteolin (**56**)	*E. farinosum*	*C. farinosa*	Triana (1995)
Sorbifolin (**57**)	*E. hirsutum*	*C. hirsuta*	Taleb-Contini et al. (2007)
	E. squalidum	*C. squalida*	Taleb-Contini et al. (2003, 2007)
7-methoxy quercetagetin (**58**)	*E. squalidum*	*C. squalida*	Taleb-Contini et al. (2003)

(continued)

Table 7.1 (continued)

Flavonoids	Basionym	Accepted name	Reference
Isalpinin (**59**)	*E. leivense*	*C. leivensis*	Torrenegra and Rodriguez (2011)
	E. tacotanum	*C. tacotana*	Rodríguez and Torrenegra (2005)
Rhamnocitrin (**60**)	*E. dictyoneurum*	*A. dictyoneura*	Eiroa et al. (2018)
	E. havanense	*A. havanensis*	Yu et al. (1987)
	E. odoratum	*C. odorata*	Wollenweber et al. (1995), Ohtsuki et al. (2009)
	E. pedale	*C. pedalis*	Lopes and Lopes (1979)
Rhamnetin (**61**)	*E. odoratum*	*C. odorata*	Wollenweber et al. (1995), Phan et al. (2001), Ling et al. (2007), Yuan et al. (2007), Ohtsuki et al. (2009), Johari et al. (2012)
	E. meridense	*C. meridensis*	Amaro-Luís and Morales-Mendez, (1983)
Sakuranetin (**62**)	*E. sternbergianum*	*A. sternbergiana*	Gonzales et al. (1984)
	E. havanense	*A. havanensis*	Dominguez and Fuente (1973), Barrio et al. (2011)
	E. espinosarum	*A. espinosarum*	Wollenweber et al. (1997)
	E. odoratum	*C. odorata*	Metwally and Ekejiuba (1981), Wollenweber et al. (1995), Pisutthanan et al. (2006), Ohtsuki et al. (2009)
	E. pedale	*C. pedalis*	Lopes and Lopes (1979)
	E. tyleri	*C. tyleri*	Perez and Espitia (1999)
	E. subscandens	*C. subscandens*	Amaro-Luis and Delgado-Mendez (1993)
	E. connivens	*C. connivens*	Tamayo-Castillo et al. (1989)
Sterubin (**63**)	*E. espinosarum*	*A. espinosarum*	Wollenweber et al. (1997)
	E. sternbergianum	*A. sternbergiana*	Gonzales et al. (1984)
	E. meridense	*C. meridensis*	Amaro-Luís and Morales-Mendez (1983)
7-methoxy aromadendrin (**64**)	*E. havanense*	*A. havanensis*	Barrio et al. (2011)
	E. illitum	*A. illita*	Castillo et al. (2015)
	E. odoratum	*C. odorata*	Pisutthanan et al. (2005, 2006), Ohtsuki et al. (2009), Heiss et al. (2014)
	E. pedale	*C. pedalis*	Lopes and Lopes (1979)
	E. tacotanum	*C. tacotana*	Sanabria-Galindo and Carrero (1995)

(continued)

Table 7.1 (continued)

Flavonoids	Basionym	Accepted name	Reference
Alpinone (**65**)	*E. leivense*	*C. leivensis*	Torrenegra and Rodriguez (2011)
	E. tacotanum	*C. tacotana*	Rodríguez and Torrenegra (2005)
Padmatin (**66**)	*E. odoratum*	*C. odorata*	Wollenweber et al. (1995), Ling et al. (2007), Johari et al. (2012)
Isosativan (**67**)	*E. odoratum*	*C. odorata*	Venkata et al. (2012)
7-methoxy kaempferol-5-*O*-glucoside (**68**)	*E. havanense*	*A. havanensis*	Yu et al. (1987)
Sakuranin (**69**)	*E. havanense*	*A. havanensis*	Yu et al. (1987)
	E. odoratum	*C. odorata*	Venkata et al. (2012)
Sakuranetin 4′-O-[glucosyl(1 → 2)-glucoside] (**70**)	*E. odoratum*	*C. odorata*	Hung et al. (2011)
Micranthoside (**71**)	*E. havanense*	*A. havanensis*	Yu et al. (1987)
	E. erythropappum[b]	*A. ligustrina*	Talapatra et al. (1985)
Genkwanin 4′-*O*-rhamnoglucoside (**72**)	*E. odoratum*	*C. odorata*	Hung et al. (2011)
Sakuranetin 4′-*O*-glucoside (**73**)	*E. havanense*	*A. havanensis*	Barrio et al. (2011)
Persinoside A (**74**)	*E. havanense*	*A. havanensis*	Barrio et al. (2011)
Hoslundin (**75**)	*E. odoratum*	*C. odorata*	Venkata et al. (2012)
Neosakuranin (**76**)	*E. havanense*	*A. havanensis*	Yu et al. (1987)
8-methoxy			
8-methoxy herbacetin (**77**)	*E. gracile*	*A. gracilis*	Torrenegra et al. (1984)
3′-methoxy			
Chryseriol (**78**)	*E. espinosarum*	*A. espinosarum*	Wollenweber et al. (1997)
Isorhamnetin (**79**)	*E. odoratum*	*C. odorata*	Na et al. (2012) Schmeda et al. (1983)
Alysifolinone (**80**)	*E. odoratum*	*C. odorata*	Pisutthanan et al. (2005, 2006), Ohtsuki et al. (2009)
Isorhamnetin 3-*O*-glucoside (**81**)	*E. tinifolium*	*A. tinifolia*	D'Agostino et al. (1990)
3′-methoxy quercetagetin-3-*O*-glucoside (**82**)	*E. tinifolium*	*A. tinifolia*	D'Agostino et al. (1990)
3′-methoxy quercetagetin 3-*O*-galactoside (**83**)	*E. tinifolium*	*A. tinifolia*	D'Agostino et al. (1990)
4′-methoxy			
Acacetin (**84**)	*E. tinifolium*	*A. tinifolia*	Moreno et al. (1980)
	E. stevioides	*A. stevioides*	Morales and Rosquete (1988)
	E. odoratum	*C. odorata*	Bose et al. (1974), Wollenweber et al. (1995), Ding et al. (2001), Suksamrarn et al. (2004), Pisutthanan et al. (2005, 2006), Yuan et al. (2005), Dat et al. (2009), Ohtsuki et al. (2009), Hung et al. (2011), Wafo et al. (2011), Na et al. (2012), Zhang et al. (2013), Heiss et al. (2014), Nath et al. (2015)

(continued)

Table 7.1 (continued)

Flavonoids	Basionym	Accepted name	Reference
	E. farinosum	*C. farinosa*	Triana (1995)
	E. hookerianum	*C. hookeriana*	Ferraro et al. (1983)
	E. congestum	*C. congesta*	Oliveira et al. (2017)
Diosmetin (**85**)	*E. altissimum*	*A. altissima*	Wollenweber et al. (1996)
	E. farinosum	*C. farinosa*	Triana (1995)
	E. odoratum	*C. odorata*	Na et al. (2012)
Kaempferide (**86**)	*E. odoratum*	*C. odorata*	Metwally and Ekejiuba (1981), Wollenweber et al. (1995), Ding et al. (2001), Phan et al. (2001), Yuan et al. (2005), Ling et al. (2007), Dat et al. (2009), Ohtsuki et al. (2009), Hung et al. (2011), Wafo et al. (2011), Johari et al. (2012), Na et al. (2012), Heiss et al. (2014), Nath et al. (2015)
Tamarixetin (**87**)	*E. odoratum*	*C. odorata*	Metwally and Ekejiuba (1981), Wollenweber et al. (1995), Shuib et al. (1999), Phan et al. (2001), Ling et al. (2007), Yuan et al. (2007), Ohtsuki et al. (2009), Johari et al. (2012), Ezenyi et al. (2014)
5,6,7-trihydroxy-4′-methoxyflavanone (**88**)	*E. odoratum*	*C. odorata*	Na et al. (2012)
Hesperitin (**89**)	*E. odoratum*	*C. odorata*	Zhang et al. (2013), Heiss et al. (2014)
	E. tyleri	*C. tyleri*	Perez and Espitia (1999)
Dihydrokaempferide (**90**)	*E. odoratum*	*C. odorata*	Pisutthanan et al. (2005, 2006), Ling et al. (2007), Yuan et al. (2007), Ohtsuki et al. (2009), Hung et al. (2011), Johari et al. (2012), Na et al. (2012), Nath et al. (2015)
Dihydrotamarixetin (**91**)	*E. odoratum*	*C. odorata*	Ling et al. (2007), Johari et al. (2012)
Isosakuranetin (**92**)	*E. sternbergianum*	*A. sternbergiana*	Gonzales et al. (1982)
	E. odoratum	*C. odorata*	Bose et al. (1973, 1974), Metwally and Ekejiuba (1981), Hai et al. (1995), Wollenweber et al. (1995), Ding et al. (2001), Suksamrarn et al. (2004), Yuan et al. (2005), Pisutthanan et al. (2006), Ling et al. (2007), Dat et al. (2009), Ohtsuki et al. (2009), Hung et al. (2011), Wafo

(continued)

Table 7.1 (continued)

Flavonoids	Basionym	Accepted name	Reference
			et al. (2011), Johari et al. (2012), Na et al. (2012), Nath et al. (2015)
6-hydroxy 4′-methoxy luteolin 7-*O*-glucoside (**93**)		*E. adenophorum*	*A. adenophora*
Li et al. (1997)			
4′-methoxy quercetagetin 7-*O*-(6″-O-E-caffeoyl)-glucoside (**94**)	*E. adenophorum*	*A. adenophora*	Wei et al. (2011)
	E. glandulosum[a]	*A. adenophora*	Nair et al. (1995)
Biochanin A (**95**)	*E. adenophorum*	*A. adenophora*	Parveen et al. (2015)
6-Methylenebis(5,7-dihydroxy-4′-methoxyflavanone) (**96**)	*E. odoratum*	*C. odorata*	Zhang et al. (2015b)
Di-methoxylated			
3,6-dimethoxy			
3,6-dimethoxy kaempferol (**97**)	*E. espinosarum*	*A. espinosarum*	Wollenweber et al. (1997)
Viscosine (**98**)	*E. deltoideum*	*A. deltoidea*	Yang et al. (1991)
	E. altissimum	*A. altissima*	Wollenweber et al. (1996)
Axillarin (**99**)	*E. saltillense*	*A. saltillensis*	Yu et al. (1986a)
	E. hirsutum	*C. hirsute*	Taleb-Contini et al. (2004, 2007)
	E. squalidum	*C. squalida*	Taleb-Contini et al. (2007)
3,7-dimethoxy			
Kumatakenin (**100**)	*E. illitum*	*A. illita*	Castillo et al. (2015)
	E. congestum	*C. congesta*	Oliveira et al. (2017)
3,7-dimethoxy quercetin (**101**)	*E. deltoideum*	*A. deltoidea*	Yang et al. (1991)
	E. dictyoneurum	*A. dictyoneura*	Eiroa et al. (2018)
6-hydroxy-3,7-dimethoxy kaempferol (**102**)	*E. deltoideum*	*A. deltoidea*	Yang et al. (1991)
Tomentin (**103**)	*E. deltoideum*	*A. deltoidea*	Yang et al. (1991)
5,7-dimethoxy			
5,7-dimethoxy apigenin (**104**)	*E. congestum*	*C. congesta*	Oliveira et al. (2017)
6,7-dimethoxy			
Skrofulein (**105**)	*E. adenophorum*	*A. adenophora*	Li et al. (1997)
	E. altissimum	*A. altissima*	Wollenweber et al. (1996)
Cirsiliol (**106**)	*E. deltoideum*	*A. deltoidea*	Yang et al. (1991)
	E. altissimum	*A. altissima*	Wollenweber et al. (1996)
Eupalitin (**107**)	*E. deltoideum*	*A. deltoidea*	Yang et al. (1991)
	E. adenophorum	*A. adenophora*	Wei et al. (2011)
	E. saltillense	*A. saltillensis*	Yu et al. (1986a)
	E. areolare	*A. areolaris*	Yu et al. (1986b)
	E. glandulosum[a]	*A. adenophora*	Nair and Sivakumar (1990)
	E. odoratum	*C. odorata*	Phan et al. (2001), Heiss et al. (2014)
	E. arnottianum	*C. arnottiana*	Clavin et al. (2007)

(continued)

Table 7.1 (continued)

Flavonoids	Basionym	Accepted name	Reference
Eupatolitin (**108**)	*E. areolare*	*A. areolaris*	Yu et al. (1986b)
	E. meridense	*C. meridensis*	Amaro-Luís and Morales-Mendez (1983)
5,4′-dihydroxy-6,7-dimethoxyflavanone (**109**)	*E. subscandens*	*C. subscandens*	Amaro-Luis and Delgado-Mendez (1993)
Eupalitin 3-*O*-glucoside (**110**)	*E. glyptophlebium*	*A. glyptophlebia*	Higuera and Perez (1989)
	E. adenophorum	*A. adenophora*	Li et al. (1997)
	E. areolare	*A. areolaris*	Yu et al. (1986b)
Eupalitin 3-O-galactoside (**111**)	*E. calophyllum*	*A. calophylla*	Fang et al. (1986)
	E. glandulosum[a]	*A. adenophora*	Nair and Sivakumar (1990)
	E. adenophorum	*A. adenophora*	Wei et al. (2011)
	E. areolare	*A. areolaris*	Yu et al. (1986b)
	E. ibaguense	*A. ibaguensis*	Triana (1995)
Eupalin (**112**)	*E. calophyllum*	*A. calophylla*	Fang et al. (1986)
	E. saltillense	*A. saltillensis*	Yu et al. (1986a)
	E. ligustrinum	*A. ligustrina*	Quijano et al. (1970)
Eupatolitin 3-*O*-glucoside (**113**)	*E. areolare*	*A. areolaris*	Yu et al. (1986b)
Eupatolitin 3-*O*-galactoside (**114**)	*E. calophyllum*	*A. calophylla*	Fang et al. (1986)
Eupatolitin 3-*O*-rhamnoside (**115**)	*E. calophyllum*	*A. calophylla*	Fang et al. (1986)
	E. ligustrinum	*A. ligustrina*	Quijano et al. (1970)
	E. saltillense	*A. saltillensis*	Yu et al. (1986a)
Eupatolitin 3-*O*-apioside (**116**)	*E. calophyllum*	*A. calophylla*	Fang et al. (1986)
7,8-dimethoxy			
7,8-dimethoxy herbacetin (**117**)	*E. gracile*	*A. gracilis*	Torrenegra et al. (1984)
3,3′-dimethoxy			
3,3′-dimethoxy quercetin (**118**)	*E. deltoideum*	*A. deltoidea*	Yang et al. (1991)
	E. espinosarum	*A. espinosarum*	Bohlmann et al. (1977)
6,3′-dimethoxy			
Jaceosidin (**119**)	*E. glyptophlebium*	*A. glyptophlebia*	Perez and Pinilla (1990)
	E. tomentellum	*A. tomentella*	Fang and Mabry (1986)
	E. arnottianum	*C. arnottiana*	Clavin et al. (2007)
	E. hirsutum	*C. hirsuta*	Taleb-Contini et al. (2004, 2007)
7,3′-dimethoxy			
Velutin (**120**)	*E. odoratum*	*C. odorata*	Arene et al. (1978)
	E. hirsutum	*C. hirsuta*	Taleb-Contini et al. (2004, 2007)
	E. laevigatum	*C. laevigata*	Flores et al. (2006)
	E. meridense	*C. meridensis*	Amaro-Luís and Morales-Mendez (1983)
5,6,4′-trihydroxy-7,3′-dimethoxyflavone (**121**)	*E. hirsutum*	*C. hirsuta*	Taleb-Contini et al. (2004)

(continued)

Table 7.1 (continued)

Flavonoids	Basionym	Accepted name	Reference
Eriodictyol 7,3′-di-O-methyl ether (**122**)	*E. espinosarum*	*A. espinosarum*	Wollenweber et al. (1997)
	E. tyleri	*C. tyleri*	Perez and Espitia (1999)
2′,5-dihydroxy-5′,7-dimethoxyflavanone (**123**)	*E. odoratum*	*C. odorata*	Hai et al. (1995)
Rhamnazin (**124**)	*E. odoratum*	*C. odorata*	Dat et al. (2009), Hung et al. (2011)
3,4′-dimethoxy			
Ermanin (**125**)	*E. espinosarum*	*A. espinosarum*	Bohlmann et al. (1977)
	E. illitum	*A. illita*	Castillo et al. (2015)
6,4′-dimethoxy			
Pectolinaringenin (**126**)	*E. angustifolium*	*A. angustifolia*	De Villarraga and De Perez (1994)
	E. glabratum	*A. glabrata*	Bustos-Brito et al. (2016)
	E. calophyllum	*A. calophylla*	Fang et al. (1986)
	E. odoratum	*C. odorata*	Wollenweber et al. (1995), Yuan et al. (2005), Zhang et al. (2013), Heiss et al. (2014)
Laciniatin (**127**)	*E. odoratum*	*C. odorata*	Wollenweber et al. (1995)
Pectolinaringenin 7-*O*-glucoside (**128**)	*E. angustifolium*	*A. angustifolia*	De Villarraga and De Perez (1994)
4′-methoxyl patulitrin (**129**)	*E. adenophorum*	*A. adenophora*	Zhang et al. (2015a)
5,7-dihydroxy-6,4′-dimethoxyflavanone (**130**)	*E. odoratum*	*C. odorata*	Pisutthanan et al. (2006), Zhang et al. (2013)
6-methoxy hesperitin (**131**)	*E. odoratum*	*C. odorata*	Heiss et al. (2014)
7,4′-dimethoxy			
7,4′-dimethoxy apigenin (**132**)	*E. fastigiatum*	*A. fastigiata*	Torrenegra et al. (1995)
	E. stevioides	*A. stevioides*	Morales and Rosquete (1988)
	E. amplum	*A. ampla*	Perez et al. (1980)
	E. angustifolium	*A. angustifolia*	De Perez and Cespedes (1984)
	E. odoratum	*C. odorata*	Pisutthanan et al. (2006)
Pilloin (**133**)	*E. farinosum*	*C. farinosa*	Triana (1995)
7,4′-dimethoxy kaempferol (**134**)	*E. odoratum*	*C. odorata*	Wollenweber et al. (1995), Zhang et al. (2013)
Ombuin (**135**)	*E. areolare*	*A. areolaris*	Yu et al. (1986b)
	E. odoratum	*C. odorata*	Wollenweber et al. (1995), Ding et al. (2001)
	E. farinosum	*C. farinosa*	Triana (1995)
	E. odoratum	*C. odorata*	Pisutthanan et al. (2005, 2006), Yuan et al. (2005), Ling et al. (2007), Hung et al. (2011), Johari et al. (2012), Zhang et al. (2012), Heiss et al. (2014)

(continued)

Table 7.1 (continued)

Flavonoids	Basionym	Accepted name	Reference
7,4′-dimethoxy naringenin (**136**)	*E. odoratum*	*C. odorata*	Arene et al. (1978), Wollenweber et al. (1995), Pisutthanan et al. (2006), Na et al. (2012), Venkata et al. (2012), Kouamé et al. (2013)
	E. tyleri	*C. tyleri*	Perez and Espitia (1999)
Persicogenin (**137**)	*E. sternbergianum*	*A. sternbergiana*	Gonzales et al. (1982), Gonzales et al. (1984)
	E. odoratum	*C. odorata*	Wollenweber et al. (1995), Suksamrarn et al. (2004), Pisutthanan et al. (2005, 2006), Ling et al. (2007), Hung et al. (2011), Johari et al. (2012)
	E. farinosum	*C. farinosa*	Triana (1995)
7,4′-dimethoxy aromadendrin (**138**)	*E. odoratum*	*C. odorata*	Wollenweber et al. (1995), Pisutthanan et al. (2006), Na et al. (2012)
Blumeatin B (**139**)	*E. odoratum*	*C. odorata*	Ohtsuki et al. (2009)
5,4′-dimethoxy			
Chromolanone (**140**)	*E. odoratum*	*C. odorata*	Dat et al. (2009)
3′,4′-dimethoxy			
3′,4′-dimethoxy luteolin (**141**)	*E. odoratum*	*C. odorata*	Wollenweber et al. (1995), Ohtsuki et al. (2009)
	E. farinosum	*C. farinosa*	Triana (1995)
Dillenetin (**142**)	*E. odoratum*	*C. odorata*	Wollenweber et al. (1995), Yuan et al. (2005), Na et al. (2012)
8′,4′-dimethoxy			
Subscandenin (**143**)	*E. subscandens*	*C. subscandens*	Amaro-Luis and Delgado-Mendez (1993)
Tri-methoxylated			
3,6,7-trimethoxy			
Penduletin (**144**)	*E. dictyoneurum*	*A. dictyoneura*	Eiroa et al. (2018)
	E. espinosarum	*A. espinosarum*	Wollenweber et al. (1997)
Chrysosplenol D (**145**)	*E. deltoideum*	*A. deltoidea*	Yang et al. (1991)
	E. hirsutum	*C. hirsuta*	Taleb-Contini et al. (2004)
5,6,7-trimethoxy			
3,4′-dihydroxy-5,6,7-trimethoxyflavone (**146**)	*E. odoratum*	*C. odorata*	Na et al. (2012)
4′-Hydroxy-5,6,7-trimethoxyflavanone (**147**)	*E. odoratum*	*C. odorata*	Barua et al. (1978), Hai et al. (1991), Suksamrarn et al. (2004), Na et al. (2012), Dhar et al. (2017)

(continued)

Table 7.1 (continued)

Flavonoids	Basionym	Accepted name	Reference
4,2′-dihydroxy-4′,5′,6′-trimethoxychalcone (**148**)	*E. odoratum*	*C. odorata*	Gupta et al. (1979), Phan et al. (2001), Suksamrarn et al. (2004), Dat et al. (2009), Zhang et al. (2012, 2013), Johari et al. (2012)
Odoratin (**149**)	*E. odoratum*	*C. odorata*	Bose et al. (1973, 1974), Arene et al. (1978), Nguyen et al. (1993), Wafo et al. (2005), Pisutthanan et al. (2006), Zhang et al. (2012), Kouamé et al. (2013)
2′,4-dihydroxy-3′,4′,6′ trimethoxychalcone (**150**)	*E. odoratum*	*C. odorata*	Dhar et al. (2017)
5,6,7-trimethoxy kaempferol-3-*O*-glucoside (**151**)	*E. tinifolium*	*A. tinifolia*	D’Agostino et al. (1991)
3,7,3′-trimethoxy			
Pachypodol (**152**)	*E. turbinatum*	*C. bigelovii*	Jakupovic et al. (1986)
3,7,4′-trimethoxy			
3,7,4′-trimethoxy kaempferol (**153**)	*E. fastigiatum*	*A. fastigiata*	Torrenegra et al. (1995)
	E. stevioides	*A. stevioides*	Morales and Rosquete (1988)
	E. tinifolium	*A. tinifolia*	Moreno et al. (1980)
Ayanin (**154**)	*E. deltoideum*	*A. deltoidea*	Yang et al. (1991), Arciniegas et al. (2018)
3,6,3′-trimethoxy			
Jaceidin (**155**)	*E. espinosarum*	*A. espinosarum*	Wollenweber et al. (1997)
	E. hirsutum	*C. hirsuta*	Taleb-Contini et al. (2004, 2007)
	E. squalidum	*C. squalida*	Taleb-Contini et al. (2007)
3,6,4′-trimethoxy			
Santin (**156**)	*E. espinosarum*	*A. espinosarum*	Wollenweber et al. (1997), Eiroa et al. (2018)
Centaureidin (**157**)	*E. saltillense*	*A. saltillensis*	Yu et al. (1986a)
	E. espinosarum	*A. espinosarum*	Wollenweber et al. (1997)
3,3′,4′-trimethoxy			
3,3′,4′-trimethoxy quercetin (**158**)	*E. deltoideum*	*A. deltoidea*	Yang et al. (1991)
7,3′,4′-trimethoxy			
7,3′,4′-trimethoxy quercetin (**159**)	*E. odoratum*	*C. odorata*	Wollenweber et al. (1995), Dat et al. (2009)
7,3′,4′-trimethoxy eriodictyol (**160**)	*E. odoratum*	*C. odorata*	Pisutthanan et al. (2006)
	E. tyleri	*C. tyleri*	Perez and Espitia (1999)
5,7,4′-trimethoxy			
5,7,4′-trimethoxy apigenin (**161**)	*E. angustifolium*	*A. angustifolia*	De Perez and Cespedes (1984)
	E. tinifolium	*A. tinifolia*	Moreno et al. (1980)
5,7,4′-trimethoxy naringenin (**162**)	*E. tyleri*	*C. tyleri*	Perez and Espitia (1999)

(continued)

Table 7.1 (continued)

Flavonoids	Basionym	Accepted name	Reference
5,7,3′-trimethoxy			
6,4′-hydroxy-5,7,3′-trimethoxyflavanone (**163**)	*E. heteroclinium*	*C. heteroclinia*	Boeker and Bohlmann (1986)
6,7,3′-trimethoxy			
Cirsilineol (**164**)	*E. tomentellum*	*A. tomentella*	Fang and Mabry (1986)
	E. hirsutum	*C. hirsuta*	Taleb-Contini et al. (2004)
6,7,4′-trimethoxy			
Salvigenin (**165**)	*E. gilbertii*	*A. gilbertii*	Herz and Gibaja (1972)
	E. jhanii	*A. jhanii*	Gonzales et al. (1979)
	E. adenophorum	*A. adenophora*	Ansari et al. (1983)
	E. stevioides	*A. stevioides*	Morales and Rosquete (1988)
	E. altissimum	*A. altissima*	Wollenweber et al. (1996)
	E. odoratum	*C. odorata*	Talapatra et al. (1974, 1977), Dat et al. (2009), Heiss et al. (2014)
5-hydroxy-6,7,4′-trimethoxyflavavone (**166**)	*E. odaratum*	*C. odorata*	Pisutthanan et al. (2006)
Eupatorin (**167**)	*E. tomentellum*	*A. tomentella*	Fang and Mabry (1986)
	E. jhanii	*A. jhanii*	Gonzales et al. (1979)
	E. altissimum	*A. altissima*	Dobberstein et al. (1977), Herz et al. (1978), Wollenweber et al. (1996)
	E. arnottianum	*C. arnottiana*	De Guttierez et al. (1995)
8-hydroxy-6,7,4′-trimethoxy flavanone (**168**)	*E. odoratum*	*C. odorata*	Ohtsuki et al. (2009)
6,3′,4′-trimethoxy			
Eupatilin (**169**)	*E. calophyllum*	*A. calophylla*	Fang et al. (1986)
	E. arnottianum	*C. arnottiana*	De Guttierez et al. (1995)
2′,4-dihydroxy-4′,5′,6′-trimethoxy chalcone (**170**)	*E. odoratum*	*C. odorata*	Barua et al. (1978), Hai et al. (1991), Nguyen et al. (1993), Ohtsuki et al. (2009)
Tetra			
Retusin (**171**)	*E. dendroides*	*A. dendroides*	Bohlmann and Grenz (1977)
3,6,3′,4′-tetramethoxy quercetagetin (**172**)	*E. espinosarum*	*A. espinosarum*	Wollenweber et al. (1997)
4′-methoxy penduletin (**173**)	*E. stevioides*	*A. stevioides*	Morales and Rosquete (1988)
	E. espinosarum	*A. espinosarum*	Wollenweber et al. (1997)
Casticin (**174**)	*E. deltoideum*	*A. deltoidea*	Yang et al. (1991)
Chrysosplenetin (**175**)	*E. squalidum*	*C. squalida*	Taleb-Contini et al. (2007)
	E. hirsutum	*C. hirsuta*	Taleb-Contini et al. (2004, 2007)

(continued)

Table 7.1 (continued)

Flavonoids	Basionym	Accepted name	Reference
Santaflavone (**176**)	*E. tomentellum*	*A. tomentella*	Fang and Mabry (1986)
	E. viscosum	*A. viscosa*	Torrenegra et al. (1990)
	E. jhanii	*A. jhanii*	Gonzales et al. (1979)
	E. altissimum	*A. altissima*	Dobberstein et al. (1977), Herz et al. (1978), Wollenweber et al. (1996)
	E. hirsutum	*C. hirsuta*	Taleb-Contini et al. (2004, 2007)
	E. odoratum	*C. odorata*	Phan et al. (2001)
7,8,3′,4′-tetramethoxy hypolaetin (**177**)	*E. viscosum*	*A. viscosa*	Torrenegra et al. (1990)
Scutellarein tetra-O-methyl ether (**178**)	*E. odoratum*	*C. odorata*	Barua et al. (1978), Vyas and Mulchandani 1986, Triratana et al. 1991, Wollenweber et al. 1995, Phan et al. (2001), Ohtsuki et al. (2009), Na et al. (2012), Venkata et al. (2012), Atindehou et al. (2013), Heiss et al. (2014)
5,6,7,4′-Tetramethoxyflavanone (**179**)	*E. odoratum*	*C. odorata*	Hai et al. (1991), Suksamrarn et al. (2004)
	E. conyzoides	*C. tunariensis*	Bohlmann et al. (1979)
5,7,3′,4′-tetramethoxyflavone (**180**)	*E. tyleri*	*C. tyleri*	Perez and Espitia (1999)
	E. heteroclinium	*C. heteroclinia*	Boeker and Bohlmann (1986)
Cystosiphonin (**181**)	*E. odoratum*	*C. odorata*	Heiss et al. (2014)
3′-hydroxy-6,7,8,4′-trimethoxy flavanone (**182**)	*E. odoratum*	*C. odorata*	Ohtsuki et al. (2009)
4′-hydroxy-5,6,7,3′-tetramethoxyflavanone (**183**)	*E. heteroclinium*	*C. heteroclinia*	Boeker and Bohlmann (1986)
2′-hydroxy-4,4′,5′,6′-tetramethoxychalcone (**184**)	*E. odoratum*	*C. odorata*	Suksamrarn et al. (2004)
2′-hydroxy-4,3′,4′,6′-tetramethoxychalcone (**185**)	*E. odoratum*	*C. odorata*	Dhar et al. (2017)
Penta			
Sinensetin (**186**)	*E. odoratum*	*C. odorata*	Barua et al. (1978), Vyas and Mulchandani (1986), Wollenweber et al. (1995), Phan et al. (2001), Atindehou et al. (2013)
	E. heteroclinium	*C. heteroclinia*	Boeker and Bohlmann (1986)
Marionol (**187**)	*E. odoratum*	*C. odorata*	Wollenweber and Roitman (1996)
Umuhengerin (**188**)	*E. arnotianum*	*C. arnottiana*	De Guttierez et al. (1995)
5,6,7,3′,4′-pentamethoxyflavanone (**189**)	*E. odoratum*	*C. odorata*	Na et al. (2012)
	E. heteroclinium	*C. heteroclinia*	Boeker and Bohlmann (1986)
2′-Hydroxy-4,3′,4′,5′,6′-pentamethoxychalcone (**190**)	*E. odoratum*	*C. odorata*	Barua et al. (1978)

[a] *Ageratina adenophora* accepted name to *Eupatorium glandulosum*
[b] *Ageratina ligustrina* accepted name to *Eupatorium erythropappum*

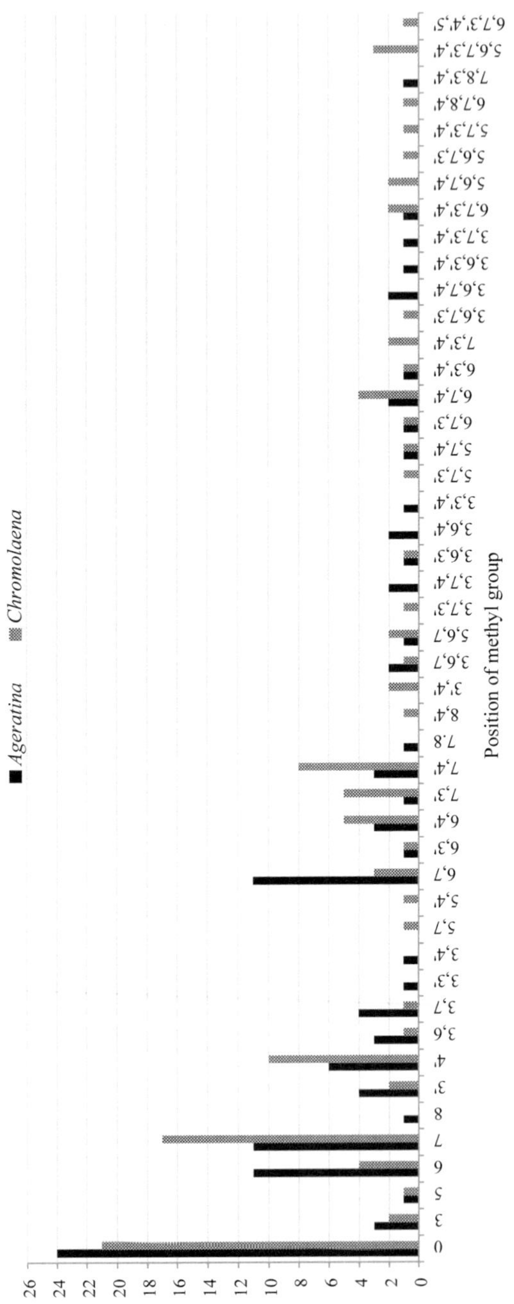

Fig. 7.2 Distribution of flavonoids in *Ageratina* and *Chromolaena* genera

One hundred and ten flavonoids have been described from *Ageratina*, classified as 24 non-methoxylated, 37 mono-methoxylated, 29 di-methoxylated, 14 tri-methoxylated and 6 tetra-methoxylated flavonoids. Already, a total of 121 flavonoids were isolated from *Chromolaena*, 21 non-methoxylated, 36 mono-methoxylated, 29 di-methoxylated, 19 tri-methoxylated, 10 tetra-methoxylated, and 6 penta-methoxylated. The most studied species is *C. odorata* with 70 different flavonoids described, including, flavone, flavonol, flavanone, and chalcone.

The distribution of flavonoids in these two species, as well as the different pattern of substitution, can be visualized in Fig. 7.2, which clearly showed that *Ageratina* accumulates mainly mono- and di-methoxy flavonoid, while flavonoids from *Chromolaena* have a more varied substitution pattern.

Furthermore, it was also observed that the methoxylated flavonoid in *Ageratina* generally has the methoxy group attached at C-6, C-7 or/and C-4′. Of the 37 mono-methoxylated flavonoids, 60% have the methoxy group attached C-6 (11 flavonoids) or C-7 (11 flavonoids), whereas in di-methoxylated, 38% are 6,7-dimethoxy flavonoid. Yu et al. (1986a) affirmed that the 6-methoxylation, 7-methoxylation, and 6,7-dimethoxylation are characteristic of the main evolutionary line in the *Ageratina* genus.

As observed for *Ageratina*, methoxylated flavonoids isolated from *Chromolaena* have the methoxyl group attached at C-6 or/and C-7 in A-ring, however, in *Chromolaena* it was observed a significant increase in the number of 5-O-methoxy flavonoids. Moreover, also it was noted an increase of 3′-O-methoxy and 4′-O-methoxy flavonoids in this genus. Another important point is the presence of highly methoxylated flavones in *Chromolaena*, such as penta-methoxychalcones isolated from *C. odorata*.

In the following sub-items, the available information on the distribution of flavonoids in *Ageratina* and *Chromolaena* is compiled. The flavonoids are ordered as non-methoxylated, mono-methoxylated, di-methoxylated, tri-methoxylated, and high methoxylated flavonoids.

7.3.1 Non-methoxylated Flavonoid

With regards to the accumulation of non-methoxylated flavonoids in *Ageratina* and *Chromolaena*, it was described 35 flavonoids, 13 of them are aglycone (**1–13**), and 22 are glycosilated flavonoids (**14–35**).

Quercetin (**5**) was the most frequent aglycone in *Ageratina*, being found in *A. calophylla* (Fang et al. 1986), *A. espinosarum* (Wollenweber et al. 1997; Bohlmann et al. 1977), *A. areolaris* (Yu et al. 1986b), *A. gracilis* (Torrenegra et al. 1984) and *A. adenophora* (Mukherjee et al. 2001). Only one aglycone flavonoid with an additional hydroxyl group in C-6 was isolated in *Ageratina*, it was quercetagetin (**6**) isolated from *A. gracilis* (Torrenegra et al. 1984). However, from *A. adenophora* it was found three glycosylated flavonoids with a hydroxyl in C-6 position, 6-hydroxykaempferol-7-*O*-glucoside (**25**) (Li et al. 1997), quercetagitrin (**27**)

(Li et al. 1997), and quercetagetin 7-*O*-(6″-*O*-acetyl)-glucoside (**28**) (Wei et al. 2011), all of them are glycosylated at C-7 position. The sugar moiety of the glycosides flavonoids described from this genus are glucose, with exception of hyperoside (**18**) isolated from *A. arbutifolia* (Perez and Molina 1995), *A. saltillensis* (Yu et al. 1986a), *A. areolaris* (Yu et al. 1986b), and *A. havanensis* (Yu et al. 1987).

On the other hand, the flavonoid aglycone most described from *Chromolaena* species was luteolin (**2**), which was found in *C. hirsuta* (Taleb-Contini et al. 2004, 2007), *C. odorata* (Suksamrarn et al. 2004; Yuan et al. 2007; Fu et al. 2007), *C. squalida* (Taleb-Contini et al. 2007), *C. moritziana* (Báez et al. 1998), *C. congesta* (Oliveira et al. 2017) and *C. chaseae* (Bohlmann et al. 1982). Flavonoids 6-OH were not described from this genus, however three flavonoids with an unsubstituted B-ring (**3**, **11**, **13**) has been found in *C. chaseae* (Bohlmann et al. 1982). Different from that observed for the genus *Ageratina*, a variety of sugar moiety has been identified from *Chromolaena* glycosylated flavonoids, including disaccharide moieties.

7.3.2 Mono-methoxylated Flavonoid

Mono-methoxylated flavonoids were described from twenty *Ageratina* species, and mostly of them have the methoxyl group attached at C-6 and C-7 in A-ring, and at C-4′ in B-ring. Only two compounds methoxylated at position C-5 and C-8 have been isolated, 5-methoxy kaempferol (**41**), isolated from *A. ligustrina* (Talapatra et al. 1985), and 8-methoxy herbacetin (**77**), isolated from *A. gracilis* (Torrenegra et al. 1984), respectively. Flavanone and isoflavone mono-methoxylated were briefly described from *Ageratina*, only one flavanone and one isoflavone, methoxylated at C-4′, have been isolated from this genus, isosakuranetin (**92**) was isolated from *A. sternbergiana* (Gonzales et al. 1982), and biochanin A (**95**) from *A. adenophora* (Parveen et al. 2015).

Glycoylated flavonoid, methoxylated at C-6, were isolated only from *Ageratina* species, and they have the sugar moiety attached at the hydroxyl group in C-3 or C-7 (**48–54**), the most frequent was 6-methoxy kaempferol 3-*O*-glucoside (**48**), that was described from *A. calophylla* (Fang et al. 1986), *A. havanensis* (Yu et al. 1987), *A. saltillensis* (Yu et al. 1986a), *A. areolaris* (Yu et al. 1986b) and *A. adenophora* (Zhang et al. 2015a).

Most of the mono-methoxylated flavonoids isolated from *Chromolaena* have the methoxy group attached at C-7 or C-4′ positions. Sixteen 7-methoxyflavonoids were described from eleven *Chromolaena* species, highlighting *C. odorata*, from which were isolated ten compounds (**60–62**, **64**, **66–67**, **69–70**, **72**, **75**). Only four 6-methoxy flavonoids were isolated from *Chromolaena*, two of them have unsubstituted B-ring, a flavonol, alnusin (**44**), isolated from *C. chaseae* (Bohlmann et al. 1982) and *C. leivensis* (Torrenegra and Rodriguez 2011; Torrenegra et al. 2016), and a flavanonol, alnustinol (**47**), from *C. chaseae* (Bohlmann et al. 1982).

Differently, of that observed for *Ageratina*, glycosylated flavonoids, methoxylated at C-6, were not described from *Chromolaena*. Only two glycosylated flavonoids, methoxylated at C-7, has been described from *C. odorata*, sakuranetin 4′-*O*-[β-D-glucoside(1→2)-βD-glucoside] (**70**) and genkwanin 4′-*O*-rhamnoglucoside (**72**) (Hung et al. 2011).

7.3.3 Di-methoxylated Flavonoid

Almost 41% of the di-methoxylated flavonoids described from *Ageratina* are 6,7-methoxy flavonoids (**105–108**, **110–116**). Eupalitin (**107**) was the most frequent, being found in *A. deltoidea* (Yang et al. 1991), *A. adenophora* (Wei et al. 2011; Nair and Sivakumar 1990), *A. saltillensis* (Yu et al. 1986a) and *A. areolaris* (Yu et al. 1986b). It is important to note that 3,7-methoxy flavonoids were isolated only from *Ageratina* species, with exception of kumatakenin (**100**), which was isolated from *A. illita* (Castillo et al. 2015) and *C. congesta* (Oliveira et al. 2017). With regards to B-ring in the di-methoxylated flavonoids, six are methoxylated at C-4′, and four at C-3′.

Glycoylated flavonoids, 6,7-methoxylated, have the sugar moiety attached at the hydroxyl at C-3 (**110–116**), and they have been described only from *Ageratina* species.

Flavonoids methoxylated at C-7 and C-4′ are the most common substitution pattern found in *Chromolaena* species, being isolated eight different flavonoids, seven of them described from *C. odorata* (**132**, **134–139**). From *C. farinosa* it was isolated pilloin (**133**), ombuin (**135**) and persigogenin (**137**) (Triana 1995). Flavonoids methoxylated at 7- and 3′-positions have been isolated only from five *Chromolaena* species, being possible to highlight velutin (**120**) which was isolated from *C. odorata* (Arene et al. 1978), *C. hirsuta* (Taleb-Contini et al. 2004, 2007), *C. laevigata* (Flores et al. 2006) and *C. meridensis* (Amaro-Luís and Morales-Mendez 1983).

7.3.4 Tri-methoxylated Flavonoid

Only fifteen tri-methoxylated flavonoids have been described from fourteen *Ageratina* species, with a wide variety of methoxylation pattern. However, almost 75% of them have a methoxyl group attached at C-4′. It is important to note that flavonoids methoxylated at 3,7,4′- (**153–154**), 3,6,4′- (**156–157**), and 3,3′,4′- (**158**) were described only from *Ageratina* species.

Already from *Chromolaena*, seventeen tri-methoxylated flavonoids have been described from seven species of this genus, ten of them were isolated from *C. odorata* (**146–150**, **159–160**, **165–166**, **168**, **170**). Aglycone flavonoid with substitution pattern of 5,6,7- (**146–150**) was described only from *C. odorata*.

Almost 78% of the tri-methoxylated flavonoids described from *Chromolaena* have a methoxyl group attached at C-7. Pachypodol (**152**), substitution pattern 3,7,3′-, was isolated from *C. bigelovii* (Jakupovic et al. 1986), and jaceidin (**155**), substitution pattern 3,6,3′-, was found in *C. hirsuta* (Taleb-Contini et al. 2004, 2007) and *C. squalida* (Taleb-Contini et al. 2007).

7.3.5 High Methoxylated Flavonoid

Highly methoxylated flavonoids have been described from *Ageratina* and *Chromolaena*, however, penta-methoxyflavonoids were obtained exclusively from *Chromolaena* genus.

Sixteen high methoxylated flavonoids, ten tetra-methoxylated flavonoids (**175–176**, **178–185**) and five penta-methoxylated flavonoid (**186–190**), have been found in *Chromolaena* species. Four of the tetra-methoxylated flavonoids belong to flavanone class, and two are chalcones, being described mainly from *C. odorata*. All penta-methoxylated flavonoids have been found in *C. odorata*, with exception of umuhengerin (**188**) that was isolated from *C. arnotiana* (De Guttierez et al. 1995).

Only five tetra-methoxylated flavonoids have been described from *Ageratina* species. Santaflavone (**176**) was the most frequent, being isolated from *A. tomenttella* (Fang and Mabry 1986), *A. viscosa* (Torrenegra et al. 1990), *A. jhanii* (Gonzales et al. 1979) and *A. altissima* (Wollenweber et al. 1996; Herz et al. 1978; Dobberstein et al. 1977).

7.4 Conclusion

In this chapter, we evaluated the flavonoid profile of *Ageratina* and *Chromolaena* species, which have *Eupatorium* as basionym. It was observed that only 21% of the flavonoids described from *Ageratina* are non-methoxylated. In *Chromolaena* this number is even lower, about 17%. Tri-methoxylated flavonoids have been found in almost 50% of *Ageratina* and *Chromolaena* species, with a miscellaneous pattern of substitution. High methoxylated flavonoids, with four or more methoxy groups, were briefly described from *Ageratina*, and it was more frequent described from *C. odorata*. So, non-, tri- and high methoxylated flavonoid were not considered as a good chemotaxonomic marker from these genera.

On the other hand, mono- and di-methoxylated flavonoids are been found in 85% of *Ageratina* species studied, and in *Chromolaena*, about 90% possess these classes of flavonoids. It was possible to observe also that 44% of the mono- and di-methoxylated flavonoids described from *Ageratina* have the methoxy attached to C-6, and 47% are methoxylated at position C-7. Already in *Chromolaena*, the methoxy group are bound at C-7 in 54% of the flavonoids mono- and di-methoxylated, and 41% have the methoxy group attached to C-4′, however, different from observed to *Ageratina*, 6-methoxy flavonoids appear in only 21%.

In conclusion, this study confirmed that flavonoids are a useful marker in *Ageratina* and *Chromolaena*. Furthermore, it can be assumed the mono- and di-methoxylated flavonoids at positions C-6 and C-7 for *Ageratina*, and C-7 and C-4′ for *Chromolaena*, can be considerate as appropriate markers for these genera.

References

Amaro-Luis JM, Delgado-Mendez P (1993) Flavonoids from the leaves of *Chromolaena subscandens*. J Prod Nat 56:610–612

Amaro-Luís JM, Morales-Mendez A (1983) Phytochemical study of Venezuelan Andean flora. II. Flavonoids of *Chromolaena meridensis*. Rev Lationoam Chem 14:86–88

Ansari S, Jain P, Tyagi RP, Joshi BC, Barar FSK (1983) Phytochemical and pharmacological studies of the aerial parts of *Eupatorium adenophorum* L. Herba Pol 29:93–96. Abstract retrieved from SciFinder (Accession No. 1985:3227)

Arciniegas A, Perez-Castorena AL, Melendez-Aguirre M, Avila JG, Garcia-Bores AM, Villasenor JL, Romo de Vivar A (2018) Chemical composition and antimicrobial activity of *Ageratina deltoidea*. Chem Biodivers 15:e1700529

Arene EO, Pettit GR, Ode RH (1978) Antineoplastic agents. Part 49. The isolation of isosakuranetin methyl ether from *Eupatorium odoratum*. Lloydia 41:186–189

Atindehou M, Lagnika L, Guérold B, Strub JM, Zhao M, Sorsselaer AV, Marchioni E, Prévost G, Haikel Y, Taddéi C, Sanni A, Metz-Boutigue MH (2013) Isolation and identification of two antibacterial agents from *Chromolaena odorata* L. active against four diarrheal strains. Adv Microbiol 3:115–121

Báez DH, Los Ríos C, Crescente O, Caserta A (1998) Antibacterial and chemical evaluation of *Chromolaena moritziana*. J Ethnopharmacol 59:203–206

Barrio G, Spengler I, Garcia T, Roque A, Alvarez AL, Calderon JS, Parra F (2011) Antiviral activity of *Ageratina havanensis* and major chemical compounds from the most active fraction. Rev Bras Farmacogn 21:915–920

Barua RN, Sharma RP, Thyagarajan G, Hertz W (1978) Flavonoids of *Chromolaena odorata*. Phytochemistry 17:1807–1808

Bauer L (1976) *Eupatorium laevigatum* Lam.—essential oil and flavonoids. Trib Farm 44:83–92. Abstract retrieved from SciFinder (Accession No. 1978:465133)

Boeker R, Bohlmann F (1986) Flavanones from *Chromolaena heteroclinium*. Rev Latinoam Quim 17:45–46. Abstract retrieved from SciFinder (Accession No. 1986:530683)

Bohlmann F, Grenz M (1977) Naturally occurring terpene derivatives. Constituents from representatives of the *Eupatorium* group. Chem Ber 110:1321–1329

Bohlmann F, Jakupovic J, Lonitz M (1977) Uber Inhaltsstoffe der *Eupatorium*-Gruppe. Chem Ber 110:301–314

Bohlmann F, Zdero C, King RM, Robinson H (1979) Naturallyoccurring terpene derivatives. Part 207. New cadinene derivatives and other constituents from*Chromolaena* species. Phytochemistry 18:1177–1179. Abstract retrieved from SciFinder (Accession No. 1980:129101)

Bohlmann F, Singh P, Jakupovic J, King RM, Robinson A (1982) Three cadinene derivatives and a prostaglandinlike acid from *Chromolaena* species. Phytochemistry 21:371–374

Bose PK, Chakrabarti P, Chakravarti S, Dutta SP, Barua AK (1973) Flavonoid constituents of *Eupatorium odoratum*. Phytochemistry 12:667–668

Bose PK, Chakrabarti P, Chakravarti S, Dutta SP, Barua AK (1974) Chemical examination of the leaves of *Eupatorium odoratum*. Trans Bose Res Inst 37:25–30. Abstract retrieved from SciFinder (Accession No. 1975:560778)

Bremer K (1994) *Asteraceae*: cladistics and classification. Timber Press, Portland, p 429

Bustos-Brito C, Sanchez-Castellanos M, Esquivel B, Calderon JS, Calzada F, Yepez-Mulia L, Joseph-Nathan P, Cuevas G, Quijano L (2015) ent-Kaurene Glycosides from *Ageratina cylindrica*. J Nat Prod 78:2580–2587

Bustos-Brito C, Vazquez-Heredia VJ, Calzada F, Yepez-Mulia L, Calderon JS, Hernandez-Ortega S, Esquivel B, Garcia-Hernandez N, Quijano L (2016) Antidiarrheal thymol derivatives from *Ageratina glabrata*. Structure and absolute configuration of 10-benzoyloxy-8, 9-epoxy-6-hydroxythymol isobutyrate. Molecules 21:1132/1–1132/13

Castillo QA, Triana J, Eiroa JL, Padron JM, Plata GB, Abel-Santos EV, Baez LA, Rodriguez DC, Jimenez MA, Perez-Pujols MF (2015) Flavonoids from *Eupatorium illitum* and their antiproliferative activities. Pharmacogn J 7:178–181

Clavin M, Gorzalczany S, Macho A, Munoz E, Ferraro G, Acevedo C, Martino V (2007) Anti-inflammatory activity of flavonoids from *Eupatorium arnottianum*. J Ethnopharmacol 112:585–589

D'Agostino M, De Simone F, Dini A, Ramundo E, Zollo F (1990) Flavonol glycosides from *Eupatorium tinifolium*. Phytochemistry 29:353–354

D'Agostino M, De Simone F, Zollo F, Pizza C (1991) Isolation of 5,7-O-dimethyl-6-methoxykaempferol-3-O-β-D-glucopyranoside from *Eupatorium tinifolium*. Fitoterapia 62:445–446. Abstract retrieved from SciFinder (Accession No. 1992:231890)

D'Agostino M, De Feo V, Pizza C, De Simone F (1994) Isolation of 3,5,7,2′,3′,4′-hexahydroxyflavone-3-*O*-β-D-glucopyranoside from *Eupatorium sternbergianum*. Fitoterapia 65:472–473. Abstract retrieved from SciFinder (Accession No. 1995:386447)

Dat NT, Lee K, Hong YS, Kim YH, Van Minh C, Lee JJ (2009) A peroxisome proliferator-activated receptor-gamma agonist and other constituents from *Chromolaena odorata*. Plant Med 75:803–807

De Guttierez AN, Catalan CAN, Diaz JG, Herz W (1995) Sesquiterpene lactones, a labdane and other constituents of *Urolepis hecatantha* and *Chromolaena arnottiana*. Phytochemistry 39:795–800

De Perez EC, Cespedes AB (1984) Isolation of 5,4′-dihydroxy-7-methoxyflavone and 5-hydroxy-7,4′-dimethoxyflavone from *Eupatorium angustifolium* (H.B.K.) Spreng (Compositae). Rev Colomb Quim 13:49–56

De Villarraga AA, De Perez EC (1994) 6-Methoxyflavonoids from flowers of *Ageratina angustifolia* R.M. King and H. Robinson (Compositae). Rev Colomb Quim 23:45–51

Dhar R, Kimseng R, Chokchaisiri R, Hiransai P, Utaipan T, Suksamram A, Chunglok W (2017) 2′,4-Dihydroxy-3′,4′,6′-trimethoxychalcone from *Chromolaena odorata* possesses anti-inflammatory effects via inhibition of NF-kB and p38 MAPK in lipopolysaccharide-activated RAW 264.7 macrophages. Immunopharmacol Immunotoxicol 40:43–51

Ding Z, Zhang X, Liu J, Ding J, Gu K (2001) Chemical constituents from *Eupatorium odoratum*. Tianran Chanwu Yanjiu Yu Kaifa 13: 22–24. Abstract retrieved from SciFinder (Accession No. 2002:20509)

Dobberstein RH, Tin-Wa M, Fong HHS, Crane FA, Farnsworth NR (1977) Flavonoid constituents from *Eupatorium altissimum* L. (Compositae). J Pharm Sci 66:600–602

Dominguez XA, Fuente ER (1973) Sakuranetin and pulcherryl acetate from *Eupatorium havanense*. Phytochemistry 12:2060

Eiroa JL, Triana J, Perez FJ, Castillo QA, Brouard I, Quintana J, Estévez F, León F (2018) Secondary metabolites from two *Hispaniola Ageratina* species and their cytotoxic activity. Med Chem Res 27:1792–1799

Elema ET, Schripsema J, Malingre TM (1989) Flavones and flavonol glycosides from *Eupatorium cannabinum* L. Pharm Weekbl Sci 11:161–164

Emerenciano VP, Barbosa KO, Scotti MT, Ferreira MJP (2007) Self-organizing maps in chemotaxonomic studies of *Asteraceae*: a classification of tribes using flavonoid data. J Braz Chem Soc 18:891–899

Ezenyi IC, Salawu OA, Kulkarni R, Emeje M (2014) Antiplasmodial activity-aided isolation and identification of quercetin-4′-methyl ether in *Chromolaena odorata* leaf fraction with high activity against chloroquine-resistant *Plasmodium falciparum*. Parasitol Res 113:4415–4422

Fang N, Mabry TJ (1986) Flavonoid aglycones from *Ageratina tomentella*. J Nat Prod 49:737–738

Fang N, Yu S, Mabry TJ (1986) Flavonoids from *Ageratina calophylla*. Phytochemistry 25:2684–2686

Ferraro G, Martino V, Coussio J (1983) Flavonoids isolated from *Eupatorium hookerianum*. An Asoc Quim Argent 71:327–329. Abstract retrieved from SciFinder (Accession No. 1984:20437)

Flores M, D'Armas H, Estaba A, Mendez B (2006) Identification of a flavonoid isolated from the leaves of *Chromolaena laevigata*. Part I. Ciencia 14:244–252. Abstract retrieved from SciFinder (Accession No. 2007:48817)

Fu Y-J, Liu W, Hou C-L, Liu X-N, Li S-M, Zu Y-G (2007) Simultaneous determination of three flavonoids in *Eupatorium odoratum* Linn by RP-HPLC/PAD. Yingyong Huaxue 24:1452–1454. Abstract retrieved from SciFinder (Accession No. 2008:79826)

Funk VA, Susanna A, Stuessy TF, Bayer RJ (2009) Systematics, evolution, and biogeography of Compositae. International Association for Plant Taxonomy, Vienna, Austria

Gonzales AG, Arteaga JM, Fraga BM, Hernnadez MG (1979) Chemistry of the compuestas. Part 40. Diterpenes, triterpenes, and flavones from *Eupatorium jhanii*. An Quim (1968–1979) 75:128–130. Abstract retrieved from SciFinder (Accession No. 1979:474736)

Gonzales AG, Fraga BM, Hernandez MG, Garcia VP (1982) Eupatarone, a 2-acetylbenzofuran from *Eupatorium sternbergianum*. Phytochemistry 21:1826–1827

Gonzales AG, Fraga BM, Garcia VP, Hernandez MG (1984) Sternbin, a new flavanone from *Eupatorium sternbergianum*. Rev Latinoam Quim 14:115–117. Abstract retrieved from SciFinder (Accession No. 1984:206488)

Gupta RK, Krishnamurti M, Parthasarathi J (1979) Synthesis of some recently isolated chalcones, their analogs and corresponding flavanones. Agric Biol Chem 23:26

Hai MA, Biswas PK, Shil KC, Ahmad MU (1991) Chemical constituents of *Eupatorium odoratum* Linn (Compositae). J Bangladesh Chem Soc 4:47–49. Abstract retrieved from SciFinder (Accession No. 1992:148159)

Hai MA, Saha K, Ahmad MU (1995) Chemical constituents of *Eupatorium odoratum* Linn (Compositae). J Bangladesh Chem Soc 8:139–142. Abstract retrieved from SciFinder (Accession No. 1996:564411)

Heiss EH, Tran TVA, Zimmermann K, Scwaiger S, Vouk C, Mayerhafer B, Malainer C, Atanasov AG, Stuppner H, Dirsch VM (2014) Identification of chromomoric acid C-I as an Nrf2 activator in *Chromolaena odorata*. J Nat Prod 77:503–508

Herz W, Gibaja S (1972) 5-Hydroxy-6,7-trimethoxyflavone from *Ageratina gilbertii*. Phytochemistry 11:2625

Herz W, Ramakrishnan G, Murari R (1978) A glycosidic germacradienolide from *Eupatorium altissimum*. Phytochemistry 17:1953–1955

Higuera BL, Perez CP (1989) Isolation of 5,4′-dihydroxy-6,7-dimethoxy-3-O-glycosylflavone of *Ageratina glyptophlebia* and characterization of its aglycon. Rev Colomb Quim 18:55–60

Hung TM, Cuong TD, Dang NH, Zhu S, Long PQ, Komatsu K, Min BS (2011) Flavonoid glycosides from *Chromolaena odorata* leaves and their *in vitro* cytotoxic activity. Chem Pharm Bull 59:129–131

Jakupovic J, Ellmauerer E, Bohlmann F, Whittemori A, Gage D (1986) Diterpenes from *Eupatorium turbinatum*. Phytochemistry 25:2677–2678

Johari SA, Kiong LS, Mohtar M, Isa MM, Man S, Mustafa S, Ali AM (2012) Efflux inhibitory activity of flavonoids from *Chromolaena odorata* against selected methicillinresistant *Staphylococcus aureus* (MRSA) isolates. Afr J Microbiol Res 6:5631–5635

Kharazian N (2014) Chemotaxonomy and flavonoid diversity of *Salvia L.* (Lamiaceae) in Iran. Acta Bot Bras 28:281–292

King RM, Robinson H (1987) The genera of the Eupatorieae (*Asteraceae*). In: Monographs in Systematic Botany. Missouri Botanical Garden, United States, pp 22, 65

Kouamé PBK, Jacques C, Bedi G, Silvestre V, Loquet D, Barillé-Nion S, Robins RJ, Tea I (2013) Phytochemicals isolated from leaves of *Chromolaena odorata*: impact on viability and clonogenicity of cancer cell lines. Phytother Res 27:835–840

Li R, Ding Z, Ding J (1997) Chemical constituents from *Eupatorium adenophorum*. Yunnan Zhiwu Yanjiu 19:196–200. Abstract retrieved from SciFinder (Accession No. 1997:683465)

Ling SK, Pisar MMd, Man S (2007) Platelet-activating factor (PAF) receptor binding antagonist activity of the methanol extracts and isolated flavonoids from *Chromolaena odorata* (L.) King and Robinson. Biol Pharm Bull 30:1150–1152

Liu P, Liu D, Li W, Zhao T, Sauriol F, Gua Y, Shi Q, Zhang M (2015) Chemical constituents of plants from the genus *Eupatorium* (1904–2014). Chem Biodivers 12:1481–1513

Lopes JLC, Lopes JNC (1979) The flavonoids of *Eupatorium* pedale DC. Ver Latinoam Quim 10:36–37. Abstract retrieved from SciFinder (Accession No. 1979:402535)

Martjnez MAP, Swain T (1985) Flavonoids and chemotaxonomy of the Commelinaceae. Bio Chem Ecol 13:391–402

Metwally AM, Ekejiuba EC (1981) Methoxylated flavonols and flavanones from *Eupatorium odoratum*. Planta Med 42:403–405

Morales MA, Rosquete C (1988) Flavonoids from *Ageratina stevioides* (Steyermark). Rev Latinoam Quim 19:47–49. Abstract retrieved from SciFinder (Accession No. 1989:92104)

Moreno B, Delle MG, Delle MF, Marini-Bettolo GB (1980) Flavones and kauranoid diterpenes from *Eupatorium tinifolium* H.B.K. Researches on plant antitumor principles. Farmaco Sci 35:457–464. Abstract retrieved from SciFinder (Accession No. 1980:466078)

Mukherjee PK, Mukherjee K, Hermans-Lokkerbol ACJ, Verpoorte R, Suresh B (2001) Flavonoid content of *Eupatorium glandulosum* and *Coolebroke oppositifolia*. J Nat Remedies 1:21–24

Na Z, Feng YL, Xu YK (2012) Chemical constituents from aerial parts of *Eupatorium odoratum*. Zhongcaoyao 43:1896–1900. Abstract retrieved from SciFinder (Accession No. 2013:182644)

Nair AGR, Sivakumar R (1990) 6-Hydroxykaempferol 6,4′-dimethyl ether 3-galactoside from *Eupatorium glandulosum*. Phytochemistry 29:1011–1012

Nair AGR, Jayaprakasam R, Gunasekaran R, Bayet C, Voirin B (1993) 6-Hydroxykaempferol 7-(6″-caffeoylglucoside) from *Eupatorium glandulosum*. Phytochemistry 33:1275–1276

Nair AGR, Gunasegaran R, Krishnan S, Bayet C, Voirin B (1995) Flavonol glycosides from leaves of *Eupatorium glandulosum*. Phytochemistry 40:283–285

Nath LR, Gorantla JN, Joseph SM, Antony J, Trankachan S, Menon DB, Sankar S, Lankalapalli RS, Anto RJ (2015) Kaempferide, the most active among the four flavonoids isolated and characterized from *Chromolaena odorata*, induces apoptosis in cervical cancer cells while being pharmacologically safe. RSC Adv 5:100912–100922

Nguyen TDT, Le VNP, Nguyen XD (1993) Chalcones of *Eupatorium odoratum* L. from Vietnam. Tap Chi Hoa Hoc 31:79–80. Abstract retrieved from SciFinder (Accession No. 1995:455219)

Ohtsuki T, Hiraka T, Kikuchi H, Koyano T, Kowithayakorn T, Sakai T, Ishibashi M (2009) Flavonoids from *Eupatorium odoratum* with death receptor 5 promoter enhancing activity. Heterocycles 77:1379–1388

Oliveira JAM, Bernardi DI, Balbinoti RB, Avincola AS, Pilau E, Carmo MRB, Sarragiotto MH, Baldoqui DC (2017) Chemotaxonomic value of flavonoids in *Chromolaena congesta* (*Asteraceae*). Biochem Syst Ecol 70:7–13

Parveen SDS, Kumar BS, Kumar SR, Khan RI, Pitchumani K (2015) Isolation of biochanin A, an isoflavone, and its selective sensing of copper (II) ion. Sens Actuators B Chem 221:75–80

Perez D, Espitia C (1999) 4′-Hydroxyflavanones from *Chromolaena tyleri*. Evidence of a characteristic EI-MS. Nat Prod Res 13:97–103

Perez CE, Molina EV (1995) Glycosidic flavonoids of *Ageratina arbutifolia*. Rev Colomb Quim 24:39–45

Perez CE, Pinilla GS (1990) Methoxyflavones of *Ageratina glyptophlebia* (Compositae). Rev Colomb Quim 19:101–105

Perez CE, Roa AM, Castellanos Y (1980) Isolation of 5-hydroxy-7,4′-dimethoxyflavone of *Eupatorium amplum* Benth (Compositae). Rev Colomb Quim 10:17–20

Phan TT, Wang L, See P, Grayer RJ, Chan SY, Lee ST (2001) Phenolic compounds of *Chromolaena odorata* protect cultured skin cells from oxidative damage: implication for cutaneous wound healing. Biol Pharm Bull 24:1373–1379

Pisutthanan N, Liawruangrath S, Bremner JB, Liawruangrath B (2005) Chemical constituents and biological activities of *Chromolaena odorata*. Chiang Mai J Sci 32:139–148. Abstract retrieved from SciFinder (Accession No. 2005:955538)

Pisutthanan N, Liawruangrath B, Liawruangrath S, Bremner JB (2006) A new flavonoid from *Chromolaena odorata*. Nat Prod Res 20:1192–1198

Quijano L, Malanco F, Rios T (1970) Structures of eupalin and eupatolin. Two new flavonol rhamnosides isolated from *Eupatorium ligustrinum*. Tetrahedron 26:1851–2859

Rodríguez OE, Torrenegra RD (2005) Flavonoids of *Chromolaena taconata* (Klatt) R. M. King and H. Rod. Actual Biol 27:113–115

Rodríguez-Cabeza BV, Díaz-Piedrahíta S, Parra-OC (2014) Chromolaena (Asteraceae: Eupatorieae). Flora de Colombia, vol 31. Instituto de Ciencias Naturales, Universidad Nacional de Colombia, Bogota, Colombia, 148 pp

Romero-Cerecero O, Zamilpa A, González-Cortazar M, Alonso-Cortés D, Jiménez-Ferrer E, Nicasio-Torres P, Aguilar-Santamaría L, Tortoriello J (2013) Pharmacological and chemical study to identify wound-healing active compounds in *Ageratina pichinchensis*. Planta Med 79:622–627

Sanabria-Galindo A, Carrero MT (1995) A flavanol with antimicrobial activity isolated from *Chromolaena taconata*. Rev Colomb Cienc Quim Farm 24:24–24. Abstract retrieved from SciFinder (Accession No. 1997:304103)

Sareedenchai V, Zidorn C (2010) Flavonoids as chemosystematic markers in the tribe Cichorieae of the *Asteraceae*. Bio Chem Ecol 38:935–957

Schmeda G, Pacheco P, Silva M (1983) Flavonoids of *Eupatorium* leave D.C. Rev Latinoam Quim 14:36–37. Abstract retrieved from SciFinder (Accession No. 1983:450287)

Shuib NS, Sam TW, Wong KC, Chinnakali K, Fun HK (1999) Tamarixetin monohydrate. Acta Crystallogr Sect C-Cryst Struct Commun 55:576–578

Singh R (2016) Chemotaxonomy: a tool for plant classification. J Med Plants Stud 4:90–93

Stevens JF, Elema ET, Wollenweber E (1995) Exudate flavonoids of *Eupatorium cannabinum*. Biochem Syst Ecol 23:451–452

Suksamrarn A, Chotipong A, Suavansri T, Boongird S, Timsuksai P, Vimuttipong S, Chuaynugul A (2004) Antimycobacterial activity and cytotoxicity of flavonoids from the flowers of *Chromolaena odorata*. Arch Pharm Res 27:507–511

Suntornsuk L, Anurukvorakun O (2005) Precision improvement for the analysis of flavonoids in selected Thai plants by capillary zone electrophoresis. Electrophoresis 26:648–660

Talapatra SK, Bhar DS, Talapatra B (1974) Flavonoid and terpenoid constituents of *Eupatorium odoratum*. Phytochemistry 1113:284–285

Talapatra SK, Bhar DS, Talapatra B (1977) Terpenoids and related compounds: part XIII. Epoxylupeol, a new triterpenoid from *Eupatorium odoratum*. Indian J Chem 15:806–807

Talapatra SK, Pal MK, Mallik AK, Talapatra B (1985) Chemical constituents of *Eupatorium erythropappum*: eupathronoside—a new flavanone glucoside and (-) 1S,2R,4S,5R-2,5-dihydroxy-p-menthane—a new monoterpene diol. J Indian Chem Soc 62:999–1002. Abstract retrieved from SciFinder (Accession No. 1988:72031)

Taleb-Contini SH, Salvador MJ, Watanabe E, Ito I, Oliveira DCR (2003) Antimicrobial activity of flavonoids and steroids isolated from two *Chromolaena* species. Rev Bras Cienc Farm 39:403–408

Taleb-Contini SH, Salvador MJ, Balanco JMF, Albuquerque S, De Oliveira DCR (2004) Antiprotozoal effect of crude extracts and flavonoids isolated from *Chromolaena hirsuta* (*Asteraceae*). Phytother Res 18:250–254

Taleb-Contini SH, Schorr K, Da Costa FB, Rodrigues ODC (2007) Detection of flavonoids in glandular trichomes of Chromolaena species (Eupatorieae, Asteraceae) by reversed-phase high-performance liquid chromatography. Rev Bras Cienc Farm 43:315–321

Tamayo-Castillo G, Jakupovic J, Bohlmann F, King RM, Robinson H (1989) ent-clerodane derivatives from *Chromolaena connivens*. Phytochemistry 28:641–642

Torrenegra RD, Rodriguez OE (2011) Chemical and biological activity of leaf extracts of *Chromolaena leivensis*. Nat Prod Commun 6:947–950

Torrenegra RD, Pedrozo JA, Escarria S (1984) Chemical study of some plants of the Bogota Savanna. Flavonoids from *Eupatorium gracile*. Rev Latinoam Quim 15:129–130. Abstract retrieved from SciFinder (Accession No. 1985:128807)

Torrenegra R, Pedrozo JA, Robles J, Fuentes O (1990) Chemical study of Colombian plants of the Compositae and Apocyanaceae. ε-Palmitolactone from *Ageratina viscosa*. Phytochemistry 29:305–306

Torrenegra R, Robles J, Pedrozo J (1995) Terpenes and flavonoids from *Ageratina fastigiata* (H.B. K.) King & Robinson. Rev Colomb Quim 24:9–16

Torrenegra RD, Rodríguez J, Rodríguez OE, Palau VE, Méndez GM (2016) Antiproliferative activity of 3,5,7-trihydroxy -6-methoxy flavone obtained from C*hromolaena leivensis* (Hieron) on cancer cell lines of breast, prostate, lung, colon and cervix. Pharmacology Online 1:7–11

Triana J (1995) Ent-norlabdanes and other constituents of four Venezuelan species previously classified in the genus. J Nat Prod 58:744–747

Triratana T, Suwannuraks R, Naengchomnong W (1991) Effect of *Eupatorium odoratum* on blood coagulation. J Med Assoc Thai 74:283–287. Abstract retrieved from SciFinder (Accession No.1992148291)

Venkata RB, Samuel LA, Pardha SM, Narashima RB, Naga VKA, Sudhakar M, Radhakrishnan TM (2012) Antibacterial, antioxidant activity and GC-MS analysis of *Eupatorium odoratum*. Asian J Pharm Clin Res 5:99–106

Viljoen AM, van Wyk BE, van Heerden FR (1998) Distribution and chemotaxonomic significance of flavonoids in Aloe (Asphodelaceae). Plant Syst Evol 211:31–42

Vyas AV, Mulchandani NB (1986) Benzene-induced proton NMR shifts of flavone protons: detection of H-3. Magn Reson Chem 24:421–423

Wafo P, Hussain H, Kouam SF, Ngadjui BT, Florke U, Krohn K (2005) 2′-Hydroxy-4,4′,5′,6′-tetramethoxychalcone. Acta Cryst 61:3017–3019

Wafo P, Kamdem RST, Ali Z, Anjum S, Begum A, Oluyemisi OO, Khan SN, Ngadjui BT, Etoa XF, Choudhary MI (2011) Kaurane-type diterpenoids from *Chromoleana odorata*, their X-ray diffraction studies and potent α-glucosidase inhibition of 16-kauren-19-oic acid. Fitoterapia 82:642–646

Wagner H, Iyengar MA, Horhammer L (1972) Flavonol-3-glucosides in eight *Eupatorium* species. Phytochemistry 2:1504–1505

Wang C, Yang R, Song L, Ning B, Ou-yang C, Cao A, He L (2016) Two new highly-oxygenated flavonoid glycosides from *Eupatorium adenophorum* Spreng. Phytochem Lett 16:245–248

Wei Y, Zhang K, Zhang G, Ito Y (2011) Isolation of five bioactive components from *Eupatorium adenophorum* Spreng using stepwise elution by high-speed countercurrent chromatography. J Liq Chromatogr Relat Technol 34:2505–2515

Wollenweber E, Roitman JN (1996) A novel methyl ether of quercetagetin from *Chromolaena odorata* leaf exudate. Biochem Syst Ecol 24:479–480

Wollenweber E, Doerr M, Muniappan R (1995) Exudate flavonoids in a tropical weed, *Chromolaena odorata* (L.) R. M. King et H. Robinson. Biochem Syst Ecol 23:873–874

Wollenweber E, Doerr M, Beyer M, Schilling E (1996) External flavonoids of 12 species of North American Eupatorieae. Z Naturforsch 51:893–896

Wollenweber E, Dorr M, Fritz H, Papendieck S, Yatskievych G, Roitman JN (1997) Exudate flavonoids in *Asteraceae* from Arizona, California, and Mexico. Z Naturforsch C J Biosci 52:301–307

Yang SL, King RA, Roberts MF (1991) The flavonoids of *Ageratina deltoidea*. Biochem Syst Ecol 18:485–486

Yu S, Fang N, Mabry TJ (1986a) Flavonoids from *Ageratina saltillensis*. J Nat Prod 49:1178–1179

Yu S, Gage DA, Fang N, Mabry TJ (1986b) Flavonols and flavonol glycosides from *Eupatorium areolare* var. *leiocarpum*. J Nat Prod 49:181–182

Yu S, Fang N, Mabry TJ (1987) Flavonoids from *Ageratina havanensis*. Rev Latinoam Quim 18:20–22. Abstract retrieved from SciFinder (Accession No. 1987:436614)

Yuan J, Yang J, Miao J (2005) Chemical constituents of *Eupatorium odoratum*. Zhongcaoyao 36:1771–1773. Abstract retrieved from SciFinder (Accession No. 2006:1290253)

Yuan J, Yang J, Miao J (2007) Flavonoids of *Eupatorium odoratum* L. Zhongyaocai 30:657–660. Abstract retrieved from SciFinder (Accession No. 2008:867611)

Zhang ML, Irwin D, Li XN, Sauriol F, Shi XN, Wang YF, Huo CH, Li LG, Gu YC, Shi QW (2012) PPARγ agonist from *Chromolaena odorata*. J Nat Prod 75:2076–2081

Zhang LK, Luo DQ, Feng YL, Dong TF (2013) Chemical constituents of *Chromolaena odorata* and their anticancer activities. ZhongChengYao 35:545–548

Zhang M, Liu W, Gao X, Tan J, Wan F (2015a) Chemical constituents from *Eupatorium adenophorum* Spreng. Redai Yaredai Zhiwu Xuebao 23:697–702. Abstract retrieved from SciFinder (Accession No. 2016:735928)

Zhang M, Liu P, Zhao T, Wang Y, Dun J, Sauriol F, Lu X, Wang R, Shi Q (2015b) A new bidihydroflavone isolated from *Chromolaena odorata*. Chem Nat Compd 51:637–639

Chapter 8
Isoquinoline Alkaloids and Chemotaxonomy

Anderson R. dos Santos and Nelissa P. Vaz

Abstract The isoquinoline alkaloids are a class of secondary metabolites classified into different groups, being the aporphinoids the most representative class. They are considered the second-largest class of alkaloids in terms of structural diversity staying behind only of indole alkaloids. The aporphines could be found in the most diverse families of the plant kingdom. Isoquinolines alkaloids have aroused great interest in chemists and pharmacists due to its wide spectrum of biological activities: highlighting dopaminergic and serotonergic, vasodilator, antiplatelet agents, antimicrobial, antiviral, and cytotoxic. Different methods are applied to improve the comprehension about the metabolic pathways of isoquinoline alkaloids biosynthesis. Those approaches contribute to chemotaxonomy: one of the most useful plant classification systems that currently exist. This class of secondary metabolites has been widely reported as chemotaxonomic markers. These technological advances concerning new methodologies and tools are used nowadays to study these chemotaxonomic relationships: since screening the chemical composition of plants extracts through analytical techniques, to the analysis of chemodiversity and chemosystematics allied through reported data in literature. In this chapter, we present the main metabolic pathways involving biosynthesis of some groups of isoquinoline alkaloids, their occurrence in different vegetable families, and a few methods used in the studies of chemotaxonomic relations.

Keywords Isoquinoline alkaloids · Biosynthesis · Occurrence · Chemotaxonomy · Biodiversity

A. R. dos Santos (✉) · N. P. Vaz
Department of Chemistry, Laboratory of Natural Products and Chemical Ecology, Federal University of Paraná, Curitiba, Paraná, Brazil
e-mail: anrogerquimica@gmail.com

N. P. Vaz
e-mail: nelissavaz@gmail.com

K. G. Ramawat (ed.), *Biodiversity and Chemotaxonomy*, Sustainable Development and Biodiversity 24, https://doi.org/10.1007/978-3-030-30746-2_8

8.1 Introduction

Plants have developed a peculiar way to defend themselves against herbivores, fungi, and other parasites. Since they are sessile organisms and cannot escape from those menaces, they involved together with their own metabolism specializing I to synthesize chemical constituents: the secondary metabolites. In order to guarantee plants survival, those compounds would not interfere in vegetables primary metabolism. In other words, regarding the first line of vegetable's defense system against pathogens and other plants, these substances show a huge variety of biological activities that can be useful if applied to pharmacology once it shows toxical, UV protection, and antioxidative activities among others (Wink 2015). Due those remarkable properties, and high structural diversity, plant secondary metabolites are target of constant investigation by researchers around the world. Latest phytochemical investigations have reported the isolation and identification of more than 21,000 alkaloids, 20,000 terpenoids, 10,000 polyphenols, 1500 polyacetylenes and fatty acids and 750 polyketides (Wink 2013) among the most abundant.

The alkaloids containing the isoquinoline nucleus are a class of substances with the second-largest structural diversity in nature getting behind only of the indole alkaloids (Da Cunha and Barbosa-Filho 2012). The aporphinoids (also known as aporphine alkaloids) are one of the major groups among the isoquinolines.

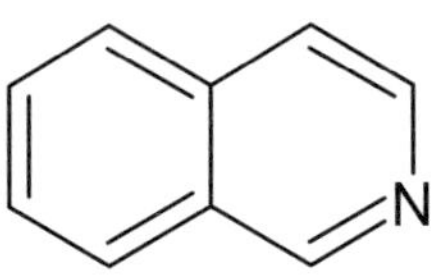

Isoquinoline nucleus

This wide variety of chemical structures is derived from the isoquinoline backbone: and it is used to define of which class (or subclass) the alkaloid belongs. The classes of isoquinoline alkaloids include: aporphines, proaporphines, secoaporphines, oxoaporphines, dehydroaporphines, 7-hidroxyaporphines, dimers of aporphines, aristolactams (Zhang et al. 2007), and 7,7-dimethylaporphines (Costa et al. 2009a). The isoquinoline alkaloids are widely distributed in many plant families like Amaryllidaceae, Ancistrocladaceae, Annonaceae, Apocynaceae, Aristolochiaceae, Atherospermataceae, Berberidaceae, Boraginaceae, Chenopodiaceae, Convolvulaceae, Cornaceae, Dioconphyllaceae, Dioscoreaceae, Euphorbiaceae, Gnetaceae, Hernandiaceae, Lauraceae, Leguminosae, Magnoliaceae, Menispermaceae, Monimiaceae, Moraceae, Nelumbonaceae, Papaveraceae, Ranunculaceae, Rubiaceae, Rutaceae, and Siparunaceae. The aporphinoids could exhibit a variety of biological properties, including dopaminergic and serotonergic, vasodilating, antiplatelet agents, antioxidant, antimicrobial, antiviral, and cytotoxic. Two aporphine alkaloids are commercially available as pharmaceuticals: boldine isolated from the leaves and bark of *Peumus boldus* species (which show free radical scavenging property) is used as medicine for the treatment of hepatobiliary dysfunction, symptomatic treatment of mild digestive disorders and as an adjunct to constipation. Apomorphine is a synthetic alkaloid that has been used in the treatment of Parkinson's disease and, more recently, on the treatment of erectile dysfunction (Stevigny et al. 2005).

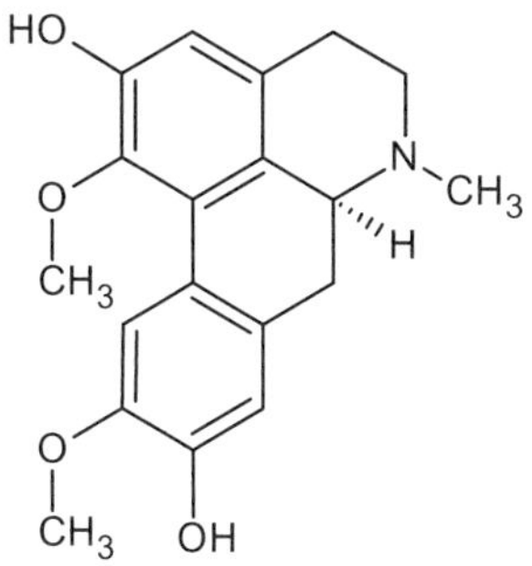

Boldine

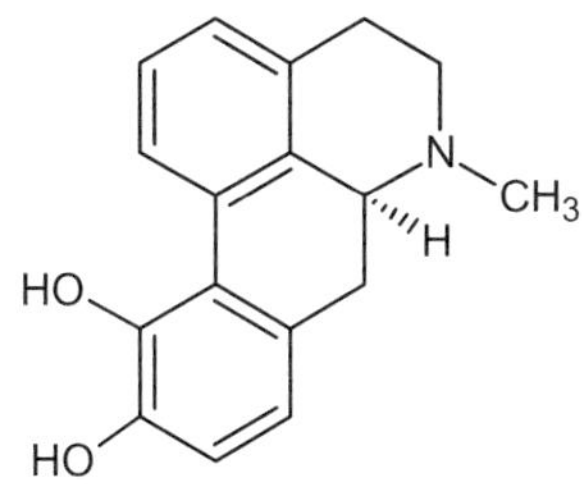

Apomorphine

With the advances in technology, plants are being recognized as living libraries containing bioactive secondary metabolites, filtered by natural selection, which have been used by humans to treat infections and health disorders, or as spices, perfumes, arrow poisons, toxins, and pesticides (Wink 2015). In this chapter, we will focus on the chemotaxonomy of isoquinoline alkaloids, reviewing about the metabolic pathways involved in biosynthesis of their main classes, as well as the current classification and the occurrence in different vegetable families and some methods used in the studies of chemotaxonomic relations.

8.2 Classification of Isoquinoline Alkaloids

The main alkaloid classes and subclasses containing an isoquinoline nucleus are summarized in Fig. 8.1. This schematic classification was made to ease the comprehension between their structural relationships, and to highlight the variety and the structural complexity of main classes and subclasses that were considered in this literature survey (Lúcio et al. 2015; Dewick 2002; Iqbal et al. 2018).

The classification presented here contains the following classes: simple isoquinolines, isoquinolones, and phenethylammonium compounds; protoberberines and tetrahydroprotoberberines; aporphinoids; aporphines; dehydroaporphines; 7-substituted aporphines; oxoaporphines; phenanthrenes; aristolactams; proaporphines; benzyltetrahydroisoquinolines; bisbenzylisoquinolines and bisbenzyltetrahydroisoquinolines; and miscellaneous isoquinoline-type alkaloids.

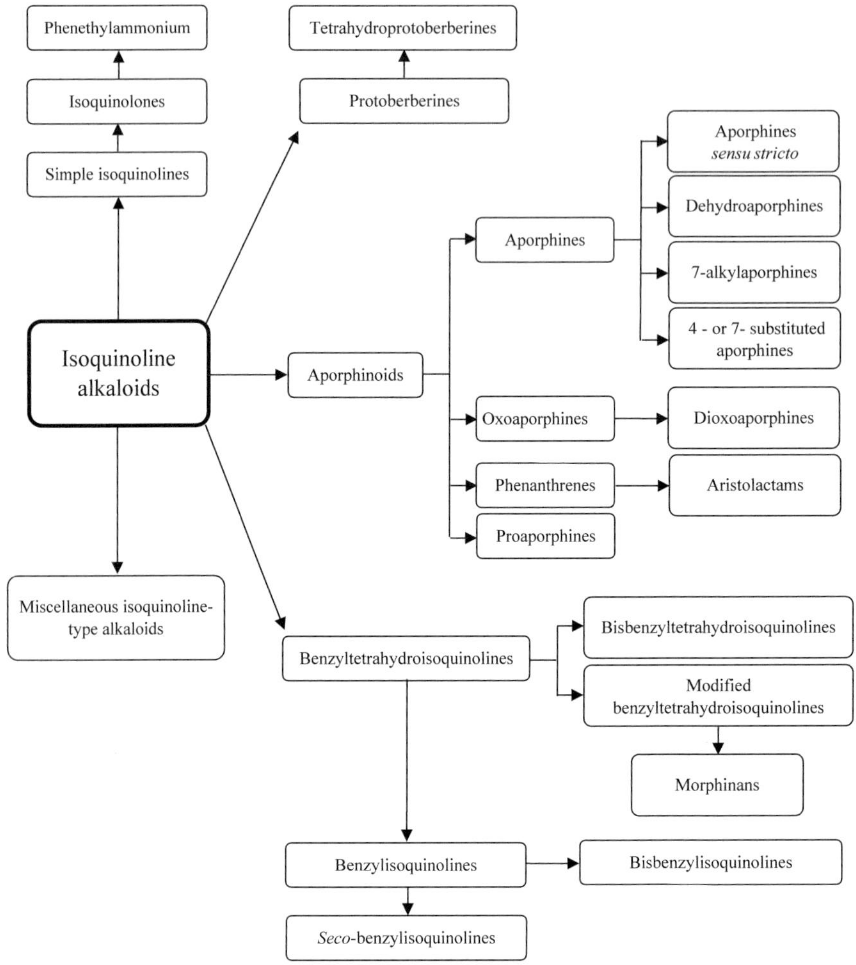

Fig. 8.1 Schematic classification of isoquinoline alkaloids

To make this variability even more evident, Table 8.1 contains the basic chemical structures of each class and subclass: Isoquinoline alkaloids are formed by the reaction of L-dopamine and 4-hydroxyphenylacetaldehyde and subsequent descarboxylation, followed by cyclization as will be properly discussed in detail later in biosynthesis section.

It is good to point out that: the bisbenzylisoquinolines occur in several structural types among plant families. There are few reports of proaporphines in the Annonacae family and to the best of our knowledge only seven representatives of this group were reported in a small number of species, mainly distributed in the genera *Annona*, *Anomianthus*, and *Uvaria*. Two alkaloids guattescidine and guattescine isolated from *Guatteria scandens* are the first members of a new class of

Table 8.1 Classes and subclasses of isoquinoline alkaloids that occur in plants

Type	Class/Subclass	Structure type
Type I	Simple isoquinolines	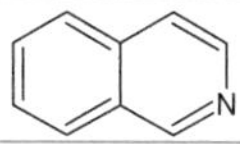
	Isoquinolones	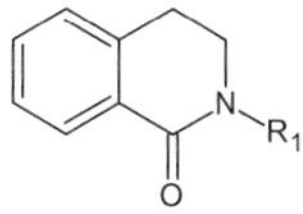
	Phenethylammonium	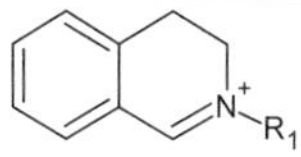
Type II	Benzyltetrahydroisoquinolines	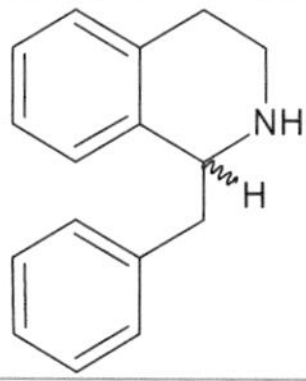
	Seco-benzylisoquinolines	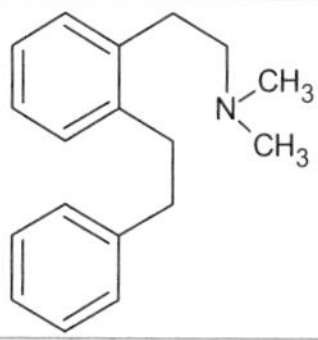
	Benzylisoquinolines	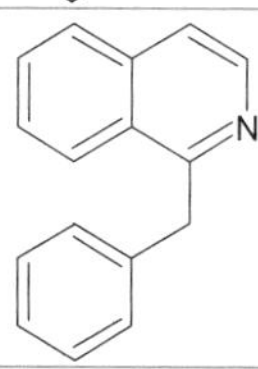
	Morphinans	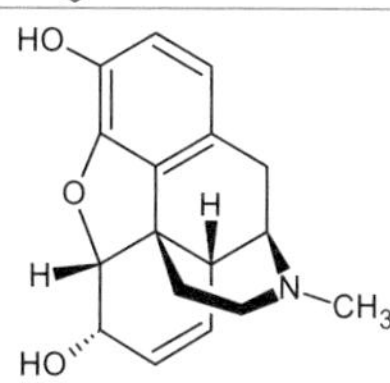
Type III	Bisbenzylisoquinolines	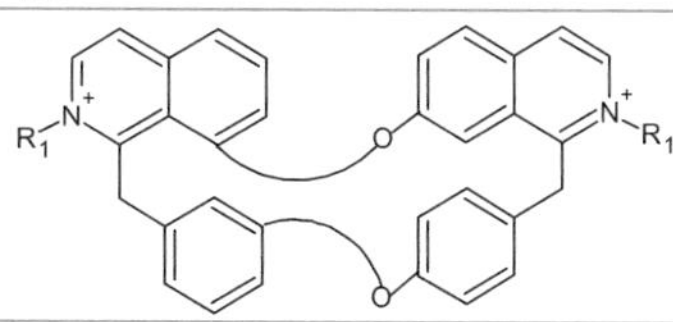

(continued)

Table 8.1 (continued)

Type	Class/Subclass	Structure type
Type IV	Bisbenzyltetrahydroisoquinolines	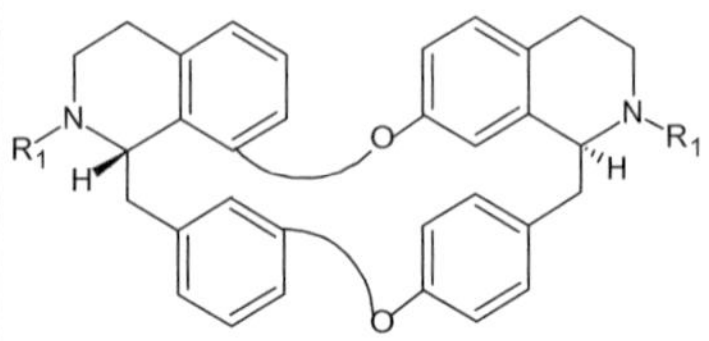
Type V	Protoberberines	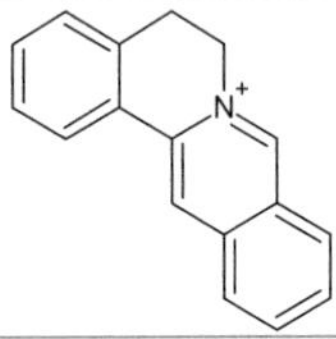
Type VI	Tetrahydroprotoberberines	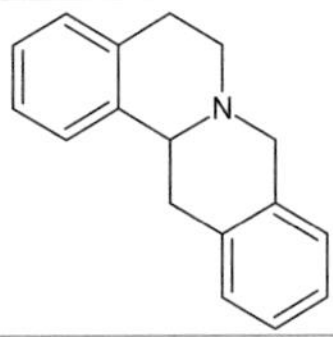
Type VII	Proaporphines	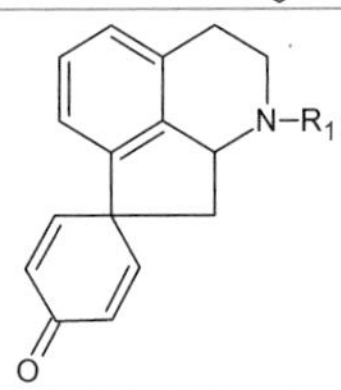
Type VIII	Aporphinoids	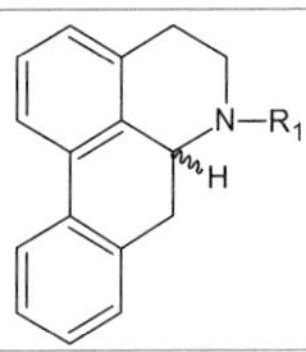
Type IX	Dehydroaporphines	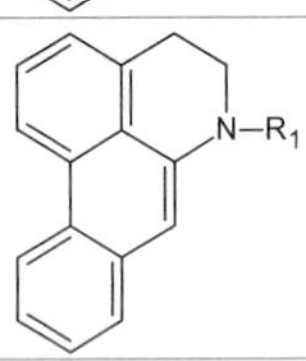
Type X	4- or 7-substituted aporphines	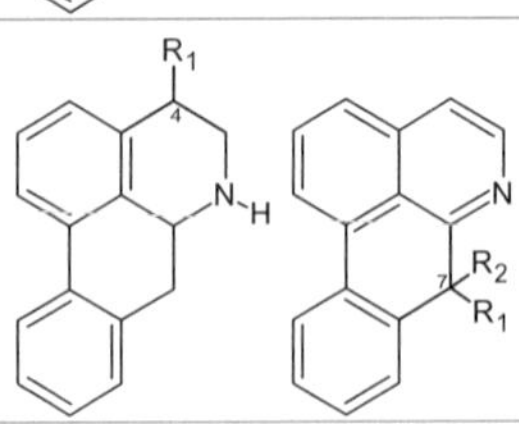

(continued)

Table 8.1 (continued)

Type	Class/Subclass	Structure type
Type XI	Oxoaporphines	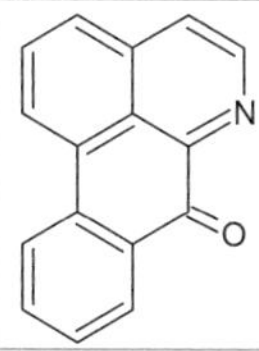
Type XII	Phenanthrenes	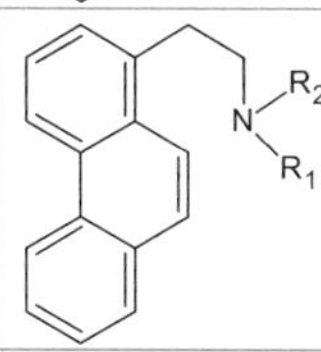
	Aristolactams	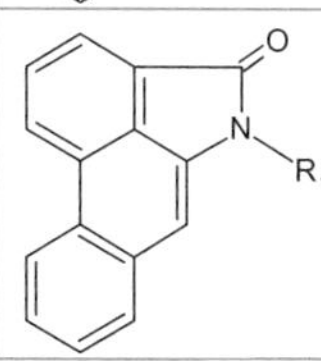
Type XIII	Miscellaneous isoquinoline-type alkaloids	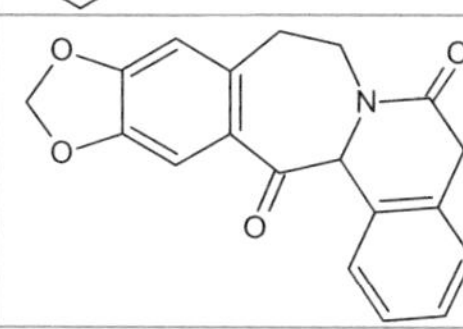

aporphinoids as they are 7-methyl substituted and cannot be regarded as oxoaporphines, since ring B is not aromatic. It may be pointed out that the 7-hydroxy or 7-methoxy aporphines are never substituted at C-11 and very rarely at C-3. The phenanthrenes (also known as "seco-aporphines") are very rare type of alkaloid. They are distributed in the families Annonaceae, Aristolochiaceae, Lauraceae, Menispermaceae, Monimiaceae, and Ranunculaceae (More detailed information could be found in Lúcio et al. 2015).

8.3 Biosynthesis

The alkaloids are a class of secondary metabolites, with low molecular mass, characteristic toxicity and a wide spectrum of pharmacological activities (De Luca and St. Pierre 2000). They are known as azo compounds derived mainly from amino acids, being present in about 20% of plants. Alkaloids play an important role in plants defense against herbivores and other pathogenic organisms. Due to the broad range of biological activities exhibited by this class of compounds, many of them are commonly exploited for their potent stimulant and pharmacological properties (Santos et al. 2017).

There are various classes of alkaloids, however, their biosynthetic origin is unique. Despite the diversity of the metabolic pathways, modern screening techniques have contributed significantly to the discovery and comprehension concerning alkaloids byosinthesis. Genome-based technologies such as express sequence tags, micro-DNA arrangement, and proteomic analysis have accelerated the discovery of new components and mechanisms involved in the synthesis of alkaloids in plants (Liscombe and Facchini 2008).

8.3.1 Benzylisoquinolines

The biosynthesis of the benzylisoquinoline alkaloids (BIA) (Fig. 8.2) begins with two molecules of L-tyrosine. One of them is decarboxylated to form tyramine that undergoes the action of phenol oxidase enzyme forming L-dopamine. The benzyl skeleton of (*S*)-norcoclaurine is formed through the transamination of the second molecule of L-tyrosine yielding 4-hidroxyphenylpiruvic acid that, after subsequent decarboxylation, leads to 4-hidroxyphenylacetaldehyde. A reaction similar to Mannich between dopamine and 4-hidroxyphenylacetaldehyde catalyzed by norcoclaurine synthase, furnishes (*S*)-norcoclaurine.

Several enzymes like norcoclaurine 6-*O*-methyltransferase (6-OMT), coclaurine *N*-methyltransferase (CNMT), (*S*)-*N*-methylcoclaurine-3-hydroxylase (NMCH), and 3-hydroxy-*N*-methylcoclaurine-4′-*O*-methyltransferase (4′OMT) produce (*S*)-reticuline from (*S*)-norcoclaurine. This precursor after a series of enzymatic reactions leads to the formation of protoberberines, aporphines, benzophenantridine alkaloids among other subclasses (Ikezawa et al. 2008). The reaction between(*S*)-reticuline and 1,2-deydroreticuline synthase (DRS) produces 1,2-dehydroreticuline which reacts with the enzyme 1,2-dehydroreticuline reductase (DRR) enzyme to form (*R*)-reticuline, the precursor of the biosynthetic route of the morphinan alkaloids (Liscombe and Facchini 2008; Dewick 2002).

8.3.2 Aporphines

Aporphine alkaloids are formed in plants (Fig. 8.3) by direct intramolecular oxidative coupling (*ortho-ortho* or *ortho-para*) of (*S*)-reticuline as the *bis*-dienone radical (Stevigny et al. 2005).

The substitution pattern of benzyltetrahydroisoquinoline precursors provides the corresponding aporphine alkaloids. Besides certain *O*-substitution positions, such as C-3 or C-7, could arise by direct oxidation of the aporphinoid nucleus. Alternatively, aporphine alkaloids may be derived from a proaporphine

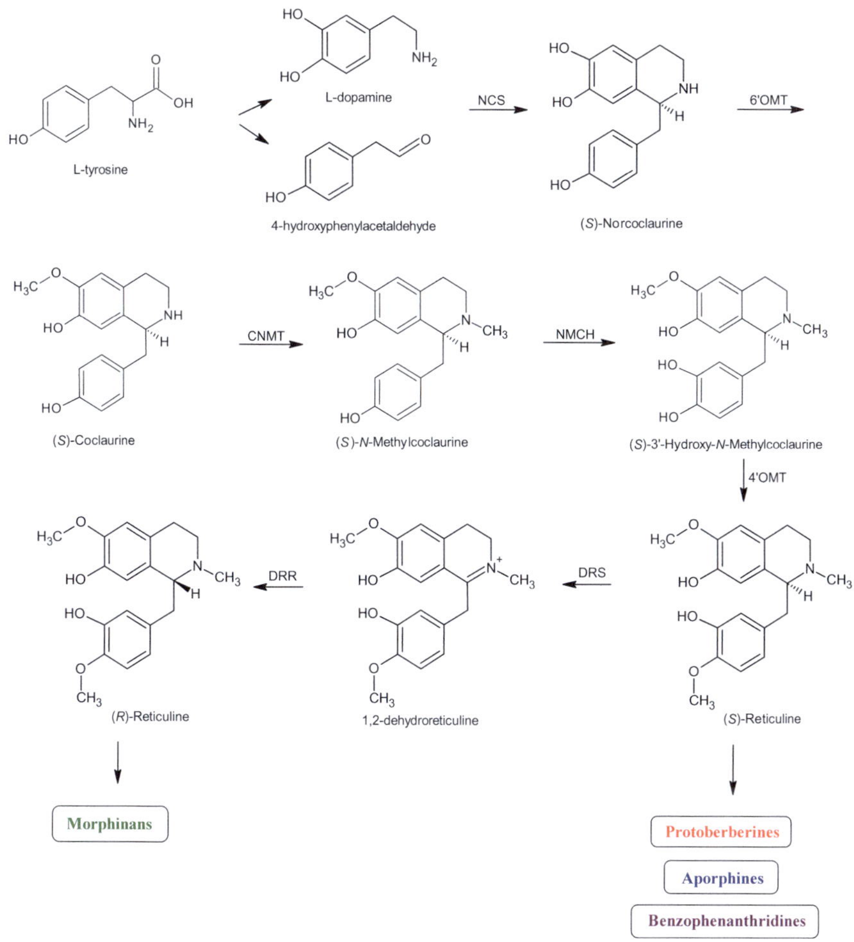

Fig. 8.2 Biosynthesis of benzylisoquinoline alkaloids

intermediate through the *ortho-para* cyclization of a tetrahydroisoquinoline diradical (Fig. 8.4), direct protonation and subsequent dienone-phenol rearrangement (Dewick 2002).

However, it was postulated previously (Lucio et al. 2015) that protoberberinium salt palmatine when biosynthetically modified by enzymes directly from polycarpine could also form aporphine alkaloids: the authors proposed a new hypothetical biogenetic route from protoberberinium salts to aporphines via benzylisoquinoline amides such as polycarpine, without involving phenolic oxidative coupling.

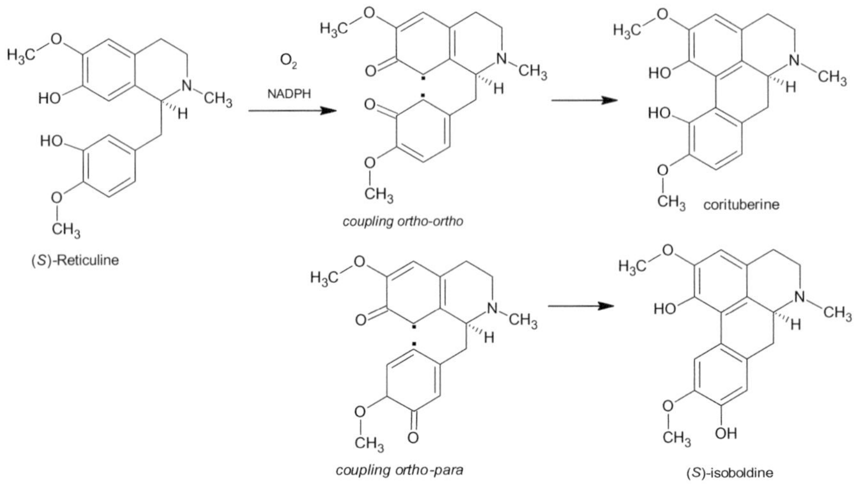

Fig. 8.3 Biosynthesis of aporphine alkaloids

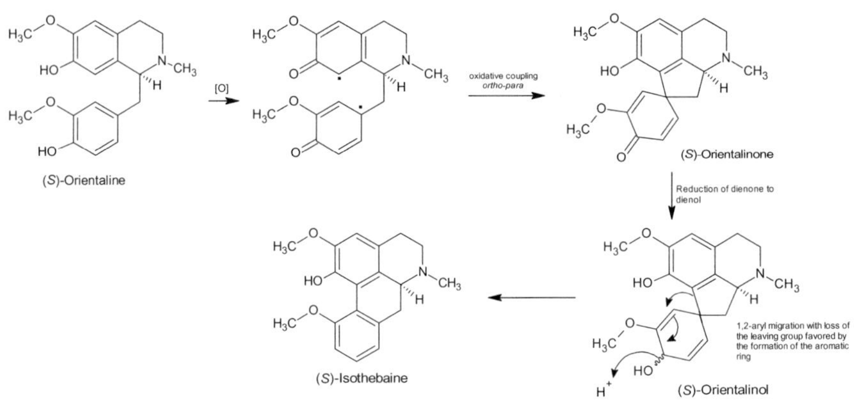

Fig. 8.4 Biosynthesis of aporphine alkaloids from proaporphine intermediate

8.3.3 Oxoaporphines

Another subclass of aporphines that deserve prominence is the oxoaporphine alkaloids. They usually have yellow, orange, or reddish-orange coloration due to the high degree of unsaturation. A rational sequence of in vivo transformations (Fig. 8.5) leads an aporphine alkaloid to its corresponding oxoaporphine via the dehydro and didehydroaporphine intermediates (Shamma and Guinaudeau 1984).

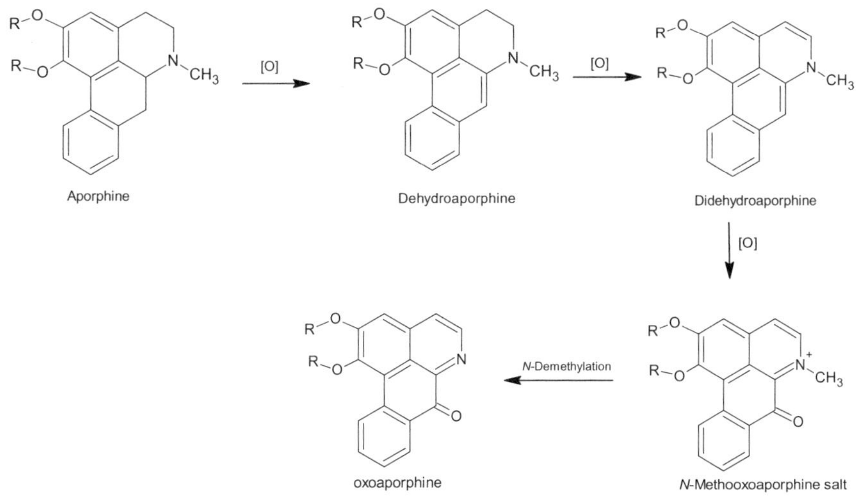

Fig. 8.5 Biosynthesis of oxoaporphine alkaloids

8.3.4 Bisbenzyltetrahydroisoquinolines

Alkaloids formed through the linkage between two benzyltetrahydroisoquinolines are known as bisbenzyltetrahydroisoquinolines. The formation of this subclass of alkaloids is explained by a phenolic oxidation mechanism. For example, the tetrandrine isolated from *Stephania tetrandra* (Menispermaceae) which is recognized as a coupling product of two (*S*)-*N*-methylcoclaurine molecules (Fig. 8.6). The two radicals formed by oxidations of an electron of a free phenol group in each

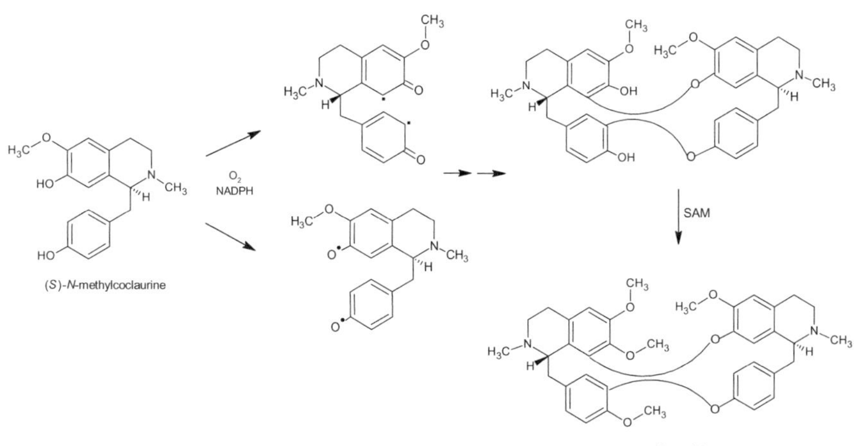

Fig. 8.6 Biosynthesis of bisbenzyltetrahydroisoquinoline alkaloids

ring couple to form ether bridges and then the product is methylated by the (*S*)-adenosyl methionine (SAM) enzyme to form tetrandrine (Dewick 2002).

8.3.5 Aristolactams

The biosynthetic pathway for the formation of aristolactams is not well known, however, Lin et al. (1997) suggest that this subclass of isoquinoline alkaloids, as well as aristolochic acid, 4,5-dioxoaporphines, 7-oxoaporphines alkaloids, and *N*-glycosylated aristolactams would be biosynthesized directly from aporphine alkaloids (Fig. 8.7). 4,5-dioxoaporphines alkaloids generated from aporphine precursors were considered as possible intermediates in the biosynthesis of aristolactams while aristolochic acid is derivatives of aristolactam.

Alkaloids are secondary metabolites known due to their toxicity to insects, slugs, and vertebrates. Besides that, they could be useful if applied as natural pesticides once they are shown to be considerably active against some fungi (mildew), bacteria, and even viruses (Wink 2018). And, this should not be different for the isoquinoline alkaloids: possibly they are important for vegetables since they act as chemical defense against herbivorous animals. Usually in nature, alkaloid-containing plants remain untouched by general herbivores, guaranteeing vegetables survival.

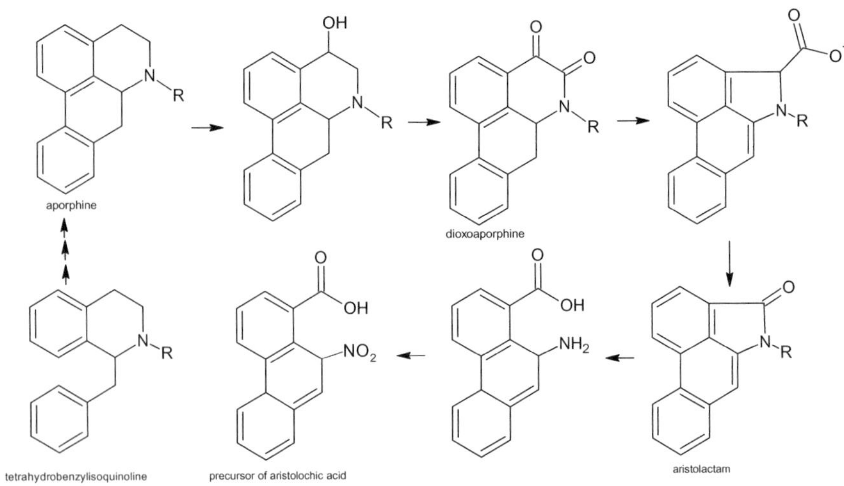

Fig. 8.7 Biosynthetic proposal for the formation of aristolactams in plants

8.4 Occurrence of Isoquinoline Alkaloids in Plants

It is estimated that plants have been surviving on this planet for more than 400 million years, being constantly attacked by herbivores and microbes (Wink 2015). During plant evolution, their metabolism has mastered in synthesized thousands of natural constituents structurally differing as means to defend themselves against herbivores and against bacteria, fungi, and viruses. Some secondary metabolites are part of plants strategy to attract pollinating and seed-dispersing animals, acting as antioxidants and UV protectants as well as mentioned before (Wink 2018).

Those metabolites are usually stored in some vegetable tissues: in the case of hydrophilic compounds, they are highly concentrated in plants' vacuole while lipophilic compounds occur more frequently in cuticle, trichomes, resin ducts, or laticifers (Schnetzler et al. 2017). Alkaloid levels are influenced and regulated by environmental factors such as heat, luminosity, humidity, high temperatures, and herbivory. These metabolites occur in higher concentrations in plants during flowering and fruit formation, once they are the first defense line in herbs in general. Most of the alkaloids in perennial species are translocated into the seeds, whereas the senescent aerial parts have very low levels (Wink 2018).

Morphine first isolated from *Papaver somniferum* (Papaveraceae), it is possibly the oldest representative of this class of alkaloids being known to humankind since 3400 B.C., when this species were grown in Mesopotamia to obtaining opium (Santos et al. 2017). Currently, many studies have been carried out with the most diverse different species reporting the presence of isoquinoline alkaloids. According to Hagel et al. (2015), BIAs are commonly found in the families Papaveraceae, Ranunculaceae, Berberidaceae, and Menispermaceae. However, the Menispermaceae family is known to possess a variety of alkaloids with the most diverse skeletons. For example, the chemical study of the *Sinomenium acutum* species resulted in the isolation of alkaloids from the morphinans, benzylisoquinoline, aporphine, oxoisoaporphine, proaporphine, and protoberberine subclasses (Li et al. 2014).

The Annonaceae family (Lucio et al. 2015) is characterized by an active metabolism specialized in biosynthetic routes for the production of alkaloids (Riley-Saldaña et al. 2017). Phytochemical investigation of some species from *Guatteriopsis*, currently reclassified as *Guatteria* genus, Annonaceae family (Erkens and Maas 2008) also revealed the presence of oxoaporphine, aporphine, and protoberberine subclasses from *G. blepharophylla* (Costa et al. 2009b) and type 7,7-dimethylaporphines from *G. friesiana* (Costa et al. 2009a).

Protoberberine-type alkaloids or berberine derivatives are found mainly in species belonging to the Berberidaceae family. The screening through analytical techniques of *Mahonia leschenaultia* and *Mahonia napaulensis* extracts showed that in addition to benzyltetrahydroisoquinolines and aporphines alkaloids, these species possesses protoberberine alkaloids as major constituents (Singh et al. 2017).

The structural diversity of isoquinoline derivatives was also observed through the chemical study of *Fumaria sepium* and *Fumaria agraria* species

(Papaveraceae). The authors reported the isolation of alkaloids with benzophenantridine, protopyne, spirobenzylisoquinoline and aporphine, isoquinolone and protoberberine skeletons (Suau et al. 2002). Representative of bisbenzylisoquinolines (thalmetine, *O*-methylthalmetine, thalicberine, *O*-methylthalicberine) and isoquinolones (thalactamine and thalflavine) subclasses were isolated from *Thalictrum minus* (Ranunculaceae) (Popovic et al. 1992).

Known by the production of alkaloids among other classes of secondary metabolites, the Rutaceae family proved to be efficient in the production of benzophenantridine, protoberberine, and benzyltetrahydroisoquinoline alkaloids in the study of the *Zanthoxylum quinduenses* species (Ladino and Suárez 2010). Although there are few reports on the isolation of isoquinoline alkaloids in Nelumbonaceae, Deng et al. (2016) report in his study the isolation of different subclasses in the *Nelumbo nucifera* as benzylisoquinoline, aporphine and bisbenzylisoquinoline alkaloids (Deng et al. 2016).

Aristolactams are a small group of aporphinoids that contain phenanthrene chromophore group and are restricted to Annonaceae, Monimiaceae, Menispermaceae, Aristolochiaceae, and Piperaceae families. Forty-five aristolactams have been isolated from 38 species of *Aristolochia* (Urzúa et al. 2013). The phytochemical investigation of *Oxandra asbeckii* species (Annonaceae) showed the presence of aristolactams AII, BII, and velutinam. In the same study, the authors verified that *Goniothalamus dumontetii* species (also belonging to the Annonaceae family) provided aristolactam AIII (Marti et al. 2013).

Finally, a study published by Chen et al. (2013) showed that hundreds of aporphine alkaloids were isolated in more than 20 families and 100 plant genera. The highest number of alkaloids in this subclass were found in the Annonaceae family (28 genera) followed by the family Menispermaceae (20 genera), Lauraceae (18 genera), Papaveraceae (13 genera), Monimiaceae (10 genera), Ranunculaceae (9 genera), Magnoliaceae (6 genera), Berberidaceae and Rhamnaceae (4 genera), Aristolochiaceae, Euphorbiaceae, Hernandiaceae, Rutaceae, Leguminosae (2 genera), Araceae, Canellaceae, Liliaceae, Piperaceae, Symplocaceae, and Sabiaceae all with a single genus.

A literature overview on the incidence of isoquinoline alkaloids in various families and plant species was performed, however, in Table 8.2 are shown some representatives on most diverse families are presented. Some plant species were reclassified within other families, e.g., the genus *Fumaria*, which was classified as belonging to Fumariaceae family and currently belongs to Papaveraceae family. The classification of the genera in their families is in agreement with that one suggested by The Plant List website that could be accessed in www.theplantlist.org.

Table 8.2 Occurrence of isoquinoline alkaloids in plants

Family	Species	Biological property	Alkaloids	Reference
Amaryllidaceae	*Galanthus nivalis* subsp. *cilicicus*	Cytotoxic	(−)-Capnoidine, (+)-bulbocapnine, (−)-lycorine, (+)-11-hydroxyvittatine and (+)-vittatine	Kaya et al. (2004)
Ancistrocladaceae	*Ancistrocladus tanzaniensis*	Malaria tropica, leishmaniasis, Chagas disease, and African sleeping sickness	Ancistrotanzanine A, ancistrotanzanine B, ancistrotectoriline A, ancistrocladidine, ancistrotectorine, ancistrotanzanine C, *O*-methylancistrocladinine and *O*,*N*-dimethylancistrocladine	Bringmann et al. (2004)
Annonaceae	*Guatteria ferruginea*, *Guatteria latifolia*, and *Guatteria sellowiana*		3-Hydroxynornuciferine, nornuciferine, lysicamine, isomoschatoline, liriodenine and *O*-methylmoschatoline	dos Santos et al. (2017)
	Annona sericea		Lysicamine, oxonantenine, nornuciferine, nornantenine, isoboldine, 3-hydroxynornuciferine, (*S*)-reticuline and (*S*)-*N*-methylcoclaurine	Campos et al. (2008)
	Xylopia laevigata	Cytotoxic	(−)-Roemerine, (+)-anonaine, (+)-glaucine, (+)-xylopine, (+)-norglaucine, asimilobine, (+)-norpurpureine, (+)-*N*-methyllaurotetanine, (+)-norpredicentrine, (+)-calycinine, (+)-laurotetanine, lanuginosine, oxoglaucine, (−)-xylopinine, (+)-discretine, (−)-corytenchine and (+)-discretamine	Menezes et al. (2016)
Apocynaceae	*Cynanchum komarovii*[a]	Antiviral	2,3-dimethoxy-6-(3-oxo-butyl)-7,9,10,11,11a,12-hexahydrobenzo[f]pyrrolo[1,2-b]isoquinoline, 7-demethoxytylophorine and 7-demethoxytylophorine N-oxide	An et al. (2001)
Aristolochiaceae	*Aristolochia constricta*	Reduction of contractions of isolated guinea-pig ileum	3,5-Di-*O*-methylconstrictosine, 5,6-dihydro-3,5-di-*O*-methylconstrictosine, 5,6-dihydroconstrictosine, constrictosine, 3-*O*-methylconstrictosine and (−)-8β-(4′-hydroxybenzyl)-2,3-dimethoxyberbin-10-ol	Rastrelli et al. (1997)

(continued)

Table 8.2 (continued)

Family	Species	Biological property	Alkaloids	Reference
Atherospermataceae	*Laureliopsis philippiana*		Fangchinoline and tetrandrine	Stanstrup et al. (2010)
Berberidaceae	*Mahonia leschenaultia and Mahonia napaulensis*		Reticuline, oblongine, magnoflorine, isoboldine, isocorydine, glaucine, 8-oxojatrorrhizine, 8-oxoberberine, demethyleneberberine, jatrorrhizine, palmatine, berberine, thalifendine, berberrubine, tetrahydropalmatine and tetrahydroberberine	Singh et al. (2017)
	Leontice altaica		Lincangenine-4-β-D-glucopyranoside	Jenis et al. (2015)
	Berberis aristata		Berberine phenoxide, ketoberberine benzoate A and ketoberberine benzoate B	Ahamad et al. (2014)
Boraginaceae	*Echium humile*		Carnegine and 7-norcarnegine	El-Shazly et al. (1996)
Chenopodiaceae	*Hammada scoparia*		*N*-Methylisosalsoline	Jarraya et al. (2008)
Convolvulaceae	*Iseia luxurians*		Iseluxine, Northalifoline, thalifoline, corydaldine, noroxyhydrastinine, oxyhydrastinine and *N*- methylcorydaldine	Schimming et al. (2000)
Cornaceae	*Alangium lamarckii*[a]		Methylisoalangiside, isoalangiside, 3-*O*-demethyl-2-*O*-methylisoalangiside, demethylneoalangiside and neoalangiside	Itoh et al. (1995)
Dioncophyllaceae	*Dioncophyllum thollonii*	Antimalarial and antitrypanosomal	Dioncophylline A, dioncophylline B, dioncophylline C,5-*O*-demethyldioncophylline A, *N*-methyldioncophylline A and dioncophylline E	Bringmann et al. (2002)
Dioscoreaceae	*Dioscorea dregeana*		Crinamine	Mulholland et al. (2002)
Euphorbiaceae	*Croton flavens*		(−)-Amuronine	Charris et al. (2000)
	Hyeronima oblonga	Toxicity	Hyeronine A and Hyeronine B	Alves and Zani (1999)

(continued)

Table 8.2 (continued)

Family	Species	Biological property	Alkaloids	Reference
Gnetaceae	*Gnetum montanum*	Antibacterial	*N*-methyllaudanosolinium trifluoroacetate, 3′-hydroxy-*N*,*N*-dimethylcoclaurinium trifluoroacetate, 1,9,10-trihydroxy-2-methoxy-6-methylaporphinium trifluoroacetate, 6a,7-didehydro-1,9,10-trihydroxy-2-methoxy-6-methylaporphinium trifluoroacetate, (−)-latifolian A and magnocurarine	Martin et al. (2011)
Hernandiaceae	*Gyrocarpus americanus*		(+)-Auroramine and (+)-maroumine	Dute et al. (1988)
Lauraceae	*Endlicheria oreocola*		*O*-Methylmoschatoline	Albarracin et al. (2017)
	Ocotea macrophylla		Dehydronantenine, (+)-nantenine, (+)-neolitsine, (+)-dicentrine, (+)-*N*-acetyl-nornantenine, (+)-cassythidine and didehydroocoteine	Barrera and Suárez (2009)
Leguminosae	*Calycotome villosa* subsp. *intermedia*		1-hydroxymethyl-6,7-dimethoxyisoquinoline-*N*-oxide	El Antri et al. (2004)
Magnoliaceae	*Michelia fuscata*		Fuscatine A, fuscatine B, northalifoline, thalifoline, corydaldine, *N*-methylcorydaldine, (6,7-dimethoxyisoquinolinyl)-(4-methoxyphenyl)-methanone, (6,7-dimethoxyisoquinolinyl)-(4-hydroxyphenyl)-methanone, liriodenine, and corydine	Li et al. (2017)
	Tsoongiodendron odorum		Lysicamine	Huang et al. (2011)
Menispermaceae	*Cissampelos capensis*		Dicentrine, glaziovine, lauroscholtzine, pronuciferine, bulbocapnine, salutaridine, cycleanine, cissacapine, crotsparine, insularine, 12-*O*-methylcurine, reticuline and insulanoline	De Wet et al. (2011)
	Stephania yunnanensis		Dehydrocrebanine, (−)-crebanine, (−)-dicentrine, (−)-cassythine, sinoacutine, (−)-corydalmine, fibrecisine (−)-sukhodianine, (−)-stephanine, (−)-roemerine, oxocrebanine, stepharanine, jatrorrhizine and palmatine	Zhang and Rao (2009)

(continued)

Table 8.2 (continued)

Family	Species	Biological property	Alkaloids	Reference
Monimiaceae	*Mollinedia costaricensis*	Antibacterial, antifungical and inhibition of cardiac arrhythmias	Molinedine	López et al. (1988)
Moraceae	*Broussonetia papirifera*	Cytotoxic	*N*-norchelerythrine, dihydrosanguinarine, oxyavicine, broussonpapyrine, nitidine, chelerythrine, and liriodenine	Pang et al. (2014)
Nelumbonaceae	*Nelumbo nucifera*	Muscle relaxant	Neoliensinine, neferine, isoliensinine, and liensinine.	Yang et al. (2018)
Papaveraceae	*Papaver clavatum and Papaver stylatum*		Berberine, meconoquintupline, amurine and isocorydine	Ünsal et al. (2008)
	Fumaria sepium[a] and *Fumaria agraria*[a]		Domesticine, (+)-isoboldine, dihydrosanguinarine, noroxyhydrastinine, oxyhidrastinine, coptsine, (−)-stilopine, protopine, densiflorine, (−)-fumaritine-*N*-oxide and (+)-parfumine	Suau et al. (2002)
Ranunculaceae	*Aconitum carmichaelii*		6-Formyl-1,2,9,10-tetramethoxy-6a,7-dehydroaporphine, glaucine and norglaucine hydrocloride	Qin et al. (2017)
	Thalictrum minus		Thalmethine, *O*-methylthalmethine, thalicberine, *O*-methylthalicberine, thaliglucine, thaliporphina, berberine, thalactamine and thalflavine	(Popovic et al. 1992)
Rubiaceae	*Pogonopus tubulosus*	Antimalarial	Tubulosine, psychotrine, and cephaeline	Sauvain et al. (1996)
	Psychotria ipecacuanha		Emetine and cephaeline	Nomura et al. (2008)

(continued)

Table 8.2 (continued)

Family	Species	Biological property	Alkaloids	Reference
Rutaceae	*Zanthoxylum simulans*		*N*-acetyldehydroanonaine, 6-methyldihydmhelerythrine and 6-methylnorchelerythine	Chen et al. (1994, 1996)
	Zanthoxylum quinduense		Nornitidine, norchelerythrine, decarine, chelerythrine, (−)-6-acetonyldihydrochelerythrine, berberine, (−)-6-carboxymethyldihydrochelerythrine, (−)-xylopinidine, *N*-methyltetrahydropalmatine, (−)-isotembetarine and *N*-methyltetrahydrocolumbamine	Ladino and Suárez (2010)
Siparunaceae	*Siparuna gilgiana*[a]		Liriodenine and cassamedine	Chiu et al. (1982)
	Siparuna apiosyce[a]		Asimilobine	Fischer et al. (1999)

[a]Plants classified according to the website: The Plant List (www.theplantlist.org)

8.5 Chemotaxonomy

The use of medicinal plants for the benefit of mankind has long been sought and has been increasing day by day. However, its categorization, that is, taxonomy is essential for understanding the communities: their origin and how they are related to each other. The plant ecosystem consists of several biotic elements such as angiosperms, fungi, mosses, lichens, and algae. Their exploitation has been increasing once they are recognized as rich sources of natural products with huge variety of chemical constituents. Among them, angiosperms containing the most developed specimens of the plant kingdom play an important role in the benefit of humankind mainly in health care (Misra and Srivastava 2016).

Over the years, plant taxonomic approaches have evolved including morphological, anatomical, and chemotaxonomic classifications, with morphological and anatomical classifications being considered traditional while chemotaxonomics are considered a modern approach to classifying plants (Singh 2016). The term chemotaxonomy emerged from chemo + taxonomy which, in itself, explains the classification of plant species through their chemical constituents (Misra and Srivastava 2016). To understand the chemical constitution in plants, it is necessary first to define some terms as primary and secondary metabolites.

Primary metabolites are macromolecular compounds that participate directly in plant metabolism and include proteins, nucleic acids, chlorophyll, and polysaccharides. All products synthesized by an organism reflect the information in the DNA or RNA and proteins. These molecules are being called semantides once they have useful information on taxonomy and phylogeny (Kumar 2017). Secondary metabolites are compounds that perform non-essential functions in plants and are used for protection and defense against predators and pathogens. They are compounds of restricted occurrence and are very useful in chemotaxonomic classification. Some classes of secondary metabolites include glycosides, alkaloids, volatile constituents, flavonoids, phenolic compounds, and terpenoids (Singh 2016).

For a precise chemotaxonomic classification, it is necessary a solid knowledge about biosynthetic pathways of secondary metabolites in plants for categorization. The junction of the morphological classification and chemical profile of plants is considered a useful tool for characterization and classification, in addition, a viable tool for researchers to determine the phylogenetic evolution of a particular species or families of plants with their chemotaxonomic profile. Another significant contribution of chemotaxonomy is to provide a rational study for the identification and quantification of specific classes of natural products, making an in-depth study of species unexplored chemically within the same genus (Misra and Srivastava 2016).

The increase in studies related to chemotaxonomy is mainly due to the advance of analytical techniques that can detect even traces of chemical compounds. Among most studied families from the chemotaxonomic point of view are: Malvaceae, Ranunculaceae, Magnoliaceae, Polygonaceae, and Solanaceae. The findings related to chemotaxonomy are useful for taxonomists, phytochemists, and pharmacologists

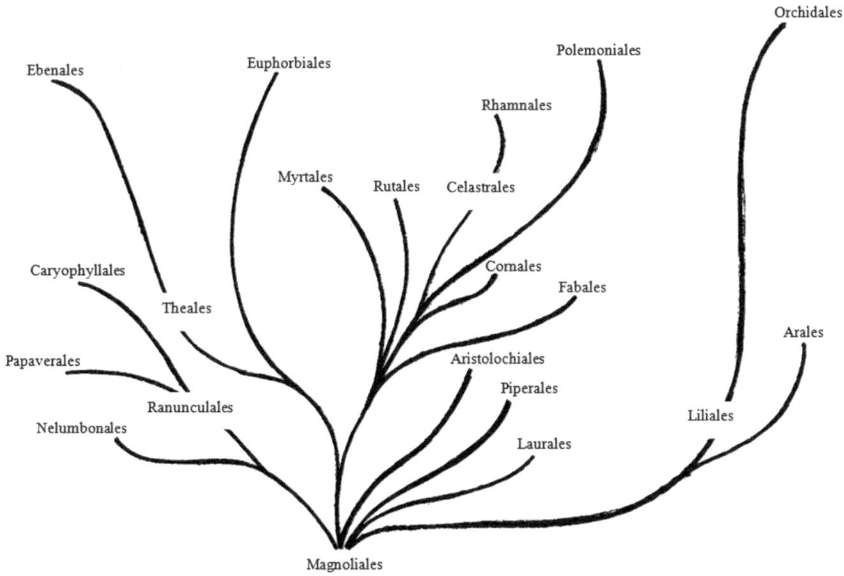

Fig. 8.8 Probable relationship between isoquinoline-synthesizing plants (based on Urzua and Cassels 1978)

to solve problems related to taxonomy (Singh 2016). An important tool for the discovery of natural products is the metabolomics (Ernst et al. 2014) that can be defined as the measurement of all metabolites in a system under particular conditions. However, when compared to animals, plants are more complex for metabolomics studies due to the great variety of metabolites (Hagel et al. 2015).

Derivatives of benzyltetrahydroisoquinolines, as well as, their biogenetic derivatives are found in about 18 orders according to the Takhtajan system (Fig. 8.8). Its distribution in angiosperms shows probable relationships among the different orders.

According to Urzúa and Cassels (1978), benzyltetrahydroisoquinolines alkaloids occur more frequently in genera that grouped around the magnoliales order, being considered as the most primitive of the angiosperms and dicotyledons. Cronquist (1977) suggests that the subclass Magnoliidae is indeed the basal phyletic group of angiosperms. The accumulation of benzylisoquinoline alkaloids has been reported in several angiosperms taxa being commonly found in the Ranunculaceae familiy and also in distant families such as Rutaceae, Lauraceae, Cornaceae, and Nelumbonaceae. In Piperales order, this class of secondary metabolites occurs very sporadically (Liscombe et al. 2005).

Although the benzylisoquinoline alkaloids are distributed in several taxa, they are more frequent in the Ranunculales order. The metabolomic study performed by Hagel et al. (2015) in Ranunculaceae order, contemplated 20 species divided into four families, Papaveraceae, Ranunculaceae, Berberidaceae, and Menispermaceae showed that many subclasses of benzylisoquinoline alkaloids occurs in Ranunculales including: simple benzylisoquinoline, protopyne, aporphine, promorphinans, morphinans, protoberberine, phthalidaisoquinolines, benzophenantridines, papaverubines, pavines, isopavines, and secoisoquinolines skeletons.

The plant species involved in this study were selected taking into account the structural diversity of the isoquinoline alkaloids and the organ of the plant where their concentration was higher. The criteria for selecting the species to be studied were based on an overview of the data available in the literature. Thus, the results obtained showed that species of *Corydalis* and *Papaver* genus (Papaveraceae) that were well known for their structural diversity in isoquinoline alkaloids really contained subclasses of the morphinans, protoberberines, benzophenantridines, pavines, and phthalidaisoquinoline. *Menispermum canadense* (Menispermaceae) produces very unusual acutumine alkaloids in addition to the bisbenzylisoquinoline alkaloids that are typical of Menispermaceae. Structural diversity was also verified within the families Ranunculaceae and Berberidaceae (Hagel et al. 2015).

Another metabolomic study, which added knowledge in chemotaxonomy, was reported by Silva et al. (2016), through the study of the leaves extracts of five different species of *Unonopsis* (*U. duckei*, *U. floribunda*, *U. stipitata*, *U. guatterioides,* and *U. rufescens*—Annonaceae family). Through this study, it was possible to verify the chemical variability (isoquinoline alkaloids) of these species, as well as the differentiation of these species through the identified metabolites. This study was made possible with the use of analytical techniques and chemometric analyses tool. Regarding the chemical variability, it was possible to verify in *Unonopsis* species, the presence of aporphine, proaporphine, and protoberberine subclasses allowing the segregation of the studied species, as well as botanical similarities among some species.

Some chemotaxonomic studies have involved chemodiversity and chemosystematic. However, more detailed and updated bibliographic survey of the micromolecules produced by species of a certain genus is necessary (Lima et al. 2018). An example of application of this chemotaxonomic tool was the study of three species of *Guatteria* (*G. latifolia*, *G. sellowiana,* and *G. ferruginea*) belonging to the Annonaceae family in which the presence of the alkaloids liriodenine and *O*-methylmoschatoline were observed in several species of *Guatteria* genus and other genera within the same subfamily. Such evidence allows the suggestion that these two oxoaporphine alkaloids could be used as chemotaxonomic markers of this subfamily (dos Santos et al. 2017).

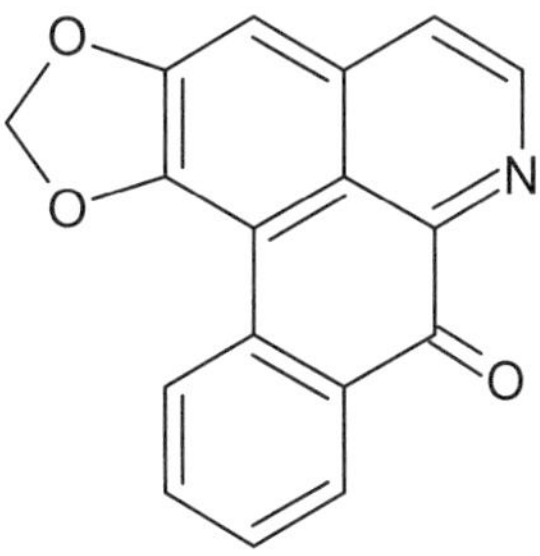

Liriodenine

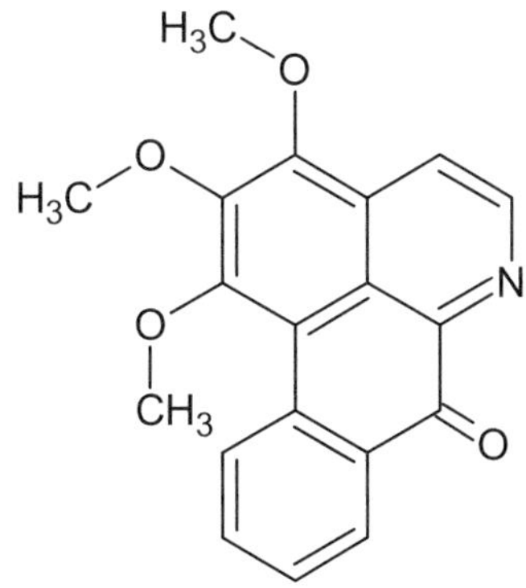

O-methylmoschatoline

8.6 Conclusions

The above survey indicates that the literature on isoquinoline alkaloids has grown considerably in the last few years and a vast field is now open to the chemist, taxonomist, and pharmacologist for in-depth investigations. A literature overview on the incidence of isoquinoline alkaloids in various families and plant species was performed. Some representatives on most diverse families were presented. However, some of the studies consulted are of early origin, very fragmentary, and incomplete. Based on this, any attempt to weave valid chemotaxonomic considerations would be futile. Nevertheless, it is apparent that isoquinoline alkaloids show strong relationships among some species of the same genus or of different genera, and mainly among Magnoliales order and some species of phylogenetically related families such as Magnoliales, Laurales, Piperales, and Aristolochiales.

For a precise chemotaxonomic classification, it is necessary to build more solid knowledge about biosynthetic pathways of those alkaloids for concise categorization. The junction of the morphological classification and chemical profile of plants is considered a useful tool for characterization and classification, in addition, a viable tool for researchers to determine the phylogenetic evolution of a particular species or families of plants with their chemotaxonomic profile. Another significant contribution of chemotaxonomy is to provide a rational study for the identification and quantification of specific classes of natural products, making an in-depth study of species unexplored chemically within the same genus. Recent research findings could also support the benefits of alkaloids and can be applied to discover and design new analogs that could be therapeutically useful for various treatments.

References

Ahamad J, Naquvi KJ, Ali M, Mir SR (2014) New isoquinoline alkaloids from the stem bark of *Berberis aristata*. Indian J Chem 53B:1237–1241

Albarracin LT, Delgado WA, Cuca LE, Ávila MC (2017) Chemical constituents of the bark of *Endlicheria oreocola* (Lauraceae) from Colombia. Biochem Syst Ecol 74:60–62

Alves T, Zani C (1999) New tetrahydroisoquinolinones from *Hyeronima oblonga* (Euphorbiaceae). Tetrahedron Lett 40:205–208

An TY, Huang RQ, Yang Z, Zhang DK, Li GR, Yao YC, Gao J (2001) Alkaloids from *Cynanchum komarovii* with inhibitory activity against the tobacco mosaic virus. Phytochemistry 58:1267–1269

Barrera EDC, Suárez LEC (2009) Aporphine alkaloids from leaves of *Ocotea macrophylla* (Kunth)(Lauraceae) from Colombia. Biochem Syst Ecol 37:522–524

Bringmann G, Messer K, Wolf K, Mühlbacher J, Grüne M, Brun R, Louis AM (2002) Dioncophylline E from *Dioncophyllum thollonii*, the first 7,3′-coupled dioncophyllaceous naphthylisoquinoline alkaloid. Phytochemistry 60:389–397

Bringmann G, Dreyer M, Faber JH, Dalsgaard PW, Stærk D, Jaroszewski JW, Ndangalasi H, Mbago F, Brun R, Christensen SB (2004) Ancistrotanzanine C and Related 5, 1′-and 7, 3′-Coupled Naphthylisoquinoline Alkaloids from *Ancistrocladus tanzaniensis*. J Nat Prod 67:743–748

Campos FR, Batista RL, Batista CL, Costa EV, Barison A, dos Santos AG, Pinheiro MLB (2008) Isoquinoline alkaloids from leaves of *Annona sericea* (Annonaceae). Biochem Syst Ecol 36:804–806

Charris J, Domínguez J, De la Rosa C, Caro C (2000) (−)-Amuronine from the leaves of *Croton flavens* L. (Euphorbiaceae). Biochem Syst Ecol 28:795–797

Chen IS, Wu SJ, Leu YL, Tsai IW, Wu TS (1996) Alkaloids from root bark of *Zanthoxylum simulans*. Phytochemistry 42:217–219

Chen IS, Wu SJ, Tsai IL, Wu TS, Pezzuto JM, Lu MC, Chai H, Suh N, Teng CM (1994) Chemical and bioactive constituents from *Zanthoxylum simulans*. J Nat Prod 57:1206–1211

Chen J, Gao K, Liu T, Zhao H, Wang J, Wu H, Liu B, Wang W (2013) Aporphine alkaloids: a kind of alkaloids' extract source, chemical constitution and pharmacological actions in different botany. Asian J Chem 25:10015–10027

Chiu S, Dobberstein R, Fong H, Farnsworth N (1982) Oxoaporphine alkaloids from *Siparuna gilgiana*. J Nat Prod 45:229–230

Costa EV, Marques FA, Pinheiro MLB, Vaz NP, Duarte MCT, Delarmelina C, Braga RM, Maia BHLS (2009a) 7,7-Dimethylaporphine alkaloids from the stem of *Guatteriopsis friesiana*. J Nat Prod 72:1516–1519

Costa EV, Pinheiro MLB, Marques FA, Braga RM, Maia BHLS (2009b) First report of alkaloids in the genus *Guatteriopsis* (Annonaceae). Biochem Syst Ecol 37:43–45

Cronquist A (1977) On the taxonomic significance of secondary metabolites in angiosperms. In: Kubitzki K (ed) Flowering plants. Springer, Vienna, pp 179–189

Da Cunha EVL, Barbosa-Filho JM (2012) Alcaloides derivados do núcleo isoquinolínico. In: Yunes RA, Cechinel-Filho V (Org.) Química de produtos naturais, novos fármacos e a moderna farmacognosia. Editora Univali, pp 291–330

De Luca V, St Pierre B (2000) The cell and developmental biology of alkaloid biosynthesis. Trends Plant Sci 5:168–173

De Wet H, Van Heerden FR, Van Wyk BE (2011) Alkaloidal variation in *Cissampelos capensis* (Menispermaceae). Molecules 16:3001–3009

Deng X, Zhu L, Fang T, Vimolmangkang S, Yang D, Ogutu C, Liu Y, Han Y (2016) Analysis of isoquinoline alkaloid composition and wound-induced variation in *Nelumbo* using HPLC-MS/MS. J Agr Food Chem 64:1130–1136

Dewick PM (2002) Medicinal natural products: a biosynthetic approach. Wiley, New York dos Santos AR, Pires C, Marques FA, Lobão AQ, Maia BHL (2017) Isoquinoline alkaloids isolated from three *Guatteria* species. Biochem Syst Ecol 73:1–2

Dute P, Chalandre MC, Cabalion P, Bruneton J (1988) (+)-auroramine and (+)-maroumine, new seco-bis-benzyl-isoquinoline dimers from *Gyrocarpus americanus*. Phytochemistry 27: 655–657

El Antri A, Messouri I, Bouktaib M, El Alami R, Bolte M, El Bali B, Lachkar M (2004) Isolation and X-ray crystal structure of a new isoquinoline-N-oxide alkaloid from *Calycotome villosa* subsp. intermedia. Fitoterapia 75:774–778

El-Shazly A, Sarg T, Ateya A, Aziz EA, El-Dahmy S, Witte L, Wink M (1996) Pyrrolizidine and tetrahydroisoquinoline alkaloids from *Echium humile*. Phytochemistry 42:225–230

Erkens RH, Maas PJ (2008) The *Guatteria* group disentangled: sinking *Guatteriopsis*, *Guatteriella*, and *Heteropetalum* into *Guatteria*. Rodriguésia 59:401–406

Ernst M, Silva DB, Silva RR, Vêncio RZN, Lopes NP (2014) Mass spectrometry in plant metabolomics strategies: from analytical platforms to data acquisition and processing. Nat Prod Rep 31:784–806

Fischer D, Gonçalves M, Oliveira F, Alvarenga M (1999) Constituents from *Siparuna apiosyce*. Fitoterapia 70:322–323

Hagel JM, Mandal R, Han B, Han J, Dinsmore DR, Borchers CH, Wishart DS, Facchini PJ (2015) Metabolome analysis of 20 taxonomically related benzylisoquinoline alkaloid-producing plants. BMC Plant Biol 15:220

Huang Y, Zeng Q, Fu JJ, Kou ZJ, Chang RJ, Jin HZ, Zhang WD (2011) Chemical constituents from *Tsoongiodendron odorum* Chun. Biochem Syst Ecol 39:209–212

Ikezawa N, Iwasa K, Sato F (2008) Molecular cloning and characterization of CYP80G2, a cytochrome P450 that catalyzes an intramolecular C-C phenol coupling of (S)-reticuline in magnoflorine biosynthesis, from cultured Coptis japonica cells. J Biol Chem 283:8810–8821

Iqbal E, Lim LBL, Salim KA, Faizi S, Ahmed A, Mohamed AJ (2018) Isolation and characterization of aristolactam alkaloids from the stem bark of *Goniothalamus velutinus* (Airy Shaw) and their biological activities. JKSUS 30:41–48

Itoh A, Tanahashi T, Nagakura N (1995) Five tetrahydroisoquinoline-monoterpene glucosides and a tetrahydro-β-carboline-monoterpene glucoside from *Alangium lamarckii*. J Nat Prod 58:1228–1239

Jarraya RM, Bouaziz A, Hamdi B, Salah A, Damak M (2008) N-methylisosalsoline from *Hammada scoparia*. Acta Crystallogr E 64:o1714

Jenis J, Nugroho AE, Hashimoto A, Deguchi J, Hirasawa Y, Wong CP, Kaneda T, Shirota O, Morita H (2015) A new benzylisoquinoline alkaloid from *Leontice altaica*. Nat Prod Commun 10:291–292

Kaya GI, Unver N, Gozler B, Bastida J (2004) (−)-Capnoidine and (+)-bulbocapnine from an Amaryllidaceae species, *Galanthus nivalis* subsp. *cilicicus*. Biochem Syst Ecol 32:1059–1062

Kumar V (2017) Study of chemotaxonomy: meaning, stages and significance. Universal Research Reports 4:6–8

Ladino OJP, Suárez LEC (2010) Isoquinoline alkaloids of *Zanthoxylum quinduense* (Rutaceae). Biochem Syst Ecol 38:853–856

Li H, Kao C, Tsai C, Li W, Chen C (2017) Isoquinoline Alkaloids from *Michelia fuscata*. Chem Nat Compd 53:504–507

Li YH, Li HM, Li Y, He J, Deng X, Peng LY, Gao LH, Zhao QS, Li RT, Wu XD (2014) New alkaloids sinomacutines A-E, and cephalonine-2-O-β-D-glucopyranoside from rhizomes of Sinomenium acutum. Tetrahedron 70:8893–8899

Lima N, Santos V, La Porta FA (2018) Quimiodiversidade, Bioatividade e Quimiossistemática do gênero Inga (FABACEAE): Uma breve Revisão. Rev Virtual Quim 10:459–473

Lin WH, Fu HZ, Hano Y, Nomura T (1997) Alkaloids from the roots of *Aristolochia triangularis* (I). J Chinese Pharm Sci 6:8–13

Liscombe DK, Facchini PJ (2008) Evolutionary and cellular webs in benzylisoquinoline alkaloid biosynthesis. Curr Opin Biotech 19:173–180

Liscombe DK, MacLeod BP, Loukanina N, Nandi OI, Facchini PJ (2005) Erratum to "Evidence for the monophyletic evolution of benzylisoquinoline alkaloid biosynthesis in angiosperms". Phytochemistry 66:2500–2520

López JA, Lin FT, Duah FK, Aly Y, Schiff PL (1988) Mollinedine, a new alkaloid from *Mollinedia costaricensis*. J Nat Prod 51:754–759

Lucio ASSC, da Silva Almeida JRG, da-Cunha EVL, Tavares JF, Barbosa Filho JM (2015) Alkaloids of the Annonaceae: occurence and a compilation of their biological activities. In: The alkaloids: chemistry and biology, vol 74, pp 233–409

Marti G, Eparvier V, Morleo B, Ven JL, Apel C, Bodo B, Amand S, Dumontet V, Lozach O, Meijer L (2013) Natural aristolactams and aporphine alkaloids as inhibitors of CDK1/Cyclin B and DYRK1A. Molecules 18:3018–3027

Martin F, Grkovic T, Sykes ML, Shelper T, Avery VM, Camp D, Quinn RJ, Davis RA (2011) Alkaloids from the Chinese vine *Gnetum montanum*. J Nat Prod 74:2425–2430

Menezes LR, Costa CO, Rodrigues ACBC, Santo FRE, Nepel A, Dutra LM, Silva F, Soares MB, Barison A, Costa EV (2016) Cytotoxic alkaloids from the stem of *Xylopia laevigata*. Molecules 21:890

Misra A, Srivastava S (2016) Chemotaxonomy: An approach for conservation & exploration of industrially potential medicinal plants. J Pharmacogn Nat Prod 2:4. https://doi.org/10.4172/2472-0992.1000e108

Mulholland DA, Crouch N, Decker B, Smith MT (2002) The isolation of the Amaryllidaceae alkaloid crinamine from *Dioscorea dregeana* (Dioscoreaceae). Biochem Syst Ecol 2:183–185

Nomura T, Quesada AL, Kutchan TM (2008) The new β-D-glucosidase in terpenoid-isoquinoline alkaloid biosynthesis in *Psychotria ipecacuanha*. J Biol Chem 283:34650–34659

Pang SQ, Wang GQ, Lin JS, Diao Y, Xu RA (2014) Cytotoxic activity of the alkaloids from *Broussonetia papyrifera* fruits. Pharm Biol 52:1315–1319

Popović M, Djurković R, Gašić O, Pal B, Dutschewska H, Kuzmanov B (1992) Chemical and cytological investigation of *Thalictrum minus* from Vojvodina Region. Biochem Syst Ecol 20:255–258

Qin XD, Yang S, Zhao Y, Gao Y, Ren FC, Zhang DY, Wang F (2017) A new Aporphine alkaloid from *Aconitum carmichaelii*. Chem Nat Compd 53:501–503

Rastrelli L, Capasso A, Pizza C, De Tommasi N, Sorrentino L (1997) New protopine and benzyltetrahydroprotoberberine alkaloids from *Aristolochia constricta* and their activity on isolated guinea-pig ileum. J Nat Prod 60:1065–1069

Riley-Saldaña CA, del Rocío Cruz-Ortega M, Vázquez MM, De-la-Cruz-Chacón I, Castro-Moreno M, González-Esquinca AR (2017) Acetogenins and alkaloids during the initial development of *Annona muricata* L. (Annonaceae). Z Naturforsch C 72:497–506

Santos CAM, Reichert CL, Santos TG (2017) Alcaloides isoquinolínicos. In: Simões CMO (ed) Farmacognosia do produto ao medicamento. Artmed, pp 331–347

Sauvain M, Moretti C, Bravo JA, Callapa J, Muñoz V, Ruiz E, Richard B, Le Men-Olivier L (1996) Antimalarial activity of alkaloids from *Pogonopus tubulosus*. Phytother Res 10:198–201

Schimming T, Jenett-Siems K, Siems K, Witte L, Gupta MP, Eich E (2000) Iseluxine: a novel isoquinolinone alkaloid from *Iseia luxurians*. Z Naturforsch C 55:1023–1025

Schnetzler BN, Teixeira SP, Marinho CR (2017) Trichomes that secrete substances of a mixed nature in the vegetative and reproductive organs of some species of Moraceae. Acta Bot Bras 31:392–402

Shamma M, Guinaudeau H (1984) Biogenetic pathways for the aporphinoid alkaloids. Tetrahedron 40:4795–4822

Silva F, Silva Filho FA, Lima BR, Almeida RA, Soares ER, Koolen HH, Souza AD, Pinheiro ML (2016) Chemotaxonomy of the Amazonian *Unonopsis* species based on leaf alkaloid fingerprint direct infusion ESI-MS and chemometric analysis. J Brazil Chem Soc 27:599–604

Singh A, Bajpai V, Kumar S, Rawat AKS, Kumar B (2017) Analysis of isoquinoline alkaloids from *Mahonia leschenaultia* and *Mahonia napaulensis* roots using UHPLC-Orbitrap-MSn and UHPLC-QqQLIT-MS/MS. J Pharm Anal 7:77–86

Singh R (2016) Chemotaxonomy: a tool for plant classification. J Med Plants 4:90–93
Stanstrup J, Schmidt JS, Rasmussen HB, Mølgaard P, Guzmán A, Staerk D (2010) Bisbenzylisoquinoline alkaloids as markers of Atherospermataceae: Tetrandrine and fangchinoline from *Laureliopsis philippiana*. Biochem Syst Ecol 38:450–453
Stevigny C, Bailly C, Quetin-Leclercq J (2005) Cytotoxic and antitumor potentialities of aporphinoid alkaloids. Curr Med Chem Anticancer Agents 5(2):173–182
Suau R, Cabezudo B, Rico R, López-Romero JM, Nájera F (2002) Alkaloids from *Fumaria sepium* and *Fumaria agraria*. Biochem Syst Ecol 30:263–265
Ünsal Ç, Eroğlu E, Şerbetçi T, Mat A, Sarıyar G (2008) Alkaloids of *Papaver clavatum* and *P. stylatum*. Biochem Syst Ecol 5:497–499
Urzua A, Cassels BK (1978) Alkaloid chemosystematics, chemotaxonomy and biogenesis in the Atherospermataceae. Lloydia 41:98–113
Urzua A, Espinoza J, Olguin A, Santander R (2013) Phenolic aristolactams from leaves and stems of *Aristolochia chilensis*. Bol Latinoam Caribe de Plantas Med Aromát 12:537–542
Wink M (2013) Evolution of secondary metabolites in legumes (Fabaceae). S Afr J Bot 89:164–175
Wink M (2015) Modes of action of herbal medicines and plant secondary metabolites. Medicines 2:251–286
Wink M (2018) Quinolizidine and pyrrolizidine alkaloid chemical ecology—a mini-review on their similarities and differences. J Chem Ecol. https://doi.org/10.1007/s10886-018-1005-6
Yang GM, Sun J, Pan Y, Zhang JL, Xiao M, Zhu MS (2018) Isolation and identification of a tribenzylisoquinoline alkaloid from *Nelumbo nucifera* Gaertn, a novel potential smooth muscle relaxant. Fitoterapia 124:58–65
Zhang A, Zhang Y, Branfman AR, Baldessarini RJ, Neumeyer JL (2007) Advances in development of dopaminergic aporphinoids. J Med Chem 50:171–181
Zhang L, Rao G (2009) Aporphine, protoberberine and morphine alkaloids from the tubers of *Stephania yunnanensis*. Biochem Syst Ecol 37:622–625

Chapter 9
Fungal Polyketides: Chemical Diversity and Their Cytotoxic Effects

Hidayat Hussain, Barbara Schulz and Ivan R. Green

Abstract Compounds isolated from different natural sources have over the years played crucial roles in the treatment of a wide range of human diseases. Over the past six decades, microorganisms have provided valuable active compounds for the treatment of various diseases. Fungi, in general, produce diverse structural classes of natural products including polyketides, a major class of secondary metabolites obtained from various natural sources, which as a group have interesting chemical diversity. In addition, polyketides are known to possess a number of biological and pharmacological effects, viz. cytotoxic, antibacterial, antifungal, antiparasitic, and immunosuppressive effects. In this chapter the focus is on describing the cytotoxic effects of polyketides isolated from fungi and in particular their potential as cancerostatic pharmaceuticals.

Keywords Polyketides · Fungi · Cytotoxicity · Chemcial diversity

9.1 Introduction

According to the analysis of Newman and Cragg (2012), microbial natural products and their mimics are and have been an important source of pharmaceuticals, for example, from the 1940s to 2010, out of a total of 175 small compound cancer

H. Hussain (✉)
Department of Bioorganic Chemistry, Leibniz Institute of Plant Biochemistry, Weinberg 3, 06120 Halle (Salle), Germany
e-mail: hussainchem3@gmail.com

B. Schulz
Institut für Mikrobiologie, Technische Universität Braunschweig, Spielmannstraße 7, 31806 Brunswick, Germany
e-mail: b.schulz@tu-bs.de

I. R. Green
Department of Chemistry and Polymer Science, University of Stellenbosch, Private Bag X1, Matieland, Stellenbosch 7600, South Africa
e-mail: irg@sun.ac.za

K. G. Ramawat (ed.), *Biodiversity and Chemotaxonomy*, Sustainable Development and Biodiversity 24, https://doi.org/10.1007/978-3-030-30746-2_9

drugs, 131 (or 75%) belong to the "other than synthetic" category. In addition, of these, 85 (or 49%) are natural products or their direct analogues (Newman and Cragg 2012). One can comfortably state that although the antibacterial activity of *Penicillium* was already observed in the nineteenth century (Mohr 2016), the microbial drug era was started in 1928 when Alexander Fleming (1929) isolated penicillin from *Penicillium notatum*, the compound which was employed as a significant antibiotic during World War II. Following this discovery, a number of other secondary metabolites, e.g., chloramphenicol and streptomycin were isolated using Fleming's method (Demain and Sanchez 2009). Fungi-derived secondary metabolites have been an inspirational and sustainable source of medicinal products. For instance, penicillins (antibacterial), lovastatin (cholesterol-lowering), echinocandin B (antifungal), and cyclosporin A (immunosuppressive) are fungal-derived marketed drugs. There are also numerous reports that natural products and their analogues have been used as anticancer agents in traditional medicine (Mohr 2016). These factors clearly demonstrate the importance of natural products and, particularly, those of fungal origin as sustainable sources for new medicines (Evidente et al. 2014).

Polyketides are reported from various fungi with considerable chemical diversity (Hussain et al. 2017). Additionally, they possess a wide range of biological effects, viz. antibiotic, antifungal, anticancer, immunosuppressive, and antiparasitic effects (Staunton and Weissman 2001). It is noteworthy that there are a good number of polyketide-derived compounds in clinical trial (Weissman and Leadlay 2005; Hussain et al. 2017). Interestingly, 20% of the best-selling drugs presently on the market are classified by pharmaceutical companies as derived from polyketides. These command some £10 billion worth per year in the UK alone (Weissman and Leadlay 2005; Hussain et al. 2017).

9.2 Cytotoxic Polyketides

9.2.1 Quinones and Xanthones

In 2008, Du et al. (2008) reported on the aromatic polyketides, viz. aspergiolide B (**1**) and (*trans*)-and (*cis*)-emodin-physcion bianthrone (**2** and **3**) (Fig. 9.1) from the fungus *Aspergillus glaucus*, based on their elucidation via advanced NMR techniques. Biological studies showed that aspergiolide B (**1**) was significantly cytotoxic towards leukaemia (HL-60; IC_{50} = 0.51 μM) and lung cancer cells (A-549; IC_{50} = 0.24 μM). In contrast, (*trans*)-emodin-physcion bianthrone **2** demonstrated only moderate cytotoxic effects towards HL-60 (IC_{50} = 7.8 μM) and A-549 (IC_{50} = 9.2 μM) cell lines, while its *cis*-isomer **3** possesses weaker effects (HL-60: IC_{50} = 44.0 μM; A-549: IC_{50} = 14.2 μM).

Lee et al. (2010) reported on the following library of xanthones, viz. sterigmatocystin (**4**), averantin (**5**), methyl-averantin (**6**), averufin (**7**), nidurufin (**8**), and

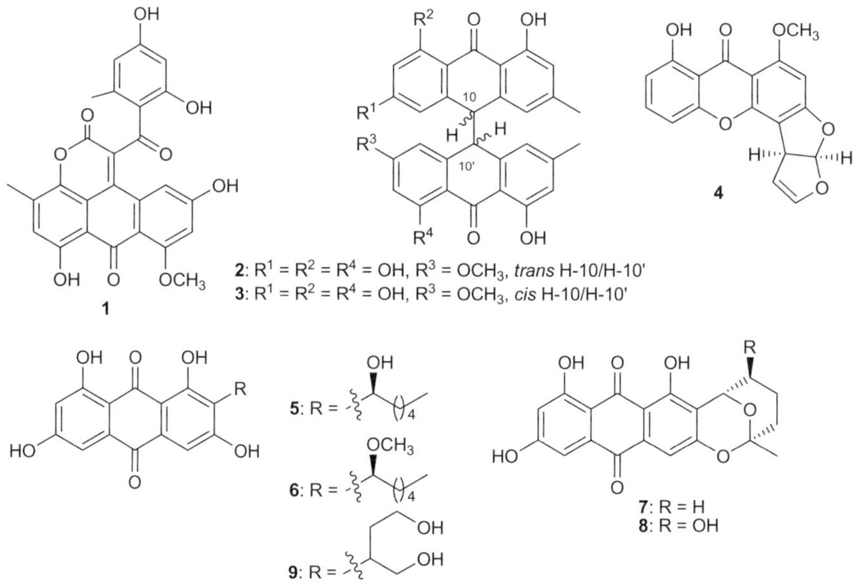

Fig. 9.1 Structures of cytotoxic quinones and xanthones **1–9**

versiconol (**9**) isolated from the fungus *Aspergillus versicolor* (Fig. 9.1) and their evaluation for cytotoxic effects. Compounds **4**, **5**, and **8** were found to possess potent activity towards lung cancer (A-549: **4**: IC_{50} = 1.86 μg/mL; **5**: IC_{50} = 3.15 μg/mL; and **8**: IC_{50} = 1.83 μg/mL), ovarian cancer (SK-OV-3: **4**: IC_{50} = 2.53 μg/mL; **5**: IC_{50} = 3.88 μg/mL; and **8**: IC_{50} = 3.39 μg/mL), skin cancer (SK-MEL-2: **4**: IC_{50} = 1.22 μg/mL; **5**: IC_{50} = 3.57 μg/mL; and **8**: IC_{50} = 3.16 μg/mL), CNS cancer (XF-498: **4**: IC_{50} = 2.75 μg/mL; **5**: IC_{50} = 3.04 μg/mL; and **8**: IC_{50} = 1.78 μg/mL), and colon cancer (HCT-15: **4**: IC_{50} = 4.61 μg/mL; **5**: IC_{50} = 3.13 μg/mL; and **8**: IC_{50} = 2.20 μg/mL). It is noteworthy that compound **5** was the most potent and showed activity towards five cancer cells lines, viz. A-549 (IC_{50} = 0.64 μg/mL), SK-OV-3 (IC_{50} = 1.17 μg/mL), SK-MEL-2 (IC_{50} = 1.10 μg/mL), XF-498 (IC_{50} = 0.41 μg/mL), and HCT-15 (IC_{50} = 0.41 μg/mL). In contrast, compounds **7** and **9** were only moderately active towards the above-mentioned five cancer cells with IC_{50s} ranging from 11.9 to 23.7 μg/mL (Lee et al. 2010).

Yao et al. (2014) isolated the xanthones, engyodontiumone H (**10**) and compound **11** (Fig. 9.2) from the fungus *Engyodontium album* and these compounds were found to possess potent cytotoxicity effects towards histiocytic lymphoma cell line (U937; **10**: IC_{50}: 4.9 μM; **11**: IC_{50}: 8.8 μM). In another study, sterigmatocystin (**4**) was isolated from the fungus *Emericella nidulans* (Kralj et al. 2006) and shown to possess cytotoxic effects towards A-549 (IC_{50}: 3.7 μM) and Neuro-2a (IC_{50}: 40.1 μM) (Bünger et al. 2004). Austocystin D (**12**), isolated from the fungus *Aspergillus* sp., was shown to possess cytotoxic effects towards a range of cancer

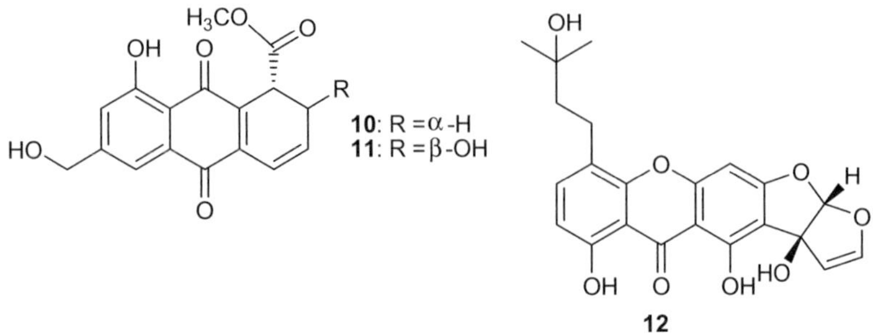

Fig. 9.2 Structures of cytotoxic quinones and xanthones **10–12**

cells, viz. brain (U-87; GI_{50}: 4946 nM), breast (MCF-7; GI_{50}: <1 nM; MDA-MB-231: GI_{50}: 549 nM), leukaemia (SR; GI_{50}: 16 nM), colon (SW-620; GI_{50}: 27 nM; HCT-15: GI_{50}: 42 nM), prostate (PC-3: GI_{50}: 3 nM), and uterine (MX-2: GI_{50}: 3358 nM) (Wang et al. 2008).

9.2.2 *Versixanthones*

Xanthone and chromanone monomers have been reported from fungi, plants, and lichens and have all been demonstrated to exhibit an interesting chemical diversity (Masters and Braese 2012; Wezeman et al. 2015; Wu et al. 2015). Moreover, xanthone–chromanone heterodimers, which are dimerized via a biaryl single bond to form a chromanone lactone and tetrahydroxanthone monomer, are very rare. Very few examples have been reported for these heterodimers and those that have include (±)-blennolide G (Zhang et al. 2008a; Cai et al. 2014, gonytolides D and E (Kikuchi et al. 2012), and blennolids I and J (El-Elimat et al. 2015). Wu et al. (2015) reported six xanthone-chromanone dimers from the fungus *Aspergillus versicolor* and named them versixanthones A–F (**13-18**) (Fig. 9.3), having different linkages of dimerization between the chromanone and tetrahydroxanthone. Interestingly, compounds **13–18** showed selective cytotoxicity towards various cancer cells, of which compound **16** possesses significant toxic effects towards HL-60 (IC_{50}: 3.1 μM), K562 (IC_{50}: 9.1 μM), A549 (IC_{50}: 12.7 μM), 803 (IC_{50}: 9.8 μM), HO-8910 (IC_{50}: 13.9 μM), and HCT-116 (IC_{50}: 6.1 μM). On the other hand, compounds **17** and **18** demonstrated cytotoxic effects towards five cancer cell lines, viz. HL-60 (**17**: IC_{50}: 1.2 μM; **18**: IC_{50}: 2.7 μM), K562 (**17**: IC_{50}: 11.1 μM; **18**: IC_{50}: 6.7 μM), A549 (**17**: not tested; **18**: IC_{50}: 10.6 μM), H1975 (**17**: IC_{50}: 2.72 μM; **18**: not tested), 803 (**17**: IC_{50}: 2.2 μM; **18**: not tested), HO-8910 (**17**: IC_{50}: 2.0 μM; **18**: IC_{50}: 20.0 μM), and HCT-116 (**17**: not tested; **18**: IC_{50}: 0.7 μM). To complete this evaluation, it was shown that compounds **13–15** were quite active towards HL-60 cell lines with IC_{50} values of 2.6, 9.9, and 7.8 μM, respectively.

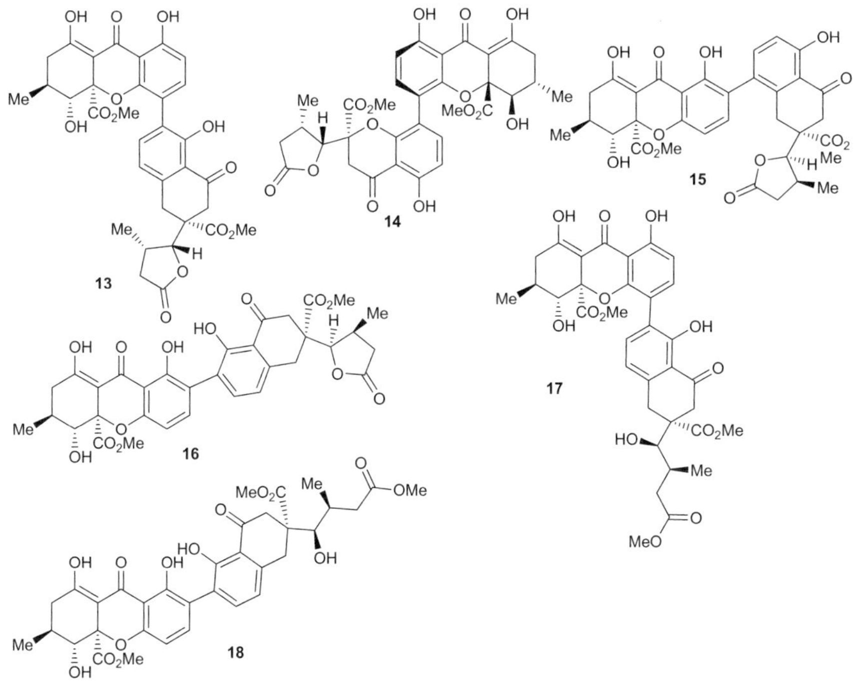

Fig. 9.3 Structures of cytotoxic versixanthones **13–18**

9.2.3 Perylenequinones

The perylenequinones are natural products isolated from fungi, aphids, and plants. These compounds are characterized by having a pentacyclic conjugated chromophore and display a number of biological effects, especially anticancer activities (Mulrooney et al. 2012). Calphostin C (**19**) (Fig. 9.4) was isolated from the fungus *Cladosporium cladosporioides* and reported to possess potent cytotoxic effects towards cervical cancer (HeLa-S_3: IC_{50}: 0.23 μM) and breast cancer (MCF-7: IC_{50}: 0.18 μM) (Takahashi et al. 1976). Additionally, hypocrellin D (**20**), which was isolated from the fungus *Shiraia bambusicola,* showed significant cytotoxicity towards liver (Bel-7721: IC_{50}: 1.8 μg/mL) and lung cancer cells (A-549: IC_{50}: 8.8 μg/mL; Anip-973: IC_{50}: 38.4 μg/mL) (Fang et al. 2006).

9.2.4 Chaetoglobosins

Chaetoglobosins and their analogues belong to the polyketide cytochalasin family present in many fungi (Hu et al. 2018). To date, over 80 chaetoglobosins have been isolated from various fungal genera, viz. *Discosia* (Donoso et al. 1997),

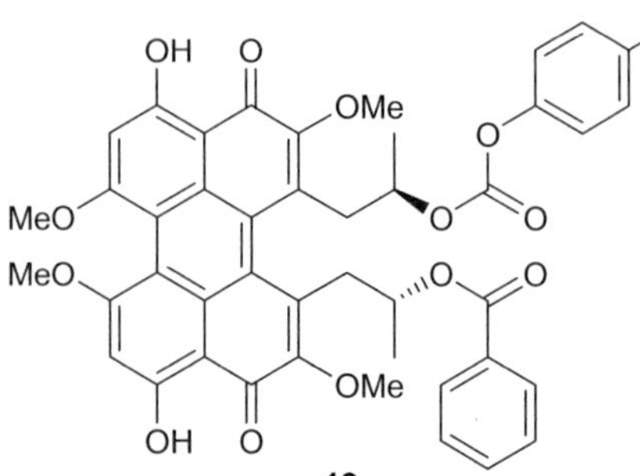
19

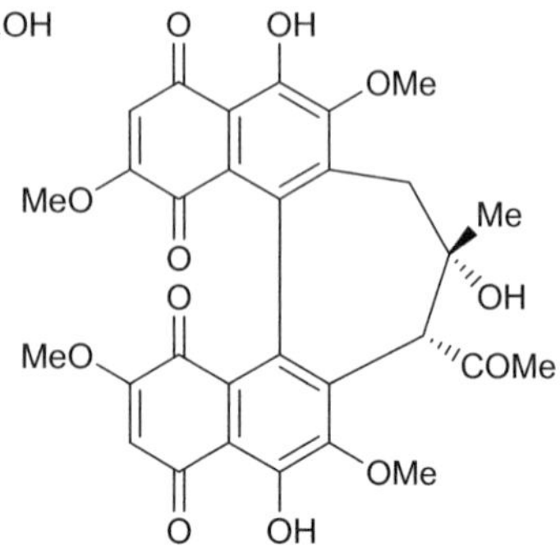
20

Fig. 9.4 Structures of cytotoxic perylenequinones **19** and **20**

Chaetomium (Zheng et al. 2014; Sekita et al. 1983; Dou et al. 2011), *Diplodia* (Springer et al. 1980), *Penicillium* (Iwamoto et al. 2001; Numafa et al. 1996), *Cylindrocladium* (Ichihara et al. 1996), *Calonectria* (Von Wallbrunn et al. 2001), and *Phomopsis* (Christian et al. 2005). Moreover, these compounds have diverse biological properties, viz. cytotoxicity (Zhang et al. 2010; Ding et al. 2006), and antimicrobial (Zhang et al. 2013a, b), anti-HIV (Chen et al. 2015) and phytotoxic activities (Li et al. 2014).

Li et al. (2014) isolated ten chaetoglobosins (**21–30**) (Fig. 9.5) from the fungus *Chaetomium globosum* and tested them for their cytotoxic effects towards various HTC116 cancer cell lines. Interestingly, among the tested compounds, chaetoglobosins **21**, **28–30** demonstrated the more potent cytotoxic properties towards HTC116 cells having IC_{50s} of 3.15, 4.43, 8.44, and 5.85 μM, respectively. On the other hand, the activity of chaetoglobosins **22–24**, **26**, and **27** was not that encouraging, having an IC_{50} > 17 μM. A subsequent SAR study showed that chaetoglobosin **1** was almost 2.5 times more active than compound **29** on HCT116 cells (**21**: IC_{50}: 3.15 μM; **29**: IC_{50}: 8.44 μM) and the authors suggested that reduction of the C-20 carbonyl group may be related to this decrease in cytotoxicity, although the latter also possesses an epoxide ring. Compound **27** is more strongly cytotoxic than chaetoglobosin **26** (**27**: IC_{50}: 17.8 μM vs. **26**: IC_{50}: >100 μM). These results lead to the suggestion that in this case, the epoxide ring at C6/C7 may be involved in the enhanced cytotoxicity. In addition, the exocyclic double bond at C-6/C-12 was crucial for activity because compound **28** (IC_{50} = 4.43 μM) was more cytotoxic than compound **26** (IC_{50}: >100 μM). Furthermore, it is well known that the presence of a fluorine atom plays a crucial role in activity as borne out by the fact that compound **30** (IC_{50} = 5.85 μM) was almost 3 times more active than compound **27** (IC_{50} = 17.8 μM) (Li et al. 2014).

Cytotoxicity towards a wide range of human cancer cell lines has been demonstrated (Zhang et al. 2012; Scherlach et al. 2010; Umeda et al. 1975; Sekita et al. 1973). Moreover, chaetoglobosins **21**, **22**, **25**, and **26** displayed cytotoxic potentials towards HeLa cells (IC_{50} = 3.2–20 μM) (Sekita et al. 1977, 1982) and in

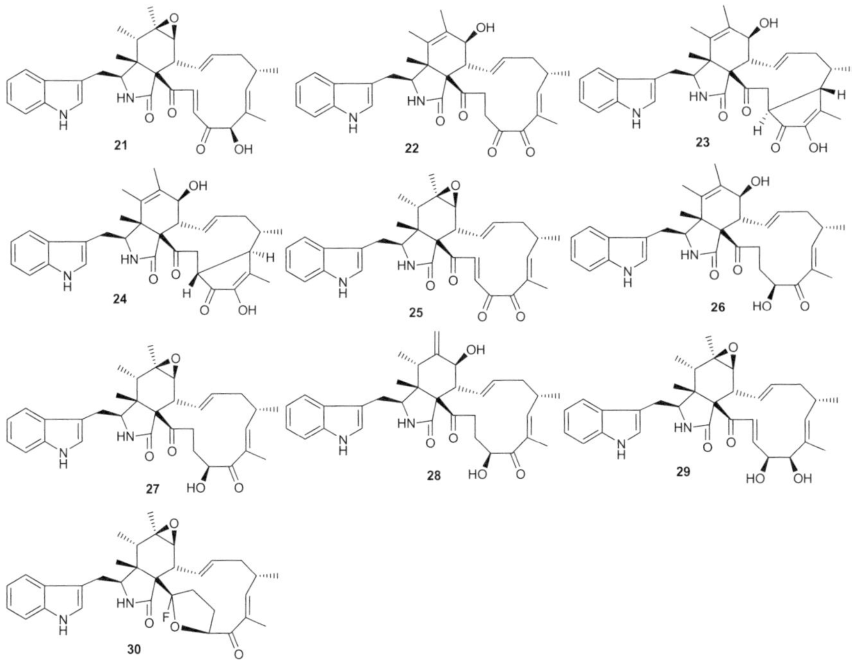

Fig. 9.5 Structures of cytotoxic chaetoglobosins **21–30**

addition, compound **21** showed potent cytotoxic effects against P388 murine leukaemia cells with IC_{50} values ranging from 1.58 to 4.92 μg/mL (Jiao et al. 2004). Additionally, compounds **22** and **25** displayed cytotoxic effects against the human breast cancer (BC1) cell lines and the IC_{50} values ranged from 6.27 to 20.0 μM (Thohinung et al. 2010). Furthermore, chaetoglobosins **21–23** and **25–28** possess promising cytotoxic effects towards the four cancer cell lines, viz. HepG2, K562, KB, and MCF-7 (Zhang et al. 2010).

9.2.5 Cytochalasans

Cytochalasans are a group of natural products with a tricyclic core comprising a macrocyclic ring (viz., carbocycle, lactone, or cyclic carbonate) fused to an isoindolone group (Zhang et al. 2013a). Presently over 80 different cytochalasans have been reported in which their macrocyclic rings constitute either 12- to 14-membered lactone rings or 11- to 13-membered carbocycles. It has been suggested that the lactone rings in some cytochalasans might be generated via a Baeyer–Villiger oxidation in which insertion of an oxygen into the carbocyclic ring has occurred (Scherlach et al. 2010; Schümann and Hertweck 2007; Zhang et al. 2013a, b). Cytochalasans have been reported to be isolated from various fungal genera (viz.,

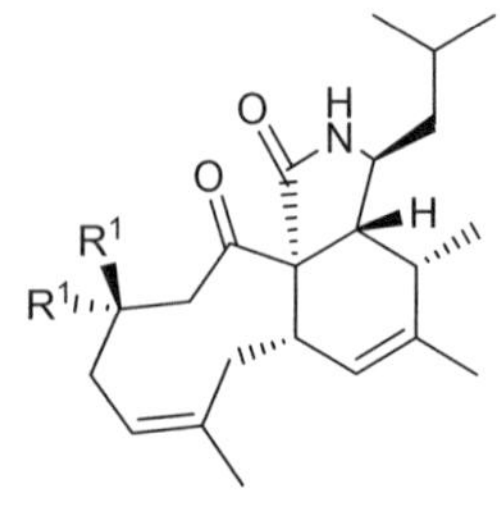

Fig. 9.6 Structures of cytotoxic cytochalasans **31** and **32**

Penicillium, *Phomopsis*, and *Aspergillus*) and possess a number of interesting biological effects: viz. cytotoxic (Jiao et al. 2004), antiviral (Zhang et al. 2008b), antimicrobial (Pongcharoen et al. 2006), and phytotoxic (Cimmino et al. 2008). Zhang et al. (2013a, b) isolated two cytochalasans A (**31**) and B (**32**) (Fig. 9.6) from the fungus *Periconia* sp. and tested them for their cytotoxic effects towards various cancer cell lines. It is noteworthy that compound **31** demonstrated potent cytotoxic effects towards the colon cancer cell HCT-8 (IC_{50}: 0.9 μM) and against gastric cancer cells BGC-823 (2.1 μM). In addition, compound **32** was shown to possess potent cytotoxic effects towards HCT-8 (IC_{50}: 0.8 μM), Bel-7402 (IC_{50}: 5.1 μM), and BGC-823 (IC_{50}: 9.4 μM).

9.2.6 Benzofuranones

Ding et al. (2009) isolated six benzofuranones named photinides A–F (**33–38**) (Fig. 9.7) from the fungus *Pestalotiopsis photiniae*. It was found that compounds **33–38** demonstrated modest but selective cytotoxic effects towards breast cancer cells (MDA-MB-231) with inhibitory rates ranging 23.1–24.6%, these compounds were not active against cervical cancer cells (HeLa). Furthermore, the photinides A–F (**33–38**) appear to be new representatives of the benzofuranone skeleton. Although, compounds **33–38** are structurally somewhat different from the already reported aurones having an oxymethylene unit connected to C-8 and a 4-methyl-γ-lactone group, while other aurones possess a para-hydroxyphenyl ring and a methyl group attached to C-8 (Ding et al. 2009).

9.2.7 Polyketide-Derived γ-Lactones

Nozawa et al. (1995, 2000) isolated waol A (**39**), which at the time was the incorrect structure (Fig. 9.8), from fungus *Myceliophthora lutea* and evaluated it for its biological activities. It is noteworthy, in this case that compound **39** was found to

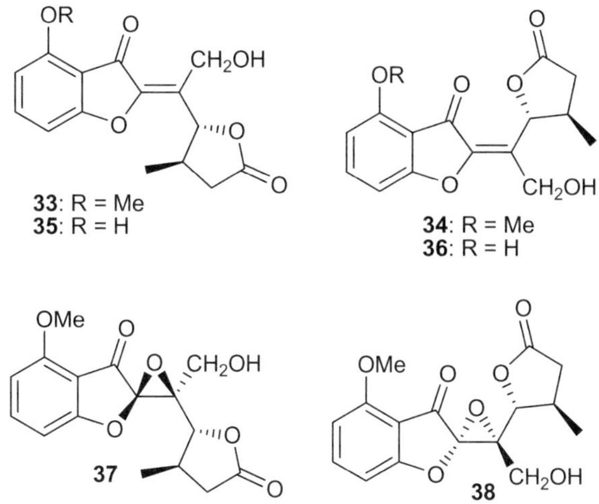

Fig. 9.7 Structures of cytotoxic benzofuranones **33–38**

be highly cytotoxic towards myeloid leukaemia (HL-60: IC_{50}: 0.2 μg/mL), Adriamycin resistant myeloid leukaemia (HL-60/ADM: IC_{50}: 0.1 μg/mL), leukaemia (P388: IC_{50}: 4.0 μg/mL), bladder cancer lines (T24: IC_{50}: 0.5 μg/mL), cervical cancer cells (HeLa: IC_{50}: 1.0 μg/mL), and lung cancer cells (A549: IC_{50}: 1.0 μg/mL). Gao and Snider (2004) published the total synthesis of waol A (**39**) and in doing so, revised its structure from **39** to **39a**. Moreover, waol A (**39a**), pestalotiopene A (**40**), and cytosporone E (**41**) were isolated from the endophytic fungus *Acremonium strictum* and evaluated for their cytotoxic effects (Hammerschmidt et al. 2014). It was additionally found that compounds **39a** and **40** were promising active agents towards the cisplatin-resistant ovarian cancer cell line A2780-Cis*R* (**39a**: IC_{50} = 12.6 μM; **40**: IC_{50} = 30.1 μM) They were, however, less active towards the sensitive ovarian cancer cell (A2780 sens (**39a**: IC_{50} = 27.1 μM; **40**: IC_{50} = 76.2 μM). On the other hand, compound **41** was more potent towards A2780 sens (IC_{50} = 8.3 μM) than towards A2780 CisR (IC_{50} = 19.0 μM). Compounds **39a** and **41** demonstrated a moderate cytotoxicity towards colorectal cancer cell HCT116 (**39a**: IC_{50} = 22.7 μM; **41**: IC_{50} = 35.4 μM).

Various natural products possessing a 5-hydroxydrimane skeleton (present in compound **40**) have been isolated from diverse plant species, viz. *Pogostemon* sp. (Paul et al. 2010)., *Zygogynum* sp. (Devkota et al. 2013), *Rhizophora mucronata* (Silva and Madureira 2012), *Canella winterana* (Ying et al. 1995a, b; Al-Said et al. 1990; Kioy et al. 1990a, b; Rastogi et al. 1998), *Cinnamodendron corticosum* (Seeram et al. 2003), *Capsicodendron dinisii* (Mahmoud et al. 1979), *Warburgia ugandensis* (Rajab and Ndegwa 2000), *W. stuhlmannii* (Kioy et al. 1990a, b), and *W. salutaris* (Mashimbye et al. 1999). In addition, fungi are also known to biosynthesize drimanes and these compounds have been isolated from various fungi, viz. *Aspergillus* sp., (Uosaki et al. 1996; Hayes et al. 1996; Grabley et al. 1996; Rahbæk et al. 1997; Belofsky et al. 1998), *Kuehneromyces* sp., (Erkel et al. 1995), *Trichopezizella barbata* (Kono et al. 2000), *Mniopetalum* sp. and *Panus* sp.,

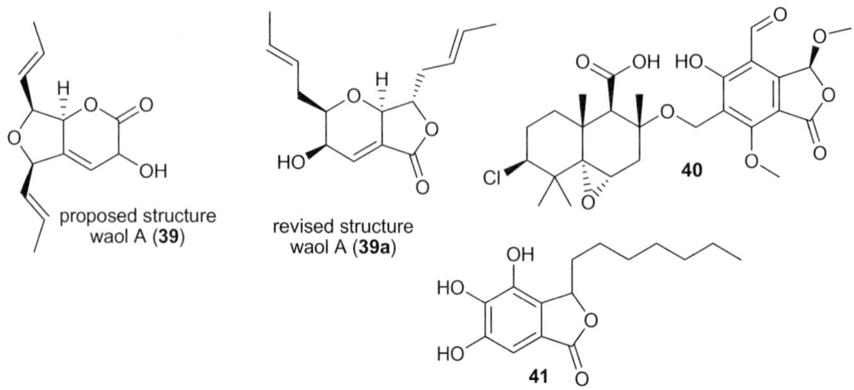

Fig. 9.8 Structures of cytotoxic polyketide-derived γ-lactones **39a–41**

(Velten et al. 1994; Lorenzen et al. 1994), *Penicillium brevicom-pactum* (McCorkindale et al. 1981), *Lactarius uvidus* (Garlaschelli et al. 1994), *Polyporus ciliates* (Cabrera et al. 2002), *Polyporusarcularius* (Cabrera et al. 2002), *Pestalotiopsis* spp., *Lepistaglaucocana*, and endophytic fungi of the yew tree (Pulici et al. 1996, 1997). This demonstrates that such molecules can be derived from both the plant and fungal kingdoms and in a sense reflect their synergetic relationship. There are even cases in which fungus and plant produce the same metabolite. This is the case with the anti-tumour metabolite taxol, which can be produced both by plant and fungus (Demain and Vaishnav 2011), albeit by the fungus, e.g. *Pestalotiopsis* spp., at low concentrations (Kumaran et al. 2010).

9.2.8 *Miscellaneous*

Kasai et al. (2005) reported on the polyketide glycoside named cladionol A (**42**), the group isolated from the fungus *Gliocladium* sp. Previously, this compound and other related polyketide glycosides were reported, e.g. the roselipins (Tomoda et al. 1999; Tabata et al. 1999) and TMC-151s (Kohno et al. 1999), which were also isolated from *Gliocladium* spp. It is noteworthy that cladionol A (**42**) possesses cytotoxicity towards two cancer cells viz. L1210 (murine leukaemia: IC_{50} values of 5 μg/mL) and KB cells epidermoid carcinoma: IC_{50}: 7 μg/mL). Mitova et al. (2006) isolated six fungal-derived polyketides named cladobotric acids A–F (**43–48**) from the fungal genus *Cladobotryum*. Moreover, compounds **43–48** (Fig. 9.9) were shown to possess cytotoxic effects towards murine P388 leukaemia cells with IC_{50} = 6.5, 27.7, 19.3, 24.9, 1.41, and 15.5 μM, respectively. Three years later, Isaka et al. (2009) reported the isolation of hypothemycin (**49**) and 4-O-demethylhypothemycin (**50**) from fungus *Aigialus parvus*. It is noteworthy in this case that compound **49** possesses potent cytotoxicity towards lung cancer

NCI-H187 (IC_{50}: 2.0 μg/mL) and Vero cells (IC_{50}: 2.1 μg/mL), while the same compound was not active towards epidermoid carcinoma (KB cells). In addition, compound **50** displays cytotoxic effects towards three cancer cells, viz. NCI-H187 (IC_{50}: 3.6 μg/mL), Vero cells (IC_{50}: 0.77 μg/mL), and breast cancer cells (BC cells: IC_{50}: 2.6 μg/mL).

Initially, Cai et al. (1999) reported on their isolation of mycoepoxydiene (**51)** from a fungus designated as OS-F66617 and demonstrated that the compound consists of δ-lactone and possessed an oxygen-bridged cyclooctadiene core (Fig. 9.10). Six years later, Lin et al. (2005) also isolated compound **51** but this time from the marine lignicolous fungus *Diaporthe* sp. and found it to be cytotoxic towards KB cells $IC_{50} < 6.25$ μg/mL. Two years later, Prachya et al. (2007) again isolated mycoepoxydiene (**51**) along with an analogue **52,** which the group named deacetylmycoepoxydiene from an endophytic fungus *Phomopsis* sp. It is noteworthy that compound **51** possesses potent effects towards 10 cancer cells with $IC_{50} < 3$ μg/mL. Interestingly, compound 51 display potent effects towards cholangiocarcinoma (HuCCA-1: IC_{50}: 0.27 ± 0.06 μg/mL), leukaemia cells (HL-60: IC_{50}: 0.79 ± 0.11 μg/mL), and P388 (IC_{50}: 0.73 ± 0.70 μg/mL). Interestingly, compound **52** was only significantly active towards HepG2 (IC_{50}: 1.05 ± 0.35 μg/mL), A549 (IC_{50}: 1.95 ± 0.21 μg/mL), and HCC-S102 (IC_{50}: 1.15 ± 0.35 μg/mL) (Prachya et al. 2007).

Yang et al. (2007) isolated three new polyketides, the tetramic acids, and myceliothermophins A (**53**), C (**54**), and E (**55**) from the fungus *Myceliophthora thermophile* and evaluated them for their cytotoxic effects. It is interesting that

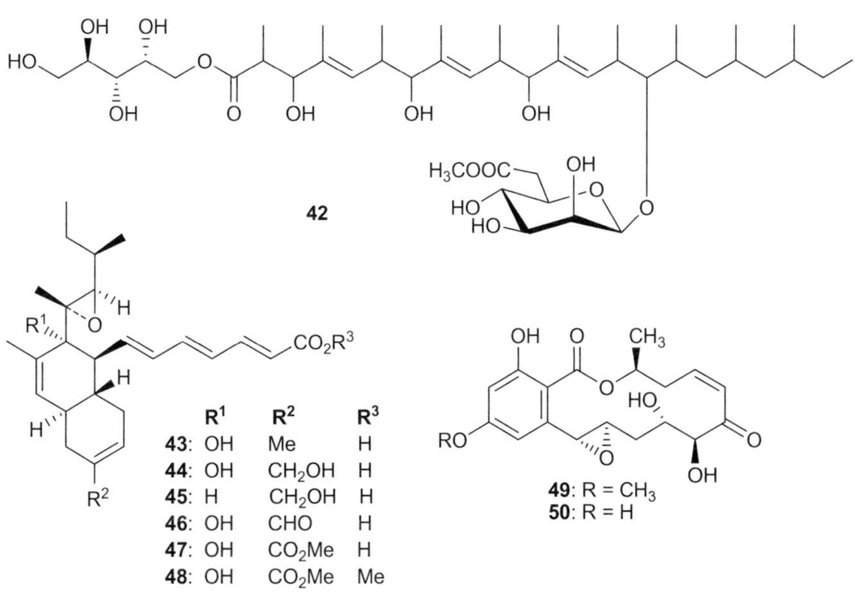

Fig. 9.9 Structures of cytotoxic polyketides **42–50**

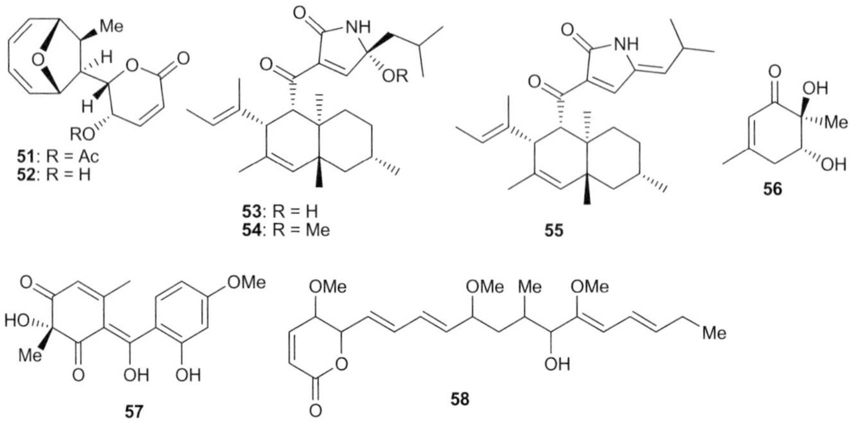

Fig. 9.10 Structures of cytotoxic polyketides **51–58**

compound **55** (Fig. 9.10) was the most potent and was active against four cancer cells, viz. A-549 (IC_{50}: 0.26 μg/mL), Hep3B (IC_{50}: 0.41 μg/mL), HepG2 (IC_{50}: 0.28 μg/mL), and MCF-7 (IC_{50}: 0.25 μg/mL). Compounds **53** and **54,** showed an almost similar level of cytotoxic effects towards the cancer cells, viz. A-549 (**53**: IC_{50}: 1.12 μg/mL; **54**: IC_{50}: 1.05 μg/mL), Hep3B (**53**: IC_{50}: 0.91 μg/mL; **54**: IC_{50}: 0.51 μg/mL), HepG2 (**53**: IC_{50}: 1.3 μg/mL; **54**: IC_{50}: 0.62 μg/mL), and MCF-7 (**53**: IC_{50}: 1.03 μg/mL; **54**: IC_{50}: 0.52 μg/mL). Moreover, natural products containing the deoxy-tetramic acid ring system have been reported, viz. myceliothermophins B and D (Yang et al. 2007), oteromycin (Singh et al. 1995), ZG-1494a, USC1025A and B (Nakai et al. 2000; Agatsuma et al. 2002), pyrrocidines A and B (He et al. 2002), ascosalipyrrolidinones A and B (Osterhage et al. 2000), and talaroconvolutins A–D (Suzuki et al. 2000).

Lin et al. (2008) isolated two polyketides, which the group named leptosphaerone C (**56**) and penicillenone (**57**) from the fungus *Penicillium* sp. and evaluated them for their cytotoxic activities. Compound **56** was shown to possess cytotoxicity towards A-549 cells having an IC_{50} = 1.45 μM and penicillenone (**57**), which was cytotoxic towards P388 cells with an IC_{50} = 1.38 μM. Pterocidin (**58**) was isolated from *Streptomyces hygroscopicus* and was cytotoxic for the cancer cell lines, viz. NCI-H522 (IC_{50}: 2.9 μM), OVCAR-3 (IC_{50}: 3.9 μM), SF539 (IC_{50}: 5.0 μM), and LOX-IMVI (IC_{50}: 7.1 μM) (Igarashi et al. 2006).

Mohamed et al. (2009) reported two new prenylated polyketides named epoxyphomalin A (**59**) and B (**60**) (Fig. 9.11) from the fungus *Phoma* sp. and evaluated them against a number of cancer cell lines. It is noteworthy that compound **59** displayed very potent and encouragingly excellent activity towards 36 cancer cell lines with IC_{50} values of below 0.5 μg/mL. In addition, compound 59 possesses potent effects towards bladder cancer cells, prostate, stomach, lung, glioblastoma, ovary, melanoma, breast, pancreas, kidney, mesothelioma, and uterus cells with $IC_{50} < 0.09$ μg/mL (Mohamed et al. 2009). On the other hand,

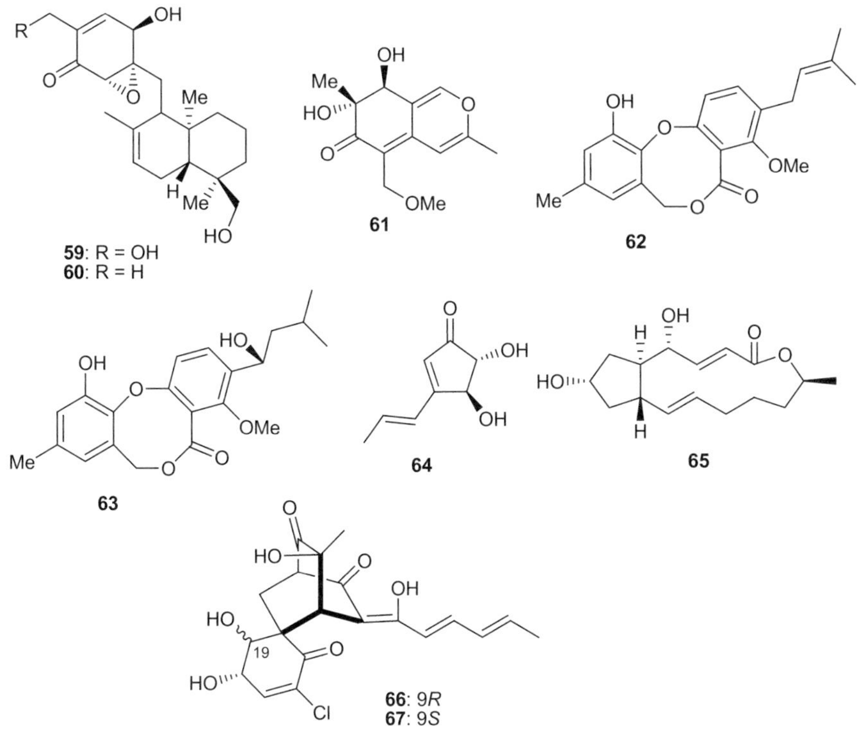

Fig. 9.11 Structures of cytotoxic polyketides **59–67**

compound **60** also significantly inhibits 36 cell lines with IC_{50} ranging from 0.2 to 11.4 µg/mL. Interestingly, this compound showed potent activity towards head, neck, bladder cancer cells, mesothelioma, glioblastoma, melanoma, breast, pancreas, ovary, prostate, and kidney with IC_{50} < 1 µg/mL (Mohamed et al. 2009).

Moreover, dothideomynone C (**61**) with one hydroxyl group less than **59** was isolated from the fungus *Dothideomycete* sp. and was cytotoxic towards HuCCA-1 (IC_{50}: 48.1 µg/mL), A549 (IC_{50}: 46.5 µg/mL), and MOLT-3 cell lines (IC_{50}: 17.4 µg/mL) (Hewage et al. 2014). In another investigation, Gao et al. (2013) isolated two penicillide derivatives, which the group named prenpenicillide (**62**) and penicillide (**63**) from the fungus *Penicillium* sp. Compounds **62** and **63** were cytotoxic for the HepG2 cell line with IC_{50} = 9.9 and 9.7 µM, respectively. Another small polyketide, known as terrein (**64**), produced by fungus *Aspergillusterreus*, demonstrated a 110-fold cytotoxicity towards MCF-7 with an IC_{50} value of 1.1 nM. In addition, terrein (**64**) possesses significant effects towards pancreatic PANC-1 (IC_{50}: 9.8 µM) and liver cancer cells HepG2 (IC_{50}: 66.8 µM) (Liao et al. 2012).

Brefeldin A (**65**) (Fig. 9.11), isolated from fungus *Penicillium brefeldianum*, was very active against HL-60 (IC_{50}: 35.7 nM), KB (IC_{50}: 32 nM), HeLa (IC_{50}:

6.4 nM), MCF-7 (IC_{50}: 7.1 nM), BC-1 (IC_{50}: 40 nM), SPC-A-1 (IC_{50}: 6.3 nM), and NCI-H187 (IC_{50}: 110 nM) (Wang et al. 2002; Chinworrungsee et al. 2008). In another investigation, Li et al. (2011) reported two sorbicillinoids, viz. chloctanspirones A (**66**) and B (**67**), both having a bicyclo[2.2.2]octane-2-spiro cyclohexane scaffold from the fungus *Penicillium terrestre*. Compound **66** was cytotoxic towards HL-60 (IC_{50}: 9.2 μM) and A-549 (IC_{50}: 39.7 μM), while compound **67** was weakly active only towards HL-60 cells (IC_{50}: 37.8 μM).

9.3 Conclusion

Numerous polyketides have been reported from various fungi and most are significantly cytotoxic towards various human cancer cells. Moreover, these polyketides are from diverse classes of natural products including quinones, xanthones, versixanthones, perylenequinones, chaetoglobosins, cytochalasans, benzofuranones, and γ-lactones. Furthermore, these cytotoxic polyketides were reported from various fungal taxa including *Aspergillus* sp., *Periconia* sp., *Phoma* sp., *Gliocladium* sp., *Dothideomycete* sp., *Cladobotryum* sp., *Phomopsis* sp., *Penicillium* sp., *Engyodontium album, Cladosporium cladosporioides, Chaetomium globosum, Pestalotiopsis photiniae, Myceliophthora lutea, Aigialus parvus,* and *Myceliophthora thermophile*.

Of particular interest are the epoxyphomalins A (**59**) and B (**60**) that have very potent activities towards bladder cancer cells, glioblastoma, head and neck, breast cells, melanoma, ovary, pancreas, prostate, mesothelioma, kidney (RXF 944 L: IC_{50}: 0.380 μg/mL) with $IC_{50} < 1$ μg/mL. Compound **51** is also a metabolite of note with significant cytotoxic effects towards lung cancer cells, cholangiocarcinoma, cervical cancer cells, breast cancer cells, and leukaemia cells with $IC_{50} < 2$ μg/mL. Austocystin D (**12**) and Averantin (**5**) are candidates for future studies with excellent inhibitions of cancer cell lines: Austocystin inhibits breast (MCF-7; GI_{50}: <1 nM), leukaemia (SR; GI_{50}: 16 nM), colon (SW-620; GI_{50}: 27 nM; HCT-15: GI_{50}: 42 nM), and prostate cancer cell lines (PC-3: GI_{50}: 3 nM), and averantin (**5**) inhibits five cancer cells: viz. SK-OV-3, HCT-15, A-549, XF-498, and SK-MEL-2 with $IC_{50} < 1$ μg/mL.

In conclusion, these results demonstrate that fungi are an excellent source of novel polyketides many of which are not only cytotoxic, but are also active against numerous cancer cell lines. Thus, considering the fact that only approximately 5% of the ca. 1.5–5.1 million fungi are presently known (Hibbett et al. 2011), fungi represent a untapped gold mine for the discovery of novel pharmaceutical drugs for treatment of cancer.

Acknowledgements The author (HH) thanks the Alexander von Humboldt Foundation for its generous support in providing the opportunity to do work in Germany, which facilitated the writing of this review. The authors want to thank Dr. Abdullah M. Al-Sadi who helped with comments and criticism by reading the manuscript.

References

Agatsuma T, Akama T, Nara S, Matsumiya S, Nakai R, Ogawa H, Otaki S, Ikeda SI, Saitoh Y, Kanda Y (2002) UCS1025A and B, new antitumor antibiotics from the fungus *Acremonium species*. Org Lett 4:4387–4390

Al-Said MS, El-Khawaja SM, El-Feraly FS, Hufford CD (1990) 9-Deoxy drimane sesquiterpenes from *Canella winterana*. Phytochemistry 29:975–977

Belofsky GN, Jensen PR, Renner MK, Fenical W (1998) New cytotoxic sesquiterpenoid nitrobenzoyl esters from a marine isolate of the fungus *Aspergillus versicolor*. Tetrahedron 54:1715–1724

Bünger J, Westphal G, Mönnich A, Hinnendahl B, Hallier E, Müller M (2004) Cytotoxicity of occupationally and environmentally relevant mycotoxins. Toxicology 202:199–211

Cabrera GM, Roberti MJ, Wright JE, Seldes AM (2002) Cryptoporic and isocryptoporic acids from the fungal cultures of *Polyporus arcularius* and *P. ciliatus*. Phytochemistry 61:189–193

Cai P, McPhail AT, Krainer E, Katz B, Pearce C, Boros C, Caceres B, Smith D, Houck DR (1999) Mycoepoxydiene represents a novel class of fungal metabolites. Tetrahedron Lett 1479–1482

Cai S, King JB, Du L, Powell DR, Cichewicz RH (2014) Bioactive sulfur-containing sulochrin dimers and other metabolites from an *Alternaria* sp. isolate from a Hawaiian soil sample. J Nat Prod 77:2280–2287

Chen C, Zhu H, Wang J, Yang J, Li XN, Wang J, Chen K, Wang Y, Luo Z, Yao G, Xue Y (2015) Armochaetoglobins K-R, anti-HIV pyrrole-based cytochalasans from Chaetomium globosum TW1-1. Eur J Org Chem 2015:3086–3094

Chinworrungsee M, Wiyakrutta S, Sriubolmas N, Chuailua P, Suksamrarn A (2008) Cytotoxic activities of trichothecenes isolated from an endophytic fungus belonging to order *Hypocreales*. Archi Pharmacal Res 31:611–616

Christian OE, Compton J, Christian KR, Mooberry SL, Valeriote FA, Crews P (2005) Using jasplakinolide to turn on pathways that enable the isolation of new chaetoglobosins from *Phomospis asparagi*. J Nat Prod 68:1592–1597

Cimmino A, Andolfi A, Berestetskiy A, Evidente A (2008) Production of phytotoxins by Phoma exigua var. exigua, a potential mycoherbicide against perennial thistles. J Agric Food Chem 56:6304–6309

Demain AL, Sanchez S (2009) Microbial drug discovery: 80 years of progress. J Antibiot 62:5–16

Demain AL, Vaishnav P (2011) Natural products for cancer chemotherapy. Microbial Biotechnol 4:687–699

Devkota KP, Covell D, Ransom T, McMahon JB, Beutler JA (2013) Growth inhibition of human colon carcinoma cells by sesquiterpenoids and tetralones of *Zygogynum calothyrsum*. J Nat Prod 76:710–714

Ding G, Song YC, Chen JR, Xu C, Ge HM, Wang XT, Tan RX (2006) Chaetoglobosin U, a cytochalasan alkaloid from endophytic *Chaetomium globosum* IFB-E019. J Nat Prod 69:302–304

Ding G, Zheng Z, Liu S, Zhang H, Guo L, Che Y (2009) Photinides A–F, cytotoxic benzofuranone-derived γ-Lactones from the plant endophytic fungus *Pestalotiopsis photiniae*. J Nat Prod 72:942–945

Dou H, Song YX, Liu XQ, Gong W, Li EG, Tan RX, Hou YY (2011) Chaetoglobosin Fex from the marine-derived endophytic fungus inhibits induction of inflammatory mediators via toll-like receptor 4 signaling in macrophages. Biol Pharm Bull 34:1864–1873

Donoso R, Rivera-sagredo A, Hueso-rodriguez JA, Elson SW (1997) A new chaetoglobosin isolated from a fungus of the genus *Discosia*. Nat Prod Lett 10:49–54

Du L, Zhu T, Liu H, Fang Y, Zhu W, Gu Q (2008) Cytotoxic polyketides from a marine-derived fungus *Aspergillus glaucus*. J Nat Prod 71:1837–1842

Erkel G, Lorenzen K, Anke T, Velten R, Gimenez A, Steglich W (1995) Kuehneromycins A and B, two new biological active compounds from a *Tasmanian kuehneromyces* sp. (Strophariaceae. Basidiomycetes Z Naturforsch C Biosci 50:1–10

El-Elimat T, Figueroa M, Raja HA, Graf TN, Swanson SM, Falkinham JO, Wani MC, Pearce CJ, Oberlies NH (2015) Biosynthetically distinct cytotoxic polyketides from *Setophoma terrestris*. Eur J Org Chem 2015:109–2021

Evidente A, Kornienko A, Cimmino A, Andolfi A, Lefranc F, Mathieu V, Kiss R (2014) Fungal metabolites with anticancer activity. Nat Prod Rep 31:617–627

Fang LZ, Qing C, Shao HJ, Yang YD, Dong ZJ, Wang F, Zhao W, Yang WQ, Liu JK (2006) Hypocrellin D, a cytotoxic fungal pigment from fruiting bodies of the ascomycete *Shiraia bambusicola*. J Antibiot 59:351–354

Fleming A (1929) On the antibacterial action of cultures of Penicillium, with special reference to their use in the isolation of *B. influenzae*. Br J Exp Pathol 10:226–236

Gao H, Zhou L, Li D, Gu Q, Zhu TJ (2013) New cytotoxic metabolites from the marine-derived fungus *Penicillium* sp. ZLN29. Helv Chim Acta 96:514–519

Garlaschelli L, Mellerio G, Vidari G, Vita-Finzi P (1994) New fatty acid esters of drimane sesquiterpenes from *Lactarius uvidus*. J Nat Prod 57:905–910

Grabley S, Thiericke R, Zerlin M, GÖHRT A, Philipps S, Zeeck A (1996) New albrassitriols from *Aspergillus* sp. (FH-A 6357). J Antibiot 49:593–595

Hayes MA, Wrigley SK, Chetland I, Reynolds EE, Ainsworth AM, Renno DV, Latif MA, Cheng XM, Hupe DJ, Charlton P, Doherty AM (1996) Novel drimane sesquiterpene esters from *Aspergillus ustus var. pseudodeflectus* with endothelin receptor binding activity. J Antibiot 49:505–512

Hammerschmidt L, Debbab A, Ngoc TD, Wray V, Hemphil CP, Lin W, Broetz-Oesterhelt H, Kassack MU, Proksch P, Aly AH (2014) Polyketides from the mangrove-derived endophytic fungus *Acremonium strictum*. Tetrahedron Lett 55:3463–3468

He H, Yang HY, Bigelis R, Solum EH, Greenstein M, Carter GT (2002) Pyrrocidines A and B, new antibiotics produced by a filamentous fungus. Tetrahedron Lett 43(9):1633–1636

Hewage RT, Aree T, Mahidol C, Ruchirawat S, Kittakoop P (2014) One strain-many compounds (OSMAC) method for production of polyketides, azaphilones, and an isochromanone using the endophytic fungus *Dothideomycete* sp. Phytochemistry 108:87–94

Hibbett DS, Ohman A, Glotzer D, Nuhn M, Kirk P, Nilsson RH (2011) Progress in molecular and morphological taxon discovery in *Fungi* and options for formal classification of environmental sequences. Fungal Biol Rev 25:35–47

Hussain H, Al-Sadi AM, Schulz B, Steinert M, Khan A, Green IR, Ahmed I (2017) A fruitful decade for fungal polyketides from 2007 to 2016: antimicrobial activity, chemotaxonomy and chemodiversity. Fut Med Chem 9:1631–1648

Hu Y, Hao X, Chen L, Akhberdi O, Yu X, Liu Y, Zhu X (2018) Gα-cAMP/PKA pathway positively regulates pigmentation, chaetoglobosin A biosynthesis and sexual development in *Chaetomium globosum*. PLoS ONE 13:e0195553

Ichihara A, Katayama K, Teshima H, Oikawa H, Sakamura S (1996) Chaetoglobosin O and other phytotoxic metabolites from *Cylindrocladium floridanum*, a causal fungus of Alfalfa black rot disease. Biosci Biotechnol Biochem 60:360–361

Igarashi Y, Miura SS, Fujita T, Furumai T (2006) Pterocidin, a Cytotoxic Compound from the Endophytic *Streptomyces hygroscopicus*. J Antibiot 59:193–195

Iwamoto C, Yamada T, Ito Y, Minoura K, Numata A (2001) Cytotoxic cytochalasans from a Penicillium species separated from a marine alga. Tetrahedron 57:2997–3004

Isaka M, Yangchum A, Intamas S, Kocharin K, Jones EG, Kongsaeree P, Prabpai S (2009) Aigialomycins and related polyketide metabolites from the mangrove fungus *Aigialus parvus* BCC 5311. Tetrahedron 65:4396–4403

Jiao W, Feng Y, Blunt JW, Cole AL, Munro MH (2004) Chaetoglobosins Q, R, and T, three further new metabolites from *Chaetomiumg lobosum*. J Nat Prod 67:1722–1725

Kasai Y, Komatsu K, Shigemori H, Tsuda M, Mikami Y, Kobayashi JI (2005) Cladionol A, a polyketide glycoside from marine-derived fungus *Gliocladium* species. J Nat Prod 68:777–779

Kikuchi H, Isobe M, Kurata S, Katou Y, Oshima Y (2012) New dimeric and monomeric chromanones, gonytolides D-G, isolated from the fungus Gonytrichum sp. Tetrahedron 68:6218–6223

Kralj A, Kehraus S, Krick A, Eguereva E, Kelter G, Maurer M, Wortmann A, Fiebig HH, Koenig GM (2006) Arugosins G and H: Prenylated polyketides from the marine-derived fungus *Emericella nidulans* var. *acristata*. J Nat Prod 69:995–1000

Kohno J, Nishio M, Sakurai M, Kawano K, Hiramatsu H, Kameda N, Kishi N, Yamashita T, Okuda T, Komatsubara S (1999) Isolation and structure determination of TMC-151s: novel polyketide antibiotics from *Gliocladium catenulatum* Gilman & Abbott TC 1280. Tetrahedron 55:7771–7786

Kono K, Tanaka M, Ogita T, Hosoya T, Kohama T (2000) F-12509A, a new sphingosine kinase inhibitor, produced by a discomycete. J Antibiot 53:459–466

Kioy D, Gray AI, Waterman PG (1990a) A comparative study of the stem-bark drimane sesquiterpenes and leaf volatile oils of *Warburgia ugandensis* and *W. stuhlmannii*. Phytochemistry 29:3535–3538

Kioy D, Gray AI, Waterman PG (1990b) 3β, 9α-Dihydroxycinnamolide: A further novel drimane sesquiterpene from the stem bark of *Canella winterana*. J Nat Prod 53:1372–1373

Kumaran RS, Kim HJ, Hur B-K (2010) Taxol promising fungal endophyte, *Pestalotiopsis* spp. from *Taxus cuspidata*. J Biosci Bioeng 110:541–546

Lee YM, Li H, Hong J, Cho HY, Bae KS, Kim MA, Kim DK, Jung JH (2010) Bioactive metabolites from the sponge-derived fungus *Aspergillus* versicolor. Arch Pharm Res 33:231–235

Li H, Xiao J, Gao YQ, Tang JJ, Zhang AL, Gao JM (2014) Chaetoglobosins from *Chaetomium globosum*, an endophytic fungus in *Ginkgo biloba*, and their phytotoxic and cytotoxic activities. J Agric Food Chem 62:3734–3741

Li D, Chen L, Zhu T, Kurtán T, Mándi A, Zhao Z, Li J, Gu Q (2011) Chloctanspirones A and B, novel chlorinated polyketides with an unprecedented skeleton, from marine sediment derived fungus *Penicillium terrestre*. Tetrahedron 67:7913–7918

Lin X, Huang Y, Fang M, Wang J, Zheng Z, Su W (2005) Cytotoxic and antimicrobial metabolites from marine lignicolous fungi, *Diaporthe* sp. FEMS Microbiol Lett 251:53–58

Lin Z, Zhu T, Fang Y, Gu Q, Zhu W (2008) Polyketides from Penicillium sp. JP-1, an endophytic fungus associated with the mangrove plant *Aegiceras corniculatum*. Phytochemistry 69:1273–1278

Liao WY, Shen CN, Lin LH, Yang YL, Han HY, Chen JW, Kuo SC, Wu SH, Liaw CC (2012) Asperjinone, a nor-neolignan, and terrein, a suppressor of ABCG2-expressing breast cancer cells, from thermophilic *Aspergillus terreus*. J Nat Prod 75:630–635

Lorenzen K, Anke T, Anders U, Hindermayr H, Hansske F (1994) Two inhibitors of platelet aggregation from a *Panus* species (Basidiomycetes). Z Naturforsch C Biosci 49:132–138

Mahmoud II, Kinghorn AD, Cordell GA, Farnsworth NR (1979) Cytotoxic drimane sesquiterpenoids from *Capsicodendron dinisii*. J Nat Prod 42(6):681

Masters K, Braese S (2012) Xanthones from fungi, lichens, and bacteria: The natural products and their synthesis. Chem Rev 112:3717–3776

Mohamed IE, Gross H, Pontius A, Kehraus S, Krick A, Kelter G, Maier A, Fiebig HH, König GM (2009) Epoxyphomalin A and B, prenylated polyketides with potent cytotoxicity from the marine-derived fungus *Phoma* sp. Org Lett 11:5014–5017

Mulrooney CA, O'Brien EM, Morgan BJ, Kozlowski MC (2012) Perylenequinones: isolation, synthesis, and biological activity. Eur J Org Chem 2012:3887–3904

McCorkindale NJ, Calzadilla CH, Hutchinson SA, Kitson DH, Ferguson G, Campbell IM (1981) The structure and chemistry of pebrolide, desacetylpebrolide and 1-deoxypebrolide, sesquiterpene benzoates from *Penicillium brevicompactum*. Tetrahedron 37(3):649–653

Mashimbye MJ, Maumela MC, Drewes SE (1999) A drimane sesquiterpenoid lactone from *Warburgia salutaris*. Phytochemistry 51:435–438

Mitova MI, Lang G, Blunt JW, Cummings NJ, Cole AL, Robinson WT, Munro MH (2006) Cladobotric Acids A–F: New cytotoxic polyketides from a New Zealand *Cladobotryum sp.* J Org Chem 71:492–497

Mohr, K. I. (2016) History of antibiotics research. In: Stadler M, Dersch P (2016) How to overcome the antibiotic crisis. Current topics in microbiology and immunology, vol 398, pp 237–272. Springer International Publishing AG

Nakai R, Ogawa H, Asai A, Ando K, Agaisuma T, Maisumiya S, Akinaga S, Yamashita Y, Mizukami T (2000) UCS1025A, a novel antibiotic produced by *Acremonium* sp. J Antibiot 53:294–296

Newman DJ, Cragg GM (2012) Natural products as sources of new drugs over the 30 years from 1981 to 2010. J Nat Prod 75:311–335

Nozawa O, Okazaki T, Sakai N, Komurasaki T, Hanada K, Morimoto S, Zx Chen, Bm He, Mizoue K (1995) A novel bioactive δ lactone FD-211. J Antibiot 48:113–118

Nozawa O, Okazaki T, Morimoto S, Chen ZX, He BM, Mizoue K (2000) Waol B, a new trihydrofuran derivative with cytocidal activity, isolated from *Myceliophthora lutea*. J Antibiot 53:1296–1300

Numafa A, Takahashi C, Ito Y, Minoura K, Yamada T, Matsuda C, Nomoto K (1996) Penochalasins, a novel class of cytotoxic cytochalasans from a *Penicillium* species separated from a marine alga: Structure determination and solution conformation. J Chem Soc Perkin Trans 1:239–245

Osterhage C, Kaminsky R, König GM, Wright AD (2000) Ascosalipyrrolidinone a, an antimicrobial alkaloid, from the obligate marine fungus *Ascochyta s alicorniae*. J Org Chem 65:6412–6417

Paul A, Thapa G, Basu A, Mazumdar P, Kalita MC, Sahoo L (2010) Rapid plant regeneration, analysis of genetic fidelity and essential aromatic oil content of micropropagated plants of Patchouli, *Pogostemon cablin* (Blanco) Benth.—An industrially important aromatic plant. Ind Crop Prod 32:366–374

Pongcharoen W, Rukachaisirikul V, Phongpaichit S, Rungjindamai N, Sakayaroj J (2006) Pimarane diterpene and cytochalasin Derivatives from the endophytic fungus *Eutypella scoparia* PSU-D44. J Nat Prod 69:856–858

Pulici M, Sugawara F, Koshino H, Uzawa J, Yoshida S, Lobkovsky E, Clardy J (1996) A new iso drimeninol from *Pestalotiopsis* sp. J Nat Prod 59:47–48

Pulici M, Sugawara F, Koshino H, Okada G, Esumi Y, Uzawa J, Yoshida S (1997) Metabolites of *Pestalotiopsis* spp., endophytic fungi of *Taxus brevifolia*. Phytochemistry 46(2):313–319

Prachya S, Wiyakrutta S, Sriubolmas N, Ngamrojanavanich N, Mahidol C, Ruchirawat S, Kittakoop P (2007) Cytotoxic mycoepoxydiene derivatives from an endophytic fungus Phomopsis sp. isolated from *Hydnocarpus anthelminthicus*. Planta Med 73:1418–1420

Rahbæk L, Christophersen C, Frisvad J, Bengaard HS, Larsen S, Rassing BR (1997) Insulicolide A: a new nitrobenzoyloxy-substituted sesquiterpene from the marine fungus *Aspergillus insulicola*. J Nat Prod 60:811–813

Rajab MS, Ndegwa JM (2000) 11α-Hydroxy muzigadiolide, a novel drimane sesquiterpene from the stem bark of *Warburgia ugandensis*. Bull Chem Soc Ethiop 14:45

Rastogi N, Abaul J, Goh KS, Devallois A, Philogène E, Bourgeois P (1998) Antimycobacterial activity of chemically defined natural substances from the *Caribbean flora* in Guadeloupe. FEMS Immunol Med Microbiol 20:267–273

Scherlach K, Boettger D, Remme N, Hertweck C (2010) The chemistry and biology of cytochalasans. Nat Prod Rep 27:869–886

Schümann J, Hertweck C (2007) Molecular basis of cytochalasan biosynthesis in fungi: gene cluster analysis and evidence for the involvement of a PKS-NRPS hybrid synthase by RNA silencing. J Am Chem Soc 129:9564–9565

Seeram NP, Francis LS, Needham OL, Jacobs H, McLean S, Reynolds WF (2003) Drimane and bisabolane sesquiterpenoids from *Cinnamodendron corticosum* (Canellaceae). Biochem Syst Ecol 6:637–640

Sekita S, Yoshihira K, Natori S, Kuwano H (1977) Chaetoglobosins G, and J, cytotoxic indol-3-yl [13]-cytochalasans from *Chaetomium globosum*. Tetrahedron Lett 32:2771–2774

Sekita S, Yoshihira K, Natori S, Kuwano H (1982) Chaetoglobosins, cytotoxic 10-(indol-3-yl)-[13]cytochalasans from *Chaetomium* spp. III. Sructures of chaetoglobosins C, E, F, G. J Chem Pharm Bull 30:1629–1638

Sekita S, Yoshihira K, Natori S, Kuwano H (1973) Structures of chaetoglobosin A and B, cytotoxic metabolites of *Chaetomium globosum*. Tetrahedron Lett 14:2109–2112

Sekita S, Yoshihira K, Natori S (1983) Chaetoglobosins, cytotoxic 10-(indol-3-yl)-[13] cytochalasans from *Chaetomium* spp. IV. ^{13}C-nuclear magnetic resonance spectra and their application to a biosynthetic study. Chem Pharm Bull 31:490–498

Silva CA, Madureira LA (2012) Source correlation of biomarkers in a mangrove ecosystem on Santa Catarina Island in southern Brazil. An Acad Bras Ciênc 84:589–604

Springer JP, Cox RH, Cutler HG, Crumley FG (1980) The structure of chaetoglobosin K. Tetrahedron Lett 21:1905–1908

Singh SB, Goetz MA, Jones ET, Bills GF, Giacobbe RA, Herranz L, Stevens-Miles S, Williams DL (1995) Oteromycin: a novel antagonist of endothelin receptor. J Org Chem 60:7040–7042

Staunton J, Weissman KJ (2001) Polyketide biosynthesis: a millennium review. Nat Prod Rep 18:380–416

Suzuki S, Hosoe T, Nozawa K, Kawai KI, Yaguchi T, Udagawa SI (2000) Antifungal substances against pathogenic fungi, talaroconvolutins, from *Talaromyces convolutus*. J Nat Prod 63:768–772

Tabata N, Ohyama Y, Tomoda H, Abe T, Namikoshi M, Omura S (1999) Structure Elucidation of Roselipins, Inhibitors of Diacylglycerol Acyltransferase Produced by *Gliodadium roseum* KF-1040. J Antibiot 52:815–826

Takahashi C, Yoshihira K, Natori S, Umeda M (1976) The structures of toxic metabolites of *Aspergillus candidus*. I. The compounds A and E, cytotoxic p-terphenyls. Chem Pharm Bull 24:613–620

Thohinung S, Kanokmedhakul S, Kanokmedhakul K, Kukongviriyapan V, Tusskorn O, Soytong K (2010) Cytotoxic 10-(indol-3-yl)-[13] cytochalasans from the fungus *Chaetomium elatum* ChE01. Arch. Pharm. Res. 33(8):1135–1141

Tomoda H, Ohyama Y, Abe T, Tabata N, Namikoshi M, Yamaguchi Y, Masuma R, Omura S (1999) Roselipins, inhibitors of diacylglycerol acyltransferase, produced by *Gliocladium roseum* KF-1040. J Antibiot 52:689–694

Umeda M, Ohtsubo K, Saito M, Sekita S, Yoshihira K, Natori S, Udagawa S, Sakabe F, Kurata H (1975) Cytotoxicity of new cytochalasans from *Chaetomium globosum*. Experientia 31:435–438

Uosaki Y, Yoshida M, Ogawa T, Saitoh Y (1996) RES-1149-1 and-2, novel non-peptidic endothelin type b receptor antagonists produced by *Aspergillm* sp. J Antibiot 49:6–12

Velten R, Klostermeyer D, Steffan B, Steglich W, Kuschel A, Anke T (1994) The mniopetals, new inhibitors of reverse transcriptases from a *Mniopetalum* species (*Basidiomycetes*). J Antibiot 47:1017–1024

von Wallbrunn C, Luftmann H, Bergander K, Meinhardt F (2001) Phytotoxic chaetoglobosins are produced by the plant pathogen *Calonectria morganii* (anamorph *Cylindrocladium scoparium*). J Gen Appl Microbiol 47:33–38

Wang J, Huang Y, Fang M, Zhang Y, Zheng Z, Zhao Y, Su W (2002) Brefeldin A, a cytotoxin produced by *Paecilomyces* sp. and *Aspergillus clavatus* isolated from *Taxus mairei* and *Torreya grandis*. FEMS Imunol Med Mic 34:51–57

Wang CCC, Chiang YM, Kuo PL, Chang JK, Hsu YL (2008) Norsolorinic acid inhibits proliferation of T24 human bladder cancer cells by arresting the cell cycle at the G0/G1 phase and inducing a Fas/membrane-bound Fas ligand-mediated apoptotic pathway. Clini Exp Pharmacol Physiol 35:1301–1308

Weissman KJ, Leadlay PF (2005) Combinatorial biosynthesis of reduced polyketides. Nat Rev Microbiol 3:925–936

Wezeman T, Bräse S, Masters KS (2015) Xanthone dimers: a compound family which is both common and privileged. Nat Prod Rep 32:6–28

Wu G, Yu G, Kurtán T, Mándi A, Peng J, Mo X, Liu M, Li H, Sun X, Li J, Zhu T, Gu Q, Li D (2015) Versixanthones A–F, cytotoxic xanthone–chromanone dimers from the marine-derived fungus *Aspergillus versicolor* HDN1009. J Nat Prod 78:2691–2698

Yang YL, Lu CP, Chen MY, Chen KY, Wu YC, Wu SH (2007) Cytotoxic polyketides containing tetramic acid moieties isolated from the fungus *Myceliophthora thermophila*: elucidation of the relationship between cytotoxicity and stereoconfiguration. Chem Eur J 13:6985–6991

Yao Q, Wang J, Zhang X, Nong X, Xu X, Qi S (2014) Cytotoxic polyketides from the deep-sea-derived fungus *Engyodontium album* DFFSCS021. Mar Drugs 12:5902–5915

Ying BP, Peiser GD, Ji YY, Mathias KM, Karasina F, Hwang YS (1995a) Structure-activity relationships of phytotoxic sesquiterpenoids from *Canella winterana*. J Agric Food Chem 43:826–829

Ying BP, Peiser G, Ji YY, Mathias K, Tutko D, Hwang YS (1995b) Phytotoxic sesquiterpenoids from *Canella winterana*. Phytochemistry 38:909–915

Zhang Y, Tian R, Liu S, Chen X, Liu X, Che Y (2008a) Alachalasins A-G, new cytochalasins from the fungus Stachybotrys charatum. Bioorg Med Chem 16:2627–2634

Zhang W, Krohn K, Flörke U, Pescitelli G, Di Bari L, Antus S, Kurtán T, Rheinheimer J, Draeger S, Schulz B (2008b) New mono-and dimeric members of the Secalonic acid family: Blennolides A–G isolated from the fungus *Blennoria* sp. Chem Eur J 14:4913–4923

Zhang J, Ge HM, Jiao RH, Li J, Peng H, Wang YR, Wu JH, Song YC, Tan RX (2010) Cytotoxic chaetoglobosins from the endophyte *Chaetomium globosum*. Planta Med 76:1910–1914

Zhang Q, Li HQ, Zong SC, Gao JM, Zhang AL (2012) Chemical and bioactive diversities of the genus *Chaetomium* secondary metabolites. Mini-Rev Med Chem 12:127–148

Zhang D, Ge H, Xie D, Chen R, Zou JH, Tao X, Dai J (2013a) Periconiasins A-C, new cytotoxic cytochalasans with an unprecedented 9/6/5 tricyclic ring system from endophytic fungus Periconia sp. Org Lett 15:1674–1677

Zhang GZ, Wang FT, Qin JC, Wang D, Zhang JY, Zhang YH, Zhang SH, Pan HY (2013b) Efficacy assessment of antifungal metabolites from *Chaetomium globosum* No. 05, a new biocontrol agent, against *Setosphaeria turcica*. Biol Control 64:90–98

Zheng QC, Kong MZ, Zhao Q, Chen GD, Tian HY, Li XX, Guo LD, Li J, Zheng YZ, Gao H (2014) Chaetoglobosin Y, a new cytochalasan from *Chaetomium globosum*. *Fitoterapia* 93:126–131

Chapter 10
The Bioactivity and Chemotaxonomy of Microalgal Carotenoids

Dónal Mc Gee and Eoin Gillespie

Abstract Microalgae are a diverse group of photosynthetic microorganisms which inhabit a wide variety of freshwater and marine environments. They play a key role in the Earth's biogeochemical cycles and hold great potential for their application in the fields of environmental remediation, biotechnology and nanotechnology. The bioactivity of carotenoids warrants their applications as nutraceuticals, cosmaceuticals or biopharmaceuticals in the treatment and prevention of chronic and age-related diseases. Their photosynthetic pigment signatures are typically taxon-specific, facilitating their application as chemotaxonomic biomarkers and provide a complementary approach to the morphogenetic taxonomy in the characterisation of new strains. This chapter provides a brief overview of carotenoids structural diversity, their bioactivity and biosynthesis. In addition, it aims to introduce the endosymbiotic theory of protist evolution which gave rise to 11 algal divisions. At present, there are 27 reported classes of photosynthetic algae representing 44 pigment types based on the distribution of their chlorophyll and carotenoid pigments.

Keywords Microalgae · Phylogeny · Chemotaxonomy · Pigment types · Carotenoids · Antioxidants · Nutraceuticals · Biopharmaceuticals

10.1 Introduction

Algae are a broad term for organisms other than higher plants that obtain energy from autotrophic photosynthesis assimilating carbon through the ribulose bisphosphate pathway (Clavin cycle). Algae are classified into divisions (phylogenetic affix: *-phyta*) based on their cellular and biochemical features such as; pigment combinations, storage products, cell wall structure (silica frustule or theca), flagella

D. Mc Gee (✉) · E. Gillespie
Department of Environmental Science, School of Science, CERIS, Centre for Environmental Research, Innovation and Sustainability, Institute of Technology Sligo, Sligo, Ireland
e-mail: dmacaoidh01@gmail.com

K. G. Ramawat (ed.), *Biodiversity and Chemotaxonomy*, Sustainable Development and Biodiversity 24, https://doi.org/10.1007/978-3-030-30746-2_10

structure and other specialised cellular structures (Tomas 1997; Jeffrey et al. 2011). As a result, they encompass an enormous diversity of organism ranging from unicellular prokaryotes (cyanobacteria and prochlorophytes) and eukaryotic protists to multicellular macroalgae (seaweeds) (Metting 1996). There are 80,000–100,000 known algae species of which currently over 200 microalgae species are culture worldwide for different sectors ranging from aquaculture, biotechnology, nanotechnology and bioremediation services (Enzing et al. 2014).

The diversity of unicellular microalgae provides an ideal source for bioprospecting novel metabolites. They have a wide distribution of habitats including; terrestrial, freshwater and marine, some are extremophiles, while others form symbiotic relationships with fungi, tropical corals and plants (Gupta et al. 2013). Their ability to adapt stressful and competitive environments is in part owed to their photosynthetic capabilities including the biosynthesis of antioxidant carotenoids with novel biopharmaceutical applications. *Chlorella* and *Arthrospira* species are commonly cultured for the production of vitamins and as health supplements. Pigments from algae have been shown to act as powerful antioxidants and have been routinely applied to food, cosmetic and pharmaceutical products. These include phycocyanin (*Spirulina*), phycoerythrin (Rhodophyta spp.), astaxanthin (*Haematococcus pluvialis*) and fucoxanthin (Phaeophyceae spp.) as well as β-carotene (*Dunaliella* spp.) (Cuellar-Bermudez et al. 2015; Chokshi et al. 2017). The following sections explore the microalgae chemotaxonomy and their bioactive carotenoid pigments.

10.2 Carotenoid Structure and Function

At present, there are over 750 carotenoids reported in nature, of which 30 have a functional role in the photosynthetic light-harvesting complexes (Takaichi 2011). Carotenoids are tetraterpenoid pigments derived from eight isoprene units, terminating in cyclic β-ionone rings (Britton et al. 2004). The central C_{40} conjugated double-bond system of the carotenoid polyene chain acts as a "chromophore" absorbing blue light (400–500 nm) for photosynthesis and giving them their characteristic yellow, orange or red colour. Carotenoids are either classified as carotenes comprising solely of hydrocarbons or oxygenate xanthophylls (Britton et al. 2004). Oxygenated functional groups present at terminal β-ionone rings of xanthophylls act to protect the cell from oxidative damage by quenching single oxygen species (Namitha and Negi 2010). Additional functional groups include allene (C=C=C), acetylene (C≡C) or acetylated ($–O–CO–CH_3$) (Fig. 10.1). The allene moiety is present in fucoxanthin associated with the Chromophyceae algae, peridinin associated with dinoflagellates and 9′-*cis*-neoxanthin predominately associated with the green algal lineage. Xanthophylls containing acetylene functional groups (C≡C) are unique to algae, which to date have been found in the Cryptophyta specific carotenoids alloxanthin, crocoxanthin and monadoxanthin in addition to the Chromophyceae xanthophyll cycle pigments diadinoxanthin and

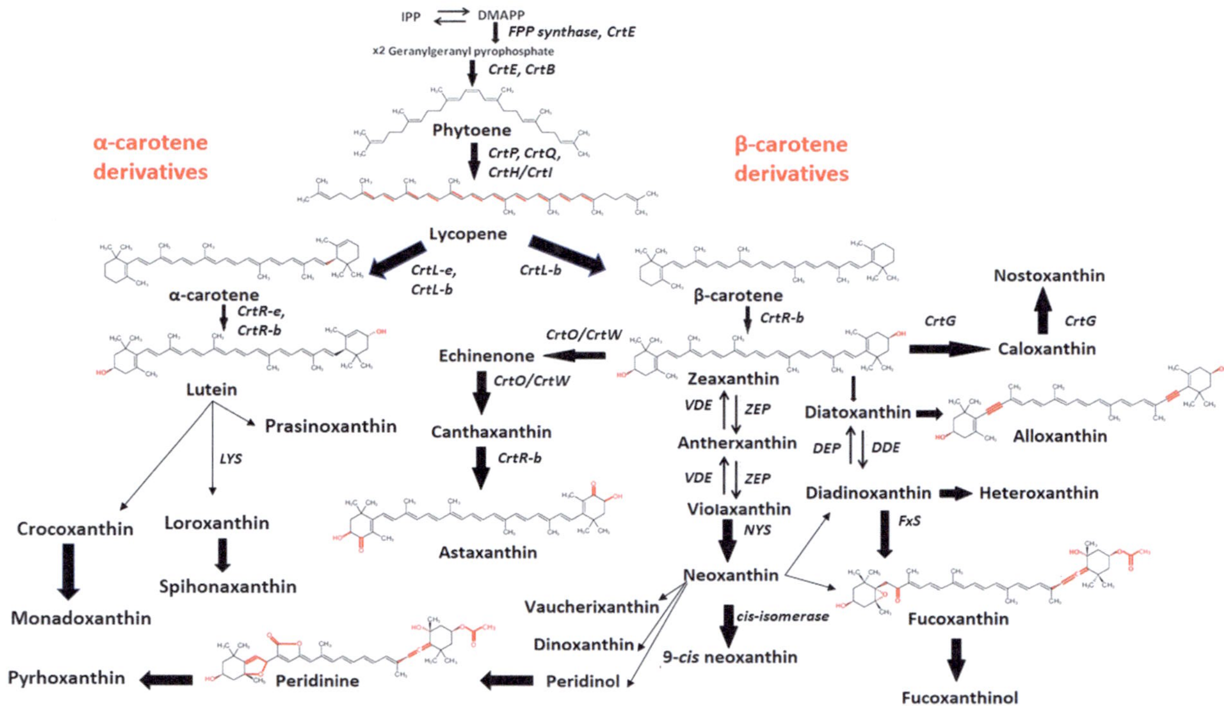

Fig. 10.1 Biosynthetic pathway of chemotaxonomically relevant and high-value carotenoids in microalgae. IPP (isopentenyl diphosphate), FPP synthase (farnesyl pyrophosphate), DMAPP (dimethylallyl diphosphate), CrtE (geranylgeranyl diphosphate synthase), CrtB (phytoene synthase), CrtP/CrtI (phytoene desaturase), CrtQ (ζ-carotene desaturase), CrtL (lycopene β-cyclase), CrtH (*cis*-to-*trans* carotene isomerase), CrtR (β-carotene hydroxylase), CrtG (2,2′-β hydroxylase), CrtO (β-carotene monoketolase), CrtW (β-carotene diketolase), VDE (violaxanthin de-epoxidase), ZEP (zeaxanthin epoxidase), DEP (diatoxanthin epoxidase), DDE (diadinoxanthin de-epoxidase), LYS (loroxanthin synthase) and FxS (fucoxanthin synthase)

diatoxanthin. Fucoxanthin, peridinin and dinoxanthin contain a novel acetylated functional group (–O–CO–CH_3) which may contribute to the bioactivity and light-harvesting capabilities of these carotenoids.

10.3 Bioactivity of Carotenoids

The bioactive potential of microalgal pigments has led to a strong market demand for their application as natural colouring agents and functional foods (Ambati et al. 2018). The global carotenoid market in 2016 had an estimated value of US$124 billion, which is projected to reach US$1.53 billion by 2021 (Markets and Markets 2016). Carotenoids commonly used as colouring agents include astaxanthin, canthaxanthin, violaxanthin, lutein, zeaxanthin and β-carotene. In vitro, in vivo and epidemiological studies have shown that consumption of the carotenoids lycopene, β-carotene, lutein, zeaxanthin, violaxanthin, astaxanthin, canthaxanthin and fucoxanthin is directly associated with the prevention of tumour progression in a range of cancers including prostate, colon, breast, lung and mammary (Chew et al. 1996; Park et al. 1998; Kotake-Nara et al. 2001; Nishino et al. 2002; Granado et al. 2003; Guerin et al. 2003; Pasquet et al. 2011; Baudelet et al. 2013; Bertram 2018). However, evidence suggests that synthetic carotenoids and the consumers lifestyle can have conflicting and detrimental health effects including raising the risk of developing cancer (Heinonen et al. 1994).

Beta-carotene is a naturally occurring primary photosynthetic carotenoid present in the reaction centre of photosystem II, where it functions to shuttle excitation energy to chlorophyll and scavenging reactive oxygen species (Telfer 2002). It is commercially marketed as the natural orange colourant E160a and is viewed as a nutraceutical due to its provitamin A and antioxidant activity (Murthy et al. 2005). Commercial production of β-carotene is predominately synthesised chemically in its *trans*-isomeric configuration. The 9′-*cis*-β-carotene isomer biosynthesised from the halotolerant microalgae *Dunaliella salina* is more favourable due to its enhanced bioavailability, antioxidant and anti-tumour activity than its *trans*-isomer (Demmig-Adams and Adams 2002; Gomez and Gonzalez 2005; García-González et al. 2005; Prieto et al. 2011). As a result, there is a strong market demand from natural 9′-*cis*-β-carotene, which is expected to reach US$334 million by 2018 (BCC Research 2015).

Similarly, the lutein supplement market is experiencing an annual growth rate of 3.6%, which is anticipated to reach US$309 million by 2018 (BCC Research 2015). Lutein is favoured over its geometric isomer zeaxanthin due to its higher percentage distribution in human serum (Alves-Rodrigues and Shao 2004). Epidemiological evidence suggests that a diet rich in lutein and zeaxanthin reduces the risk of developing cataracts and age-related macular degeneration (Moeller et al. 2008). This is a consequence of their accumulation within the macula and lens of the eye, where they act to filter out short-wavelength light and quench reactive oxygen species (Seddon et al. 1994; Krinsky et al. 2003). In addition to their function as macular pigments, lutein and zeaxanthin are also believed to play a critical role in the prevention of lung cancer and in protecting the skin from damaging UV-B radiation (Chew et al. 1996; Park et al. 1998; Nishino et al. 2002; Granado et al. 2003).

Traditionally, lutein and zeaxanthin have been extracted from the marigold flower *Tagetes erecta* as an oleoresin marketed under E161b (Delgado-Vargas et al. 2000; Krinsky et al. 2003). Microalgae such as *Murellopsis* and *Scenedesmus* spp. offer a more sustainable source of lutein due to their fast growth rates and production capacity, which is tenfold higher than marigold plants (Sanchez et al. 2008a, b).

Astaxanthin is a secondary keto-carotenoid which accumulates as lipid vesicles in some *Chlorophyceae* microalgae as a survival strategy in response to unfavourable conditions (Table 10.1). The antioxidant capacity of astaxanthin has been reported to be 10-fold higher than that of β-carotene (Guerin et al. 2003). The esterification of astaxanthin in fatty acids protects lipids from reactive oxygen species (ROS)-induced peroxidation and enhances its bioavailability (Saw et al. 2013; Zuluaga et al. 2018). It is widely applied as a colouring agent for the characteristic pink/orange colour in farmed salmon and rainbow trout (Choubert et al. 2006; Chitchumroonchokchai and Failla 2017). Commercial astaxanthin production is predominated by synthetic racemic mixture of *3S, 3′S*; *3R, 3′S* and *3R, 3′ R* stereoisomers and it is marketed at a value of over US$240 million per annum (Han et al. 2013).

Naturally, derived astaxanthin is preferred due to increasing concerns over the safety and consumer demand for natural products. Currently, natural astaxanthin is obtained from *Haematococcus pluvalis* under controlled stress conditions, where it undergoes transition from green vegetative cells to red palmelloid cysts, accumulating up to 2–5% dry weight astaxanthin as a single chiral 3S, 3′S stereoisomer.

Table 10.1 Astaxanthin accumulating microalgae strains

Strain	Astaxanthin content[a]	References
Haematococcus pluvalis	4–5%	Han et al. (2013)
Botryococcus braunii	3–8%	Rao et al. (2010)
Chlamydocapsa spp.	0.04%	Leya et al. (2009)
Chlorococcum spp.	0.7%	Ma and Chen (2001)
Chlorella zofingiensis	0.7%	Orosa et al. (2000)
Chlamydomonas nivalis	0.004%	Leya et al. (2009)
Neochloris wimmeri	1.9%	Orosa et al. (2000)
Protosiphon botryoides	1.4%	Orosa et al. (2000)
Scenedesmus spp.	0.3%	Qin et al. (2008)
Scotiellopsis oocystiformis	1.1%	Orosa et al. (2000)
Tovellia sanguinea	7 pg/cell	Frassanito et al. (2006)
Euglena sanguinea	0.53%	Grung and Liaaen-Jensen (1993)

[a]% DW astaxanthin

However, production of astaxanthin from *Haematococcus pluvalis* is costly due to its fastidious growth rates and difficulties with the extraction of the carotenoid from encysted cells (Hata et al. 2001). In order to remain competitive with synthetic market, alternative strains, cultivation conditions and biorefinery technologies are sought for economical and sustainable astaxanthin production (Hanagata and Dubinsky 1999; Orosa et al. 2001; Ahmed et al. 2014; Mao et al. 2018).

Fucoxanthin is an abundant pigment in nature responsible for the golden-brown colour in *Heterokontophyta* and *Haptophyta* (Jeffrey et al. 2011). Among the carotenoids, fucoxanthin exhibits pronounced antioxidant and anti-inflammatory activity when compare to lycopene, β-carotene, lutein, zeaxanthin, β-cryptoxanthin and astaxanthin (Nomura et al. 1997; Ishikawa et al. 2008). Its potent bioactivity is attributed to the presence of an unusual allelic bond and numerous functional moieties (hydroxyl, epoxy, carbonyl and carboxyl) (Nomura et al. 1997; Hosokawa et al. 1999; Das et al. 2008; Nakazawa et al. 2009; Woo et al. 2010; Kim et al. 2010). It holds considerable promise as a nutraceutical and medical adjuvant for the treatment of diabetes, obesity, malaria and cancer (Afolayan et al. 2008; Woo et al. 2009, 2010; Hosokawa et al. 2010; Park et al. 2011; Wang et al. 2012; Rengarajan et al. 2013).

Abidov et al. (2010) carried out a 16-week randomised double-blind study on the effect of Xanthigen™ (brown marine algae fucoxanthin and pomegranate seed oil) on body weight, body fat, liver lipids and blood biochemistry on 151 non-diabetic, obese premenopausal women with non-alcoholic fatty liver disease. The study found that Xanthigen™ reduced body and liver fat content and improved liver function tests.

Fucoxanthin induces anti-proliferative and apoptotic effects on tumours through a range of molecular mechanisms including the stimulation of gap-junction cellular communication, immunomodulation of the JAK/STAT pathway, activation of capases-3, ROS generation, cell cycle arrest and anti-angiogenic effects (Zhang et al. 1991; Kim et al. 2010; Wang et al. 2012; Rengarajan et al. 2013). Its non-toxic pharmacokinetics warrants its application to alleviate the side effects of chemotherapy (Iio et al. 2011a, b). However, full-scale clinical trials are required to prove its safety and efficiency against a range of cancer therapies.

The highest reported fucoxanthin containing microalgae includes *Mallomonas* sp. SBV13, *Odentella aurita*, *Isochrysis galbana* and *Phaeodactylum tricornutum* which are capable of biosynthesising up to 26.6, 18.5, 18.2 and 8.6 mg g^{-1} DW fucoxanthin, respectively (Kim et al. 2012; Xia et al. 2013; Gómez-Loredo et al. 2016; Petrushkina et al. 2017). Current commercial sources of fucoxanthin are derived from macroalgae. However, the fucoxanthin content in microalgae is significantly higher than that reported for macroalgae (Xiao et al. 2012; Jaswir et al. 2013; Schmid et al. 2017). Cultivation of *Phaeodactylum tricornutum* under low light conditions with enriched nitrate medium led to significant increase in fucoxanthin content from 9.9 ± 4.2 mg g^{-1} to 59.2 ± 22.8 mg g^{-1} (McClure et al. 2018).

Other high-value carotenoids associated with brown algae (Chromophyceae) include violaxanthin and canthaxanthin which are commonly used as food colouring additive (E161g). Violaxanthin is a primary light-harvesting carotenoid and is the major xanthophyll pigment associated with Eustigmatophyceae (Owens et al. 2004). Violaxanthin isolated by bioassay-guided fractionation possessed strong anti-proliferative activity against human mammary cancer cells (Pasquet et al. 2011). *Nannochloropsis gaditana* is also capable of biosynthesising the high-value keto-carotenoids canthaxanthin and astaxanthin representing 0.7% of dry weight biomass (Lubián et al. 2000).

Another source of microalgae derived natural colouring agents are phycobiliproteins. These water-soluble fluorescent pigments function as accessory or antenna pigments to harvest light for photosynthesis in *Cyanobacteria*, *Cryptophyta* and *Rhodophyta* spp. (Sekar and Chandramohan 2008). They are linear tetrapyrrole chromophores (bilins) covalently bound to apoproteins via thioether links to the cysteine residues forming phycobiliproteins (Stadnichuk et al. 2015).

C-phycocyanin is a blue fluorescent pigment protein with a range of applications ranging from natural colouring in food and drink, fluorescent marker in molecular diagnostic or as a nutraceutical due to its antioxidant, anti-inflammatory, hepatoprotective and cholesterol-lowering effects (Eriksen 2008). Similarly, phycoerythrin is a fluorescent phycobiliprotein which possesses anti-inflammatory activity and has applications in anticancer photodynamic therapy as well as in the field of immunodiagnostics (Oi et al. 1982; Cian et al. 2012; Cai et al. 2014). Due to its novel properties, patent applications have been filled for its application in immunoassays and flow cytometry (Chiueh 2003; Chiueh et al. 2003).

10.4 Carotenoid Biosynthesis

Carotenogenesis in microalgae is undertaken via either the mevalonate (MVA) pathway or the 1-deoxy-D-xylulose-5-phosphate (DOXP) pathway initiating from the precursor isopentenyl pyrophosphate (IPP). IPP is enzymatically condensed and elongated to geranylgeranyl diphosphate (GGDP) farnesyl pyrophosphate via geranylgeranyl diphosphate synthase. The head–head condensation of two GGDP by phytoene synthase yields the C_{40} molecule, phytoene. Subsequent desaturation of phytoene forms the red coloured carotenoid, lycopene. This, in turn, acts as the template for synthesis of the dicyclic carotenoids, α-carotene and β-carotene, and their oxygenated xanthophylls derivatives (Takaichi 2011; Huang et al. 2017). Alpha- and beta-carotene undergo hydroxylation via β-carotene hydroxylase forming the xanthophylls lutein and zeaxanthin, respectively (Fig. 10.1). While the epoxidation of zeaxanthin at position 5,6 or 5′,6′ yields antheraxanthin and violaxanthin, respectively. Violaxanthin is converted to neoxanthin by the rearrangement of an ionone ring in produces two allenic double bonds. Neoxanthin subsequently acts as a precursor for the synthesis of acetylenic xanthophylls vaucheriaxanthin, peridinin, fucoxanthin and its 19′-acyloxy derivatives, diadinoxanthin, heteroxanthin, diatoxanthin and alloxanthin (Fig. 10.1). Lutein acts as a precursor for the acetylenic xanthophylls crocoxanthin, monadoxanthin and loroxanthin (Bhosale and Bernstein 2005; Takaichi 2011).

Peridinin and fucoxanthin are unique in that they contain an additional conjugated carbonyl functional group (Liaaen-Jensen 1979). In the case of lutein, the addition of carbonyl functional groups yields prasinoxanthin and siphonaxanthin. The addition of endo-cyclic keto functional groups to β-carotene yields the secondary carotenoids echinenone, canthaxanthin and astaxanthin (Bhosale and Bernstein 2005; Takaichi 2011).

10.5 Analysis of Microalgae Carotenoids

At present, high-performance liquid chromatography (HPLC) represents the gold standard analytical technique for the separation and quantification of photosynthetic pigments in microalgal cultures and phytoplankton community samples. Since the 1980s, various intercalibration initiatives carried out by NASA and the SORC/UNSECO have paved the way towards developing standardised protocols for the analysis of phytoplankton pigments (Wright et al. 1991; Claustre et al. 2004). Advances in HPLC methodologies for the analysis of different chlorophyll and carotenoid derivatives have been extensively reviewed by Garrido et al. (2011). Developments in the field include the application of mobile phase modifier (Zapata et al. 2000), a C_{16}-amide column (Jayaraman et al. 2011) or a pentafluorophenyloctadecyl silica column (Sanz et al. 2015) for the enhanced resolution of chemotaxonomically important monovinyl and divinyl chlorophyll pigment pairs in *Prochlorococcus* spp. High-throughput analysis has been achieved through the application of a monolithic column (Mc Gee et al. 2017) or using UHPLC systems (Suzuki et al. 2015).

In order to gain reliable pigment data, the SCOR working group set out guidelines for the "Minimum identification criteria for phytoplankton pigments". These include the use of standards, biological reference material and the application of LC–MS to accurately identify chlorophylls and carotenoids present in extracts (Airs and Garrido 2011; Egeland 2011). The application of HPLC to the analyses of phytoplankton pigments has rapidly expanded our knowledge of pigment chemodiversity providing insights into species-specific chemotaxonomic biomarkers or class-specific pigments types (Zapata et al. 2004; Laza-Martinez et al. 2007; Paliwal et al. 2016; Serive et al. 2017). Various multivariant statistical analyses can be applied to the HPLC derived pigment matrices in order to determine phytoplankton biodiversity, estimate abundances and to characterise new strains (Mackey et al. 1996; Laza-Martinez et al. 2007; Paliwal et al. 2016; Serive et al. 2017; Mc Gee et al. 2017).

10.6 Chemotaxonomy

Current phylogenetic reconstructions propose that eukaryotic microalgae lineages have evolved through a series of endosymbiotic events (Fig. 10.2). A primary endosymbiotic event involving the phagocytosis of a primitive cyanobacterium by a heterotrophic eukaryote host gave rise to the Archaeplastida lineage, comprising the divisions Glaucophyta, Rhodophyta and Chlorophyta (Keeling 2013). Subsequent multiple secondary and tertiary eukaryotic–eukaryotic endosymbiotic events led to the diversification of microalgae plastids and the evolution of the hacrobians, SAR supergroup, Excavates lineages (Keeling 2013).

The unique distribution of photosynthetic plastids reflects the evolutionary history of the different photosynthetic microalgal lineages and is largely supported

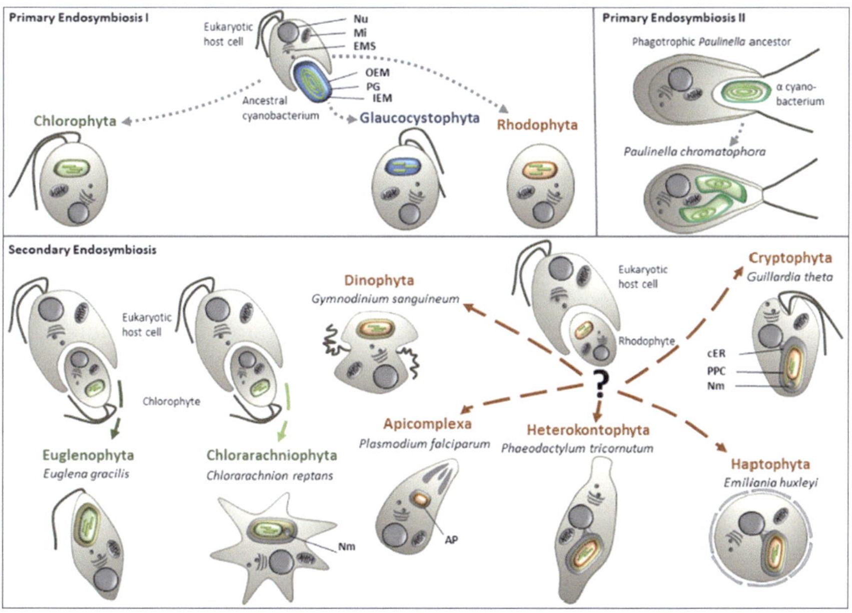

Fig. 10.2 Schematic of microalgae plastic evolution through endosymbiosis. Primary plastids originated from the phagocytotic engulfing of a cyanobacterium, resulting in two-membrane plastids in the Archaeplastida lineage (glaucophytes, red algae and green algae). The glaucocystophytes retained the peptidoglycan, which was lost in the other groups. An additional independent primary endosymbiotic event occurred between a α-cyanobacterial and *Paulinella chromatophora*. At least three independent eukaryote-eukaryote secondary endosymbiotic events occurred in which the plastid originated from either green or red algae. These subsequent events yielded plastids surrounded by three (euglenophytes, dinoflagellates) or four membranes (chlorarachniophytes, apicomplexans, heterokontophytes, haptophytes and cryptophytes). Nucleus (Nu), nucleomorph (Nm), mitochondrion (Mi), endomembrane system (EMS), outer envelope membrane (OEM), inner envelope membrane (IEM), peptidoglycan (PG), chloroplast ER (cER), apicoplast (AP), periplastidal compartment (PPC). Reprint from Gentil et al. (2017) with permission from Springer Publishing Group

Table 10.2 Chloroplast light-harvesting complexes in microalgae

	Light-harvesting complex (LHC)	Microalgal class
1.	*Chlorophyll a/ phycobiliprotein*	Cyanobacteria, Glaucophyceae, Rhodophyceae and Cryptophyceae
2.	*Chlorophyll a/c*	Haptophyceae, Heterokontophyceae, Cryptophyceae and Alevolates
3.	*Chlorophyll a/b*	Chlorophyceae and Euglenophyceae

by molecular phylogenetics (Keeling 2013). Microalgae can be classified into three hierarchical groups based on their light-harvesting complex (LHC) pigment signatures (Table 10.2).

These photosynthetic light-harvesting complexes (LHCs) harness solar light energy within the photosynthetically active radiation (PAR) spectrum ranging from 400 to 700 nm (Schulze et al. 2014). The carotenoids bound within these LHCs function in light harvesting, stabilising the photosystems and provide photoprotection from reactive oxygen species (Campbell 1996; Lohr and Wilhelm 1999; MacIntyre et al. 2002; Jahns and Holzwarth 2012; Depauw et al. 2012; Polimene et al. 2012; Musser et al. 2015).

The distribution of different carotenoids within these three hierarchical lineages is typically taxon-specific, facilitating their application as chemotaxonomic biomarkers. Certain pigments are class-specific diagnostic markers such as α-carotene and the acetylene xanthophylls alloxanthin, crocoxanthin and monadoxanthin in *Cryptophyceae* or the acetylated carotenoid, peridinin, in *Dinoflagellates.* Some pigments can also act as species-specific chemotaxonomic tracers; for example, the divinyl chlorophyll *a* for the marine cyanobacteria, *Prochlorococcus marinus* (Jeffrey et al. 2011). However, this distinction is not always so clearly defined, limiting the application of the single pigment biomarker approach (Zapata 2005; Laza-Martinez et al. 2007). This is evident in the distribution of fucoxanthin in the classes; *Heterokontophyta, Haptophyta and Dinoflagellate* pigment type DINO 2 and 3 (Jeffrey et al. 2011). To overcome this limitation, the total plastid pigment signature can be used to define class-specific pigment types (Zapata 2005; Jeffrey et al. 2011). At present, there are 44 pigment types representing 27 microalgal classes from 11 divisions (Jeffrey et al. 2011).

10.6.1 Cyanobacteria Pigment Signatures

Cyanobacteria are prokaryotic organisms which represent the oldest phototrophic life forms on earth, originating between 2600 and 3500 million years ago (Hedges et al. 2001). They comprise a diverse group of unicellular, filamentous and colonial microorganisms. Cyanobacteria LHC are comprised of chlorophyll and phycobiliproteins (phycocyanin, allophycocyanin and phycoerythrin). Cyanobacteria regulate the composition of these pigments by complementary chromatic adaptation and the distribution of light energy through the photosystems via state transitions to maximise their photosynthetic capacity in fluctuating light environments (Campbell 1996; McConnell 2002). There are five pigment types defining cyanobacteria and Prochlorophytes. The CYANO-2 pigment type comprises the basic cyanobacterial pigment signature (chlorophyll *a*, phycobillin's, zeaxanthin and β-carotene) and is predominately associated with picocyanobacterium species. Filamentous cyanobacteria are classified as CYANO-1 which contains the additional myxoxanthophyll, oscillaxanthin, nostoxanthin, aphanizophyll and 4-keto-myxoxanthophyll. The Prochlorophyceae species are classified into CYANO-3-5 pigment types defined by the presence of cryptoxanthin (α and β), divinyl chlorophylls and chlorophyll *d*, respectively (Jeffrey et al. 2011).

10.6.2 Primary Endosymbiont Pigment Signatures

Plastids of the lineage Archaeplastida are composed of the light-harvesting pigments chlorophyll *a* and phycobiliproteins (phycocyanin, phycoerythrin and allophycocyanin) in addition to accessory carotenoids. The presence of phycobiliproteins enables them to utilise light within the red, yellow and green electromagnetic spectrum (Schulze et al. 2014).

10.6.2.1 Glaucophyta

Glaucophyceae represent primitive primary endosymbionts which have been suggested to be an evolutionary connection between cyanobacteria, Rhodophyceae and Chlorophyceae (John et al. 2011). The phyla comprising of three main groups; *Cyanophora*, *Glaucocystis* and *Gloeochaete* based on cellular morphology, pigment composition and nuclear and plastid-encoded protein phylogeny (Reyes-Prieto and Bhattacharya 2007; John et al. 2011). They share features with cyanobacteria including peptidoglycan layer, carboxysomes, an ancestral form of fructose-1,6-bisphosphate aldolase and the phycobiliproteins. The GLAUCO-1 pigment type is similar to CYANO-2 comprising of chlorophyll *a*, phycobillin's, β-cryptoxanthin and zeaxanthin (Jeffrey et al. 2011).

10.6.2.2 Rhodophyta

Red algae lineage is comprised of 32 orders and 90 families and contains high species diversity with up to 6000 species of macroalgae reported to date, of which 97% are found in marine and benthic environments (Norton et al. 1996; Schneider and Wynne 2007). Microalgae within the red microalgae contain less biodiversity falling within three classes; Cyanidiales, Rhodellophyceae and Porphyridiophyceae.

Zeaxanthin is the predominant carotenoid in unicellular Rhodophyceae with trace levels of violaxanthin, antheraxanthin and β-cryptoxanthin (Schubert et al. 2006). In contrast to the cyanobacterium pigment type, phycoerythrin is the dominant accessory phycobiliprotein pigment in the light-harvesting antenna complex of red algae, bestowing upon them their characteristic dark red colouration. Rhodophyceae are commonly found in highly exposed open ocean sites and it has been proposed that phycoerythrin provides red algae with an advantage by enabling them to utilise the blue–green light spectrum more efficiently (Boney and Corner 1960).

10.6.2.3 Chlorophyta

The green algae comprise a diverse group of filamentous, colonial, motile and solitary cells within the classes Chlorophyceae, Charophyceae, Trebouxiophyceae, Ulvophyceae, Prasinophyceae, Mesostigmatophyceae (John et al. 2011). Microalgae within this lineage commonly used in research and industry include; *Chlamydomonas* spp., *Tetraselmis* spp., *Dunaliella* spp., *Lobosphaera* spp., *Botryococcus* spp., *Haematococcus* spp., *Coccomonas* spp., *Senedesmus* spp. and *Coccomyxa* spp. (Vanessa et al. 2012; Ambati et al. 2018). The accessory phycobiliproteins have been lost in Chlorophyta and replaced with chlorophyll *b*. In addition, the plastids of the *Chlorophyta* lineage contain the most diverse set of carotenoids. The CHLORO-1 pigment type comprises the basic set of carotenoids for the division; 9′-*cis*-neoxanthin, violaxanthin, lutein, zeaxanthin and β-carotene. Lutein serves as a precursor for the Prasinophyceae carotenoids; loroxanthin and its esters (PRASINO-2A), siphonaxanthin and its esters (PRASINO-2B), prasinoxanthin (PRASINO-3A) and uriolide, micromonol and micromonal (PRASINO-3B). The Trebouxiophyceae class contains CHLORO-1 plus vaucheriaxanthin esters (TREBOUX-1), while the Mesostigmatophyceae class contains CHLORO-1 plus *trans*-neoxanthin, siphonaxanthin and its ester derivatives (MESOTIG-1) (Jeffrey et al. 2011).

10.6.3 Secondary and Tertiary Endosymbiont Pigment Signatures

Eukaryotic algae descending from secondary endosymbiotic events between a primitive red algae cell and a heterotrophic host are assigned to three major supergroups: HACROBIA (Cryptophyceae and Haptophyceae), SAR (Heterokontophyceae, dinoflagellates and Chlorarachniophyceae) and euglenozoa. Microalgae within these lineages have undergone numerous gains and losses of genetic and plastid components, generating a high diversity of eukaryotic photosynthetic microalgae (Kim et al. 2014). The red algae lineage acquired their plastid through a series of secondary and tertiary endosymbiotic events. Their chemotaxonomic pigment types are defined based on the differences of chlorophyll *c* in the classes Prymnesiophyceae, Bacillariophyceae, Pelagophyceae and fucoxanthin and its derivatives within the Dinophyceae, Pelagophyceae, Dictyophyceae, Chrysophyceae, Raphidophyceae and Phaeophyceae macroalgae.

10.6.3.1 Cryptophyta

It has been proposed that Cryptophyceae and Haptophyceae monophyletic lineage defined as HACROBIA (Sakaguchi et al. 2009). However, further genome analysis has confirmed that Haptophyceae are a sister group of SAR and revealed that Cryptophyceae may represent an early diverging lineage with close relationship to plastid-lacking katablepharids with close phylogeny to Archaeplastida (Burki et al. 2012). Cryptophyceae are biflagellate flattened asymmetrical cells with a distinctive offset close to the atypical end. They are a common component of marine and freshwater phytoplankton communities. The cryptomonad pigment type CRYPT-1 comprises the definitive class-specific pigments chlorophyll *a*, chlorophyll c_2, α-carotene and the acetylene xanthophylls crocoxanthin and monadoxanthin and alloxanthin. The presence of phycobiliproteins within the thylakoid lumen gives *Chroomonas* spp. and *Rhodomonas* spp. their distinctive green and red colours, respectively.

10.6.3.2 Haptophyceae

Haptophyceae are composed of predominately marine algae known to form extensive blooms visible from space by satellite reflectance spectroscopy. They are classified by the presence of calcified scales (coccoliths) and a haptonema, although genetic analysis now also includes some strains which have lost their haptonema. Haptophyceae are divided into the classes Pavlovophyceae and Prymnesiophyceae of which commonly studied members include; *Isochrysis* spp., *Pavlova* spp. and *Emiliania huxleyi* (Andersen 2004).

Their pigment chemotaxonomy is defined based on the presence of derivatives of chlorophyll *c* and fucoxanthin resulting in 8 pigment types. HAPTO-1 and 2 are predominately associated with the class Pavlovophyceae. HAPTO-1 comprises the basic pigment signature of the group containing chlorophyll *a*, chlorophyll c_1 and c_2, fucoxanthin, diadinoxanthin, diatoxanthin and β-carotene, while the HAPTO-2 contains the additional chlorophyll c_2 *P. gyrans* pigment.

The pigments types HAPTO 3–8 are affiliated with the Prymnesiophyceae. They are defined by the variable distribution of six different chlorophyll *c* derivatives (Chl c_3, c_2, c_1, MV Chl c_3, Chl c_2-MGDG (18:4/14:0) and Chl c_2-MGDG (14:0/14:0)) and four fucoxanthin derivatives (4-ketofucoxanthin, 19′-butanoyloxyfucoxanthin, 19′-hexanoyloxyfucoxanthin and 4-keto-19′-hexanoyloxyfucoxanthin) (Zapata et al. 2004; Jeffrey et al. 2011).

10.6.3.3 Heterokontophyceae

Heterokontophyceae are a monophyletic group of microalgae comprising of 17 classes defined based on their morphology, chloroplast pigments, ultrastructural features and genomics. This large protist group contains heterotrophic and photoautotrophic microorganisms. The photoautotrophic lineage (Orchophyta) is comprised of the classes Bacillariophyceae (DIATO 1-3), Bolidophyceae (BOLIDO-1), Chrysophyceae (CHRYSO-1), Dictyochophyceae (DICTYO-1), Eustigmatophyceae (EUSTIG-1), Pelagophyceae (PELAGO-1), Raphidophyceae (RAHIDO-1), Synurophyceae (SYNURO-1) and Xanthophyceae (XANTHO-1) (Andersen 2004). As in the haptophyta lineage, the chemotaxonomy of Heterokontophyta is defined based on the presence of chlorophyll *c* and fucoxanthin derivatives with low levels of violaxanthin and zeaxanthin. The Eustigmatophyceae and Xanthophyceae classes lack fucoxanthin but contain vaucheriaxanthin and its esters.

10.6.3.4 Dinoflagellates

Dinoflagellates are a diverse group of microalgae occurring in a wide range of habitats including open ocean, marine sediments, freshwater and in symbiotic relationships with corals. They have both photoautotrophic, heterotrophic or mixotrophic lifestyles and can form cyst's under suboptimal conditions. The majority of dinoflagellates are characterised by their theca cell wall and "whirling" swimming motility. They classified into four phylogenetic orders comprising the Prorocentrales, Dinophysiales, Peridiniales and the Gymnodiniales (Tomas 1997). Species such as *Dinophysis* and *Alexandrium* biosynthesise potent biotoxins and responsible for harmful algal bloom events while others are responsible for oceanic bioluminescence (*Pyrocystis* spp. and *Gonyaulax* spp.) (Knaust et al. 1998; Dees et al. 2017). Their LHC is composed of chlorophyll *a*/*c* and the primary light-harvesting xanthophyll pigment, peridinin (DINO-1) (Jeffrey et al. 2011). Some species of dinoflagellates have acquired additional accessory pigments through tertiary endosymbiotic events between a haptophyta (*Karlodinium* spp., DINO-2), a diatom (*Durinskia* spp., DINO-3), a Cryptophyta (*Dinophysis* spp., DINO-4) or a Chlorophyta (*Lepidodinium* spp., DINO-5) (Figs. 10.2 and 10.3).

10.6.3.5 Chlorarachniophyceae and Euglenozoa

The Chlorarachniophyceae are marine amoeboid organisms who have acquired their plastids from green algae and contain a pigment signature similar to PRASINO-2A. Euglenoids are flagellate microalgae found in eutrophic aquatic environments and marine sediments. They are classified within the most basal eukaryotic supergroup, Excavata, which originated from a secondary endosymbiotic event between a heterotrophic protist and a Chlorophyta (Keeling 2013). Their

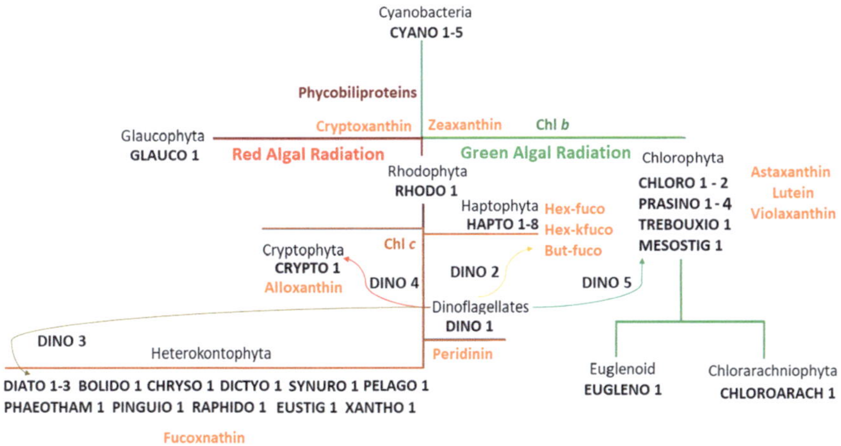

Fig. 10.3 Chemotaxonomic distribution of microalgal pigment types and their high-value pigments

pigment signature reflects their complex evolutionary history comprising of both Chromophyceae pigments (diadinoxanthin, diatoxanthin and heteroxanthin) and green algal pigments (chlorophyll *b*, 9′-*cis*-neoxanthin, loroxanthin and siphonaxanthin). They also contain species-specific xanthophylls such as eutreptiellanone associated with *Eutreptiella gymnastica* and hexadehydro-β,β-caroten-3-ol and octadehydro-β,β-carotene associated with *Euglena viridis* (Fiksdahl et al. 1984; Fiksdahl and Liaaen-Jensen 1988). *Euglena sanguinea* is capable of biosynthesising the secondary keto-carotenoids adonirubin (3%), diesters of (3S, 3′R)-adonixanthin (13%) and diesters of (3S, 3′S)-astaxanthin (75%); however, the presence of ichthyotoxins limits the commercial development of these species (Grung and Liaaen-Jensen 1993; Triemer et al. 2003).

10.7 Conclusions

Microalgal pigment chemotaxonomy provides a key diagnostic tool as part of a polyphasic taxonomic approach in the characterisation of new isolates. In particular, plastid pigment types can facilitate the identification of pico-algae which lack molecular data and can be notoriously difficult to identify by microscopy alone. Advances in HPLC techniques have increased the resolution of chemotaxonomic pigment pairs and shone a light on the diversity of pigments in microalgae. These include the distribution in the Chl-*c* and fucoxanthin derivatives within red algal radiation and lutein derivatives within the green algae radiation. The application of LC–MS to pigments analysis continues to lead to the identification of new carotenoids and chlorophylls. Similarly, molecular tools are providing new insight into carotenoid biosynthesis which could lead to improved strains for the industrial production of high-value carotenoids.

References

Abidov M, Ramazanov Z, Seifulla R, Grachev S (2010) The effects of Xanthigen™ in the weight management of obese premenopausal women with non-alcoholic fatty liver disease and normal liver fat. Diabetes Obes Metab 12:72–81. https://doi.org/10.1111/j.1463-1326.2009.01132.x

Afolayan AF, Bolton JJ, Lategan CA, Smith PJ, Beukes DR (2008) Fucoxanthin, tetraprenylated toluquinone and toluhydroquinone metabolites from *Sargassum heterophyllum* inhibit the in vitro growth of the malaria parasite *Plasmodium falciparum*. Zeitschrift fur Naturforschung C J Biosci 63:848–852. https://doi.org/10.1515/znc-2008-11-1211

Ahmed F, Fanning K, Netzel M, Turner W, Li Y, Schenk PM (2014) Profiling of carotenoids and antioxidant capacity of microalgae from subtropical coastal and brackish waters. Food Chem 165:300–306. https://doi.org/10.1016/j.foodchem.2014.05.107

Airs RL, Garrido JL (2011) Liquid chromatography-mass spectrometry for pigment analysis. In: Roy S, Llewellyn C, Egeland ES, Johnsen G (eds) Phytoplankton pigments. Cambridge University Press, Cambridge, pp 314–342

Alves-Rodrigues A, Shao A (2004) The science behind lutein. Toxicol Lett 150:57–83. https://doi.org/10.1016/J.TOXLET.2003.10.031

Ambati RR, Gogisetty D, Aswathanarayana RG, Ravi S, Bikkina PN, Bo L, Yuepeng S (2018) Industrial potential of carotenoid pigments from microalgae: current trends and future prospects. Crit Rev Food Sci Nutr 1–22. https://doi.org/10.1080/10408398.2018.1432561

Andersen RA (2004) Biology and systematics of heterokont and haptophyte algae. Am J Bot 91:1508–1522. https://doi.org/10.3732/ajb.91.10.1508

Baudelet P-H, Gagez A-L, Berard J-B, Juin C, Bridiau N, Kaas R, Picot L (2013) Antiproliferative activity of *Cyanophora paradoxa* pigments in melanoma, breast and lung cancer cells. Marine Drugs 11:4390–4406. https://doi.org/10.3390/md11114390

BCC Research (2015) The global market for carotenoids—FOD025C. http://www.bccresearch.com/market-research/food-and-beverage/carotenoids-market-fod025c.html. Accessed 27 Jan 2016

Bertram JS (2018) Cancer chemoprevention by retinoids and carotenoids: proposed role of gap junctional communication. In Jacobs MM (ed) Vitamins and minerals in the prevention and treatment of cancer, 1st edn. CRC Press, pp 46–65. https://doi.org/10.1016/j.bbadis.2005.01.003

Bhosale P, Bernstein PS (2005) Microbial xanthophylls. Appl Microbiol Biotechnol 68:445–455. https://doi.org/10.1007/s00253-005-0032-8

Boney AD, Corner EDS (1960) A possible function of phycoerythrin in intertidal red algae. Nature 188:1042–1043. https://doi.org/10.1038/1881042a0

Britton G, Liaaen-Jensen S, Pfander H (eds) (2004) Carotenoids—handbook. Springer, Birkhäuser Basel, p 647

Burki F, Okamoto N, Pombert J-F, Keeling PJ (2012) The evolutionary history of haptophytes and cryptophytes: phylogenomic evidence for separate origins. Proc Biol Sci 279:2246–2254. https://doi.org/10.1098/rspb.2011.2301

Cai C, Wang Y, Li C, Guo Z, Jia R, Wu W, He P (2014) Purification and photodynamic bioactivity of phycoerythrin and phycocyanin from *Porphyra yezoensis* Ueda. J Ocean Univ China 13:479–484. https://doi.org/10.1007/s11802-014-2148-x

Campbell D (1996) Complementary chromatic adaptation alters photosynthetic strategies in the cyanobacterium *Calothrix*. Microbiology 142:1255–1263. https://doi.org/10.1099/13500872-142-5-1255

Chew BP, Wong MW, Wong TS (1996) Effects of lutein from marigold extract on immunity and growth of mammary tumors in mice. Anticancer Res 16:3689–3694 PMID:9042242

Chitchumroonchokchai C, Failla ML (2017) Bioaccessibility and intestinal cell uptake of astaxanthin from salmon and commercial supplements. Food Res Int 99:936–943. https://doi.org/10.1016/J.FOODRES.2016.10.010

Chiueh C-C (2003) Fluorescent labeling reagents and its kit fabricated by B-phycoerythrin and its derivatives. US patent US2004142380

Chiueh C-C, Chan E-C, Shen C-R (2003) Diagnostic kit and method using phycoerythrin to label tumor tissues and cells. US patent US2005009117

Chokshi K, Pancha I, Ghosh A, Mishra S (2017) Oxidative stress-induced bioprospecting of microalgae. In: Systems biology of marine ecosystems. Springer International Publishing, Cham, pp 251–276. https://doi.org/10.1007/978-3-319-62094-7_13

Choubert G, Mendes-Pinto MM, Morais R (2006) Pigmenting efficacy of astaxanthin fed to rainbow trout *Oncorhynchus mykiss*: effect of dietary astaxanthin and lipid sources. Aquaculture 257:429–436. https://doi.org/10.1016/j.aquaculture.2006.02.055

Cian RE, López-Posadas R, Drago SR, de Medina FS, Martínez-Augustin O (2012) Immunomodulatory properties of the protein fraction from *Phorphyra columbina*. J Agric Food Chem 60:8146–8154. https://doi.org/10.1021/jf300928j

Claustre H, Hooker SB, Van Heukelem L, Berthon J-F, Barlow R, Ras J, Marty J-C (2004) An intercomparison of HPLC phytoplankton pigment methods using in situ samples: application to remote sensing and database activities. Mar Chem 85:41–61. https://doi.org/10.1016/J.MARCHEM.2003.09.002

Cuellar-Bermudez SP, Aguilar-Hernandez I, Cardenas-Chavez DL, Ornelas-Soto N, Romero-Ogawa MA, Parra-Saldivar R (2015) Extraction and purification of high-value metabolites from microalgae: essential lipids, astaxanthin and phycobiliproteins. Microb Biotechnol 8:190–209. https://doi.org/10.1111/1751-7915.12167

Das SK, Hashimoto T, Kanazawa K (2008) Growth inhibition of human hepatic carcinoma HepG2 cells by fucoxanthin is associated with down-regulation of cyclin D. Biochem Biophys Acta 1780:743–749. https://doi.org/10.1016/j.bbagen.2008.01.003

Dees P, Bresnan E, Dale AC, Edwards M, Johns D, Mouat B, Davidson K (2017) Harmful algal blooms in the Eastern North Atlantic Ocean. Proc Natl Acad Sci USA 114:E9763–E9764. https://doi.org/10.1073/pnas.1715499114

Delgado-Vargas F, Jimenez AR, Paredes-Lopez O (2000) Natural pigments: carotenoids, anthocyanins, and betalains—characteristics, biosynthesis, processing, and stability. Crit Rev Food Sci Nutr 40:173–289. https://doi.org/10.1080/10408690091189257

Demmig-Adams B, Adams WW (2002) Antioxidants in photosynthesis and human nutrition. Science 298:2149–2153. https://doi.org/10.1126/science.1078002

Depauw FA, Rogato A, Ribera d'Alcala M, Falciatore A (2012) Exploring the molecular basis of responses to light in marine diatoms. J Exp Bot 63:1575–1591. https://doi.org/10.1093/jxb/ers005

Egeland ES (2011) Minimum identification criteria for phytoplankton pigments. In: Roy S, Llewellyn C, Egeland ES, Johnsen G (eds) Phytoplankton pigments: characterization, chemotaxonomy and applications in oceanography. Cambridge University Press, Cambridge, pp 650–652

Enzing C, Ploeg M, Barbosa M, Sijtsma L, Vigani M, Parisi C, Cerezo ER (2014) Microalgae-based products for the food and feed sector: an outlook for Europe. European Commission JRC Scientific and Policy Reports, pp 1–75. https://doi.org/10.2791/3339

Eriksen NT (2008) Production of phycocyanin—a pigment with applications in biology, biotechnology, foods and medicine. Appl Microbiol Biotechnol 80:1–14. https://doi.org/10.1007/s00253-008-1542-y

Fiksdahl A, Liaaen-Jensen S (1988) Diacetylenic carotenoids from *Euglena viridis*. Phytochemistry 27:1447–1450. https://doi.org/10.1016/0031-9422(88)80213-9

Fiksdahl A, Bjørnland T, Liaaen-Jensen S (1984) Algal carotenoids with novel end groups. Phytochemistry 23:649–655. https://doi.org/10.1016/S0031-9422(00)80400-8

Frassanito R, Flaim G, Mancini I, Guella G (2006) High production of unexpected carotenoids in Dinophyceae. Astaxanthin esters from the freshwater dinoflagellate *Tovellia sanguinea*. Biochem Syst Ecol 34:843–853. https://doi.org/10.1016/j.bse.2006.05.013

García-González M, Moreno J, Manzano JC, Florencio FJ, Guerrero MG (2005) Production of *Dunaliella salina* biomass rich in 9-*cis*-beta-carotene and lutein in a closed tubular photobioreactor. J Biotechnol 115:81–90. https://doi.org/10.1016/j.jbiotec.2004.07.010

Garrido JL, Airs RL, Rodríguez F, Van Heukelem L, Zapata M (2011) New HPLC separation techniques. In: E.S.E., Suzanne Roy GJ, Llewellyn Carole A (eds) Phytoplankton pigments: characterization, chemotaxonomy and applications in oceanography. Cambridge University Press, Cambridge, pp 165–194

Gentil J, Hempel F, Moog D, Zauner S, Maier UG (2017) Review: origin of complex algae by secondary endosymbiosis: a journey through time. Protoplasma 254:1835–1843. https://doi.org/10.1007/s00709-017-1098-8

Gomez PI, Gonzalez MA (2005) The effect of temperature and irradiance on the growth and carotenogenic capacity of seven strains of *Dunaliella salina* (Chlorophyta) cultivated under laboratory conditions. Biol Res 38:151–162. https://doi.org/10.4067/S0716-97602005000200005

Gómez-Loredo A, Benavides J, Rito-Palomares M (2016) Growth kinetics and fucoxanthin production of *Phaeodactylum tricornutum* and *Isochrysis galbana* cultures at different light and agitation conditions. J Appl Phycol 28:849–860. https://doi.org/10.1007/s10811-015-0635-0

Granado F, Olmedilla B, Blanco I (2003) Nutritional and clinical relevance of lutein in human health. Br J Nutr 90:487–502. https://doi.org/10.1079/bjn2003927

Grung M, Liaaen-Jensen S (1993) Algal carotenoids 52; secondary carotenoids of algae 3; carotenoids in a natural bloom of *Euglena sanguinea*. Biochem Syst Ecol 21:757–763. https://doi.org/10.1016/0305-1978(93)90088-9

Guerin M, Huntley ME, Olaizola M (2003) *Haematococcus* astaxanthin: applications for human health and nutrition. Trends Biotechnol 21:210–216. https://doi.org/10.1016/s0167-7799(03)00078-7

Gupta Vishal, Ratha Sachitra Kumar, Sood Anjuli, Vidhi Chaudhary RP (2013) New insights into the biodiversity and applications of cyanobacteria (blue-green algae)—prospects and challenges. Algal Res 2:79–97. https://doi.org/10.1016/J.ALGAL.2013.01.006

Han D, Li Y, Hu Q (2013) Astaxanthin in microalgae: pathways, functions and biotechnological implications. ALGAE 28:131–147. https://doi.org/10.4490/algae.2013.28.2.131

Hanagata N, Dubinsky Z (1999) Secondary carotenoid accumulation in *Scenedesmus komarekii* (Chlorophycea, Chlorophyta). J Phycol 35:960–966

Hata N, Ogbonna JC, Hasegawa Y, Taroda H, Tanaka H (2001) Production of astaxanthin by *Haematococcus pluvialis* in a sequential heterotrophic-photoautotrophic culture. J Appl Phycol 13:395–402. https://doi.org/10.1046/j.1529-8817.1999.3550960.x

Hedges SB, Chen H, Kumar S, Wang DY-C, Thompson AS, Watanabe H (2001) A genomic timescale for the origin of eukaryotes. BMC Evol Biol 1:4. https://doi.org/10.1186/1471-2148-1-4

Heinonen OP, Huttunen JK, Albanes D, Haapakoski J, Palmgren J, Pietinen P, Pikkarainen J, Rautalahti M, Virtamo J, Edwards BK, Greenwald P et al (1994) The effect of vitamin e and beta carotene on the incidence of lung cancer and other cancers in male smokers. New Engl J Med 330:1029–1035. https://doi.org/10.1056/NEJM199404143301501

Hosokawa M, Wanezaki S, Miyauchi K, Kurihara H, Kohno H, Kawabata J, Takahashi K (1999) Apoptosis-inducing effect of fucoxanthin on human leukemia cell line HL-60. Food Sci Technol Res 5:243–246. https://doi.org/10.3136/fstr.5.243

Hosokawa M, Miyashita T, Nishikawa S, Emi S, Tsukui T, Beppu F, Miyashita K (2010) Fucoxanthin regulates adipocytokine mRNA expression in white adipose tissue of diabetic/obese KK-Ay mice. Arch Biochem Biophys 504:17–25. https://doi.org/10.1016/j.abb.2010.05.031

Huang JJ, Lin S, Xu W, Cheung PCK (2017) Occurrence and biosynthesis of carotenoids in phytoplankton. Biotechnol Adv 35(5):597–618. https://doi.org/10.1016/j.biotechadv.2017.05.001

Iio K, Okada Y, Ishikura M (2011a) Single and 13-week oral toxicity study of fucoxanthin oil from microalgae in rats. Shokuhin eiseigaku zasshi (J Food Hygienic Soc Jpn) 52:183–189 PMID: 21720124

Iio K, Okada Y, Ishikura M (2011b) Bacterial reverse mutation test and micronucleus test of fucoxanthin oil from microalgae. Food Hyg Safety Sci (Shokuhin Eiseigaku Zasshi) 52:190–193. https://doi.org/10.3358/shokueishi.52.190

Ishikawa C, Tafuku S, Kadekaru T, Sawada S, Tomita M, Okudaira T, Mori N (2008) Anti-adult T-cell leukemia effects of brown algae fucoxanthin and its deacetylated product, fucoxanthinol. Int J Cancer (J Int du cancer) 123:2702–2712. https://doi.org/10.1002/ijc.23860

Jahns P, Holzwarth AR (2012) The role of the xanthophyll cycle and of lutein in photoprotection of photosystem II. Biochimica et Biophysica Acta (BBA) - Bioenergetics 1817, 182–193. https://doi.org/10.1016/j.bbabio.2011.04.012

Jaswir I, Noviendri D, Salleh HM, Taher M, Miyashita K, Ramli N (2013) Analysis of fucoxanthin content and purification of all-*trans*-fucoxanthin from *Turbinaria turbinata* and *Sargassum plagyophyllum* by SiO_2 open column chromatography and reversed phase-HPLC. J Liq Chromatogr Relat Technol 36:1340–1354. https://doi.org/10.1080/10826076.2012.691435

Jayaraman S, Knuth ML, Cantwell M, Santos A (2011) High performance liquid chromatographic analysis of phytoplankton pigments using a C16-Amide column. J Chromatogr A 1218:3432–3438. https://doi.org/10.1016/J.CHROMA.2011.03.058

Jeffrey SW, Wright SW, Zapata M (2011) Microalgal classes and their signature pigments. In: Roy S, Llewellyn C, Egeland ES, Johnsen G (eds) Phytoplankton pigments. Cambridge environmental chemistry series. Cambridge University Press, Cambridge, pp 3–77

John DM, Whitton BA, Brook AJ, British Phycological Society, Natural History Museum, London E (2011) The freshwater algal flora of the British Isles : an identification guide to freshwater and terrestrial algae, 2nd edn. Cambridge University Press, Cambridge

Keeling PJ (2013) The number, speed, and impact of plastid endosymbioses in eukaryotic evolution. Annu Rev Plant Biol 64:583–607. https://doi.org/10.1098/rspb.2011.2301

Kim K-N, Heo S-J, Kang S-M, Ahn G, Jeon Y-J (2010) Fucoxanthin induces apoptosis in human leukemia HL-60 cells through a ROS-mediated Bcl-xL pathway. Toxicol In Vitro 24:1648–1654. https://doi.org/10.1016/J.TIV.2010.05.023

Kim SM, Kang S-W, Kwon O-N, Chung D, Pan C-H (2012) Fucoxanthin as a major carotenoid in *Isochrysis aff. galbana*: characterization of extraction for commercial application. J Korean Soc Appl Biol Chem 55:477–483. https://doi.org/10.1007/s13765-012-2108-3

Kim KM, Park J-H, Bhattacharya D, Yoon HS (2014) Applications of next-generation sequencing to unravelling the evolutionary history of algae. Int J Syst Evol Microbiol 64:333–345. https://doi.org/10.1099/ijs.0.054221-0

Knaust R, Urbig T, Li L, Taylor W, Hastings JW (1998) The circadian rhythm of bioluminescence in pyrocystis is not due to differences in the amount of luciferase: a comparative study of three bioluminescent marine dinoflagellates. J Phycol 34:167–172. https://doi.org/10.1046/j.1529-8817.1998.340167.x

Kotake-Nara E, Kushiro M, Zhang H, Sugawara T, Miyashita K, Nagao A (2001) Carotenoids affect proliferation of human prostate cancer cells. J Nutr 131:3303–3306. https://doi.org/10.1093/jn/131.12.3303

Krinsky NI, Landrum JT, Bone RA (2003) Biologic mechanisms of the protective role of lutein and zeaxanthin in the eye. Annu Rev Nutr 23:171–201. https://doi.org/10.1146/annurev.nutr.23.011702.073307

Laza-Martinez A, Seoane S, Zapata M, Orive E (2007) Phytoplankton pigment patterns in a temperate estuary: from unialgal cultures to natural assemblages. J Plankton Res 913–929. https://doi.org/10.1093/plankt/fbm069

Leya T, Rahn A, Lutz C, Remias D (2009) Response of arctic snow and permafrost algae to high light and nitrogen stress by changes in pigment composition and applied aspects for biotechnology. FEMS Microbiol Ecol 67:432–443. https://doi.org/10.1111/j.1574-6941.2008.00641.x

Liaaen-Jensen S (1979) Carotenoids—a chemosystematic approach. Pure Appl Chem 51:661–675. https://doi.org/10.1351/pac197951030661

Lohr M, Wilhelm C (1999) Algae displaying the diadinoxanthin cycle also possess the violaxanthin cycle. Proc Natl Acad Sci USA 96:8784–8789. https://doi.org/10.1073/PNAS.96.15.8784

Lubián LM, Montero O, Moreno-Garrido I, Huertas IE, Sobrino C, González-del Valle M, Parés G (2000) *Nannochloropsis* (Eustigmatophyceae) as source of commercially valuable pigments. J Appl Phycol 12:249–255. https://doi.org/10.1023/A:1008170915932

Ma RY-N, Chen F (2001) Enhanced production of free *trans*-astaxanthin by oxidative stress in the cultures of the green microalga *Chlorococcum* sp. Process Biochem 36:1175–1179. https://doi.org/10.1016/S0032-9592(01)00157-1

MacIntyre HL, Kana TM, Anning T, Geider RJ (2002) Photoacclimation of photosynthesis irradiance response curves and photosynthetic pigments in microalgae and cyanobacteria. J Phycol 38:17–38. https://doi.org/10.1046/j.1529-8817.2002.00094.x

Mackey MD, Mackey DJ, Higgins HW, Wright SW (1996) CHEMTAX—a program for estimating class abundances from chemical markers: application to HPLC measurements of phytoplankton. Mar Ecol Prog Ser 144:265–283. https://doi.org/10.3354/meps144265

Mao X, Wu T, Sun D, Zhang Z, Chen F (2018) Differential responses of the green microalga *Chlorella zofingiensis* to the starvation of various nutrients for oil and astaxanthin production. Biores Technol 249:791–798. https://doi.org/10.1016/J.BIORTECH.2017.10.090

Markets and Markets (2016) Carotenoids market by type & application. Global Forecast 2021

Mc Gee D, Archer L, Paskuliakova A, Mc Coy GR, Fleming GTA, Gillespie E, Touzet N (2017) Rapid chemotaxonomic profiling for the identification of high-value carotenoids in microalgae. J Appl Phycol 1–15. https://doi.org/10.1007/s10811-017-1247-7

McClure DD, Luiz A, Gerber B, Barton GW, Kavanagh JM (2018) An investigation into the effect of culture conditions on fucoxanthin production using the marine microalgae *Phaeodactylum tricornutum*. Algal Res 29:41–48. https://doi.org/10.1016/J.ALGAL.2017.11.015

McConnell MD (2002) Regulation of the distribution of chlorophyll and phycobilin-absorbed excitation energy in cyanobacteria. A structure-based model for the light state transition. Plant Physiol 130:1201–1212. https://doi.org/10.1104/pp.009845

Metting FB (1996) Biodiversity and application of microalgae. J Ind Microbiol Biotechnol 17:477–489. https://doi.org/10.1007/BF01574779

Moeller SM, Voland R, Tinker L, Blodi BA, Klein ML, Gehrs KM, Grp CS (2008) Associations between age-related nuclear cataract and lutein and zeaxanthin in the diet and serum in the Carotenoids in the Age-Related Eye Disease Study (CAREDS), an ancillary study of the women's health initiative. Arch Ophthalmol 126:354–364. https://doi.org/10.1001/archopht.126.3.354

Murthy KNC, Vanitha A, Rajesha J, Swamy MM, Sowmya PR, Ravishankar GA (2005) In vivo antioxidant activity of carotenoids from *Dunaliella salina*—a green microalga. Life Sci 76:1381–1390. https://doi.org/10.1016/j.lfs.2004.10.015

Musser AJ, Maiuri M, Brida D, Cerullo G, Friend RH, Clark J (2015) The nature of singlet exciton fission in carotenoid aggregates. J Am Chem Soc 137:5130–5139. https://doi.org/10.1021/jacs.5b01130

Namitha KK, Negi PS (2010) Chemistry and biotechnology of carotenoids. Crit Rev Food Sci Nutr 50(8):728–760

Nakazawa Y, Sashima T, Hosokawa M, Miyashita K (2009) Comparative evaluation of growth inhibitory effect of stereoisomers of fucoxanthin in human cancer cell lines. J Funct Foods 1:88–97. https://doi.org/10.1016/j.jff.2008.09.015

Nishino H, Murakoshi M, Ii T, Takemura M, Kuchide M, Kanazawa M, Jinno K (2002) Carotenoids in cancer chemoprevention. Cancer Metastasis Rev 21:257–264. https://doi.org/10.1023/a:1021206826750

Nomura T, Kikuchi M, Kubodera A, Kawakami Y (1997) Proton-donative antioxidant activity of fucoxanthin with 1,1-Diphenyl-2-Picrylhydrazyl (DPPH). IUBMB Life 42:361–370. https://doi.org/10.1080/15216549700202761

Norton TA, Melkonian M, Andersen RA (1996) Algal biodiversity. Phycologia 35:308–326. https://doi.org/10.2216/i0031-8884-35-4-308.1

Oi VT, Glazer AN, Stryer L (1982) Fluorescent phycobiliprotein conjugates for analyses of cells and molecules. J Cell Biol 93:981–986 PMID: 6749865

Orosa M, Torres E, Fidalgo P, Abalde J (2000) Production and analysis of secondary carotenoids in green algae. J Appl Phycol 12:553–556. https://doi.org/10.1023/A:1008173807143

Orosa M, Valero JF, Herrero C, Abalde J (2001) Comparison of the accumulation of astaxanthin in *Haematococcus pluvialis* and other green microalgae under N-starvation and high light conditions. Biotech Lett 23:1079–1085. https://doi.org/10.1023/A:1010510508384

Owens TG, Gallagher JC, Alberte RS (2004) Photosynthetic light-harvestign function of violaxanthin in *Nannochloropsis* spp. (Eustigmatophyceae)1. J Phycol 23:79–85. https://doi.org/10.1111/j.0022-3646.1987.00079.x

Paliwal C, Ghosh T, George B, Pancha I, Maurya R, Chokshi K, Mishra S (2016) Microalgal carotenoids: potential nutraceutical compounds with chemotaxonomic importance. Algal Res 15:24–31. https://doi.org/10.1016/j.algal.2016.01.017

Park JS, Chew BP, Wong TS (1998) Dietary lutein from marigold extract inhibits mammary tumor development in BALB/c mice. J Nutr 128:1650–1656. https://doi.org/10.1093/jn/128.10.1650

Park HJ, Lee MK, Park YB, Shin YC, Choi MS (2011) Beneficial effects of *Undaria pinnatifida* ethanol extract on diet-induced-insulin resistance in C57BL/6J mice. Food Chem Toxicol 49:727–733. https://doi.org/10.1016/J.FCT.2010.11.032

Pasquet V, Morisset P, Ihammouine S, Chepied A, Aumailley L, Berard J-B, Picot L (2011) Antiproliferative activity of violaxanthin isolated from bioguided fractionation of *Dunaliella tertiolecta* extracts. Marine Drugs 9:819–831. https://doi.org/10.3390/md9050819

Petrushkina M, Gusev E, Sorokin B, Zotko N, Mamaeva A, Filimonova A, Kuzmin D (2017) Fucoxanthin production by heterokont microalgae. Algal Res 24:387–393. https://doi.org/10.1016/J.ALGAL.2017.03.016

Polimene L, Brunet C, Allen JI, Butenschon M, White DA, Llewellyn CA (2012) Modelling xanthophyll photoprotective activity in phytoplankton. J Plankton Res 34:196–207. https://doi.org/10.1093/plankt/fbr102

Prieto A, Pedro Canavate J, Garcia-Gonzalez M (2011) Assessment of carotenoid production by *Dunaliella salina* in different culture systems and operation regimes. J Biotechnol 151:180–185. https://doi.org/10.1016/j.jbiotec.2010.11.011

Qin S, Liu G-X, Hu Z-Y (2008) The accumulation and metabolism of astaxanthin in *Scenedesmus obliquus* (Chlorophyceae). Process Biochem 43:795–802. https://doi.org/10.1016/J.PROCBIO.2008.03.010

Rao R, Sarada A, Ravishankar R, Aswathanarayana G (2010) Enhancement of carotenoids in green alga-*Botryococcus braunii* in various autotrophic media under stress conditions. Int J Biomed Pharm Sci 4:87–92

Rengarajan T, Rajendran P, Nandakumar N, Balasubramanian MP, Nishigaki I (2013) Cancer preventive efficacy of marine carotenoid fucoxanthin: cell cycle arrest and apoptosis. Nutrients 5:4978–4989. https://doi.org/10.3390/nu5124978

Reyes-Prieto A, Bhattacharya D (2007) Phylogeny of nuclear-encoded plastid-targeted proteins supports an early divergence of glaucophytes within plantae. Mol Biol Evol 24:2358–2361. https://doi.org/10.1093/molbev/msm186

Sakaguchi M, Takishita K, Matsumoto T, Hashimoto T, Inagaki Y (2009) Tracing back EFL gene evolution in the cryptomonads–haptophytes assemblage: Separate origins of EFL genes in haptophytes, photosynthetic cryptomonads, and goniomonads. Gene 441:126–131. https://doi.org/10.1016/J.GENE.2008.05.010

Sanchez JF, Fernandez-Sevilla JM, Acien FG, Ceron MC, Perez-Parra J, Molina-Grima E (2008a) Biomass and lutein productivity of *Scenedesmus almeriensis*: influence of irradiance, dilution rate and temperature. Appl Microbiol Biotechnol 79:719–729. https://doi.org/10.1007/s00253-008-1494-2

Sanchez JF, Fernandez JM, Acien FG, Rueda A, Perez-Parra J, Molina E (2008b) Influence of culture conditions on the productivity and lutein content of the new strain *Scenedesmus almeriensis*. Process Biochem 43:398–405. https://doi.org/10.1016/j.procbio.2008.01.004

Sanz N, García-Blanco A, Gavalás-Olea A, Loures P, Garrido JL (2015) Phytoplankton pigment biomarkers: HPLC separation using a pentafluorophenyloctadecyl silica column. Methods Ecol Evol 6:1199–1209. https://doi.org/10.1111/2041-210X.12406

Saw CLL, Yang AY, Guo Y, Kong A-NT (2013) Astaxanthin and omega-3 fatty acids individually and in combination protect against oxidative stress via the Nrf2–ARE pathway. Food Chem Toxicol 62:869–875. https://doi.org/10.1016/J.FCT.2013.10.023

Schmid M, Guihéneuf F, Stengel DB (2017) Ecological and commercial implications of temporal and spatial variability in the composition of pigments and fatty acids in five Irish macroalgae. Mar Biol 164:158. https://doi.org/10.1007/s00227-017-3188-8

Schneider CW, Wynne MJ (2007) A synoptic review of the classification of red algal genera a half century after Kylin's "Die Gattungen der Rhodophyceen". Bot Mar 50:197–249. https://doi.org/10.1515/BOT.2007.025

Schubert N, García-Mendoza E, Pacheco-Ruiz I (2006) Carotenoid Composition of Marine Red Algae. J Phycol 42:1208–1216. https://doi.org/10.1111/j.1529-8817.2006.00274.x

Schulze PSC, Barreira LA, Pereira HGC, Perales JA, Varela JCS (2014) Light emitting diodes (LEDs) applied to microalgal production. Trends Biotechnol 32:422–430. https://doi.org/10.1016/j.tibtec

Seddon JM, Ajani UA, Sperduto RD, Hiller R, Blair N, Burton TC, Willett W (1994) Dietry carotenoids, vitamin-A, vitamin-C, and vitamin-E, and advanced age-related macular degeneration. Jama-J A M Assoc 272:1413–1420. https://doi.org/10.1001/jama.272.18.1413

Sekar S, Chandramohan M (2008) Phycobiliproteins as a commodity: trends in applied research, patents and commercialization. J Appl Phycol 20:113. https://doi.org/10.1007/s10811.007.9188.1

Serive B, Nicolau E, Bérard J-B, Kaas R, Pasquet V, Picot L, Cadoret J-P (2017) Community analysis of pigment patterns from 37 microalgae strains reveals new carotenoids and porphyrins characteristic of distinct strains and taxonomic groups. PLoS ONE 12:e0171872. https://doi.org/10.1371/journal.pone.0171872

Stadnichuk IN, Krasil'nikov PM, Zlenko DV (2015) Cyanobacterial phycobilisomes and phycobiliproteins. Mikrobiologiia 84:131–143. https://doi.org/10.1134/S0026261715020150

Suzuki K, Kamimura A, Hooker SB (2015) Rapid and highly sensitive analysis of chlorophylls and carotenoids from marine phytoplankton using ultra-high performance liquid chromatography (UHPLC) with the first derivative spectrum chromatogram (FDSC) technique. Mar Chem 176:96–109. https://doi.org/10.1016/j.marchem.2015.07.010

Takaichi S (2011) Distributions, biosyntheses and functions of carotenoids in algae. Marine drugs 9:1101–1118 PMID: 21747749

Telfer A (2002) What is beta-carotene doing in the photosystem II reaction centre? Philos Trans R Soc London Ser B-Biol Sci 357:1431–1439. https://doi.org/10.1098/rstb.2002.1139

Tomas CR (1997) Identifying marine phytoplankton (Chaps. 1–3). Academic Press, pp 1–584

Triemer RE, Zimba PV, Rowan M (2003) 163 identification of euglenoids that produce ichthyotoxin(s) (Euglenophyta). J Phycol 39:56–57. https://doi.org/10.1111/j.0022-3646.2003.03906001_163.x

Vanessa B-V, Rocio O-B, Ruth R-S, Paola T-M, Adelaida H-G, Edgar C-E (2012) Microalgae of the Chlorophyceae class: potential nutraceuticals reducing oxidative stress intensity and cellular damage. In: Oxidative stress and diseases. InTech. https://doi.org/10.5772/32550

Wang J, Chen S, Xu S, Yu X, Ma D, Hu X, Cao X (2012) In vivo induction of apoptosis by fucoxanthin, a marine carotenoid, associated with down-regulating STAT3/EGFR signaling in sarcoma 180 (S180) xenografts-bearing mice. Marine Drugs 10:2055–2068. https://doi.org/10.3390/md10092055

Woo M-N, Jeon S-M, Shin YC, Lee M-K, Kang MA, Choi M-S (2009) Anti-obese property of fucoxanthin is partly mediated by altering lipid-regulating enzymes and uncoupling proteins of visceral adipose tissue in mice. Mol Nutr Food Res 53:1603–1611. https://doi.org/10.1002/mnfr.200900079

Woo M-N, Jeon S-M, Kim H-J, Lee M-K, Shin S-K, Shin YC, Choi M-S (2010) Fucoxanthin supplementation improves plasma and hepatic lipid metabolism and blood glucose concentration in high-fat fed C57BL/6N mice. Chem Biol Interact 186:316–322. https://doi.org/10.1016/J.CBI.2010.05.006

Wright SW, Jeffery SW, Mantoura RFCEN (1991) Improved HPLC method for the analysis of chlorophylls and carotenoids from marine phytoplankton. Mar Ecol Prog Ser 77:183–196. https://doi.org/10.3354/meps077183

Xia S, Wang K, Wan L, Li A, Hu Q, Zhang C (2013) Production, characterization, and antioxidant activity of fucoxanthin from the marine diatom *Odontella aurita*. Marine drugs 11:2667–2681. https://doi.org/10.3390/md11072667

Xiao X, Si X, Yuan Z, Xu X, Li G (2012) Isolation of fucoxanthin from edible brown algae by microwave-assisted extraction coupled with high-speed countercurrent chromatography. J Sep Sci 35:2313–2317. https://doi.org/10.1002/jssc.201200231

Zapata M (2005) Recent advances in pigment analysis as applied to picophytoplankton. Vie et Milieu 55:233–248

Zapata M, Rodríguez F, Garrido JL (2000) Separation of chlorophylls and carotenoids from marine phytoplankton: a new HPLC method using a reversed phase C8 column and pyridine-containing mobile phases. Mar Ecol Prog Ser 195:29–45. https://doi.org/10.3354/meps195029

Zapata M, Jeffrey S, Wright S, Rodríguez F, Garrido J, Clementson L (2004) Photosynthetic pigments in 37 species (65 strains) of Haptophyta: implications for oceanography and chemotaxonomy. Mar Ecol Prog Ser 270:83–102. https://doi.org/10.3354/meps270083

Zhang LX, Cooney RV, Bertram JS (1991) Carotenoids enhance gap junctional communication and inhibit lipid peroxidation in C3H/10T1/2 cells: relationship to their cancer chemopreventive action. Carcinogenesis 12:2109–2114. https://doi.org/10.1093/carcin/12.11.2109

Zuluaga M, Gueguen V, Letourneur D, Pavon-Djavid G (2018) Astaxanthin-antioxidant impact on excessive reactive oxygen species generation induced by ischemia and reperfusion injury. Chem Biol Interact 279:145–158. https://doi.org/10.1016/J.CBI.2017.11.012

Chapter 11
Combined Phylogenetic Analysis in *Echinocereus* (Cactaceae), the Use of Morphology, and Taxonomic Implications

Daniel Sánchez, Salvador Arias, Monserrat Vázquez-Sánchez and Teresa Terrazas

Abstract Phylogenies based on molecular characters has dominated publications rather than those based on morphological characters. Some authors have defended the use of morphology in phylogenetic reconstruction. In Cactaceae few studies have been made combining molecular and morphological characters. A good example about the use of morphology in phylogenetic analysis has been addressed in *Echinocereus*. *Echinocereus* is a morphologically diverse genus including 67 species that have been grouped into eight taxonomic sections based on morphological traits. Previous molecular phylogenetic analyses did not show entirely the relationships in *Echinocereus* species, and did not provide useful characters to recognize clades. Therefore, we performed a combined phylogenetic analysis with a set of 44 morphological characters and six chloroplast DNA sequences. Topologies from parsimony and Bayesian analyses resulted mostly congruent. However, relationships of *E. poselgeri* did not agree between analyses. A second bayesian analysis using long-branch extraction test resulted in a topology with a morphologically congruent position of *E. poselgeri*. Parsimony and Bayesian analyses corroborated the monophyly of *Echinocereus*, which included eight monophyletic

D. Sánchez
CONACYT—Laboratorio Nacional de Identificación y Caracterización Vegetal, Centro Universitario de Ciencias Biológicas y Agropecuarias, Universidad de Guadalajara, Zapopan, Jalisco, Mexico

S. Arias (✉)
Jardín Botánico, Instituto de Biología, Universidad Nacional Autónoma de México, Coyoacán, Ciudad de México, Mexico
e-mail: sarias@ib.unam.mx

M. Vázquez-Sánchez
Departamento de Botánica, Instituto de Biología, Universidad Nacional Autónoma de México, Coyoacán, Ciudad de México, Mexico

T. Terrazas
Colegio de Postgraduados, Carretera México-Texcoco Km 36.5, Montecillo, 56230 Texcoco, Estado de México, Mexico

K. G. Ramawat (ed.), *Biodiversity and Chemotaxonomy*, Sustainable Development and Biodiversity 24, https://doi.org/10.1007/978-3-030-30746-2_11

groups. The clades did not recover the recent infrageneric classification. As a consequence, a new sectional classification for *Echinocereus* is proposed based on the eight recovered clades, which are supported by a combination of morphological and molecular characters. An identification key for sections in the genus is included.

Keywords Combined analyses · *Echinocereus poselgeri* · Echinocereeae · Long-branch attraction · Long-branch extraction · Morphology · Parsimony · Taxonomy

11.1 Introduction

The phylogenies based on molecular characters has dominated publications during last decades rather than those based on morphological characters. Despite this tendency, there are several essays that advocated the using of morphology in phylogenetic reconstruction (Jenner 2004; Smith and Turner 2005; Wheeler 2004). It has been demonstrated that inclusion of morphological characters can modify the topologies from molecular phylogenies (Bergsten 2005), and it has been seated the importance of simultaneous or combined phylogenetics analyses using different sources of evidence (Nixon and Carpenter 1996). In several families, including Cactaceae, morphology has been mainly included as few traits mapped on molecular phylogenies (e.g. Demaio et al. 2011; Hernández-Hernández et al. 2011). In this chapter, we disclose the importance to include categorical characters as structural changes in sequences as well as morphological characters in phylogenetic analyses, as a way to improve accuracy in topology, using as an example the diverse genus *Echinocereus*.

Molecular versus morphological data. Scotland et al. (2003) indicate that the advantages of molecular characters are the quantity of data and that they can be obtained easily and can be analysed using models of molecular evolution. This idea was adopted by some systematists in Cactaceae, who also argued that morphological traits are dominated by phenotypic plasticity due to ambient conditions and parallel evolution (Wallace and Gibson 2002). However, those authors did not realize a deep and comparative morphological revision to correctly code characters and character states to probe through a phylogenetic analysis and corroborate the uselessness of morphology for grouping taxa. The qualification and designation of the similarities as homology or homoplasy are not possible a priori, and they should be explained under a phylogenetic analysis in congruence with other character states (De Pinna 1991; Luna and Mishler 1996). Also, Smith and Turner (2005) suggest that both molecular and morphological characters can be homoplastic. In Cactaceae, for example, the phylogeny of the tribe Cacteae (Butterworth et al. 2002) based on the chloroplast marker *rpl16* showed a consistency index (CI) of 0.494 and homoplasy index (HI) of 0.516, so half of the informative sites of the marker sequence are homoplastic in the most parsimonious tree.

Taxonomic scope of the molecular phylogenies in Cactaceae. The initial changes in the taxonomy of Cactaceae from molecular phylogenies occurred at subfamily and tribe level, but with poor resolution in a generic level (Nyffeler 2002; Wallace 1995; Wallace and Gibson 2002). Few years later, analyses with a greater sampling and more molecular markers allowed to delimit genera with high support as in *Gymnocalycium* Pfeiff. (Demaio et al. 2011); *Opuntia* Mill (Majure et al. 2012); some genera in tribe Rhipsalideae (Korotkova et al. 2011; Calvente et al. 2011); genera in tribe Cacteae (Vázquez-Sánchez et al. 2013); and several genera in Echinocereeae (Arias et al. 2005; Sánchez et al. 2014) and Hylocereeae (Cruz et al. 2016; Korotkova et al. 2017). The results of those studies have allowed to construct a natural classification as is preferred in phylogenetic systematics (Wheeler 2004). In a natural classification, each taxon contains information about characters and the hierarchy allows to describe the distribution of the characters, so that the taxa names construct a system referring to diagnosis of those taxa (Alves and Machado 2007; Nixon and Carpenter 2000). However, a classification based only on molecular phylogeny will not be informative until the morphological characters are included. In Cactaceae, although the molecular phylogenies have been integrated into the family's classification (Hunt 2012, 2016; Nyffeler and Eggli 2010), they miss of the analyses of morphological characters, and the proposed taxa do not file synapomorphies or diagnostic characters.

The role of morphology in phylogenetic reconstruction. Scotland et al. (2003) propose that morphology may be integrated into phylogenetics through the study of few characters, which should be examined in the context of the molecular phylogeny. Character mapping has been supported for other authors (Bollback 2006; Cunningam et al. 1998), who argue that this approach provides a historical framework to understand evolution of morphology. Morphological characters in Cactaceae have been mapped on molecular phylogenies (*Echinopsis* Zucc., Schlumberger and Renner 2012; *Rebutia* K. Schum., Ritz et al. 2007) and have been used in the reconstruction of putative ancestral states on the molecular phylogenies of some genera (*Copiapoa*, Larridon et al. 2015; *Gymnocalycium* Pfeiff., Demaio et al. 2011; *Pereskia* Mill., Edwards et al. 2005). However, mapping morphological characters does not allow a test for congruence of characters and does not determine whether some morphological characters are strictly synapomorphies (Assis 2009; De Pinna 1991; Patterson 1982). Studies in Cactaceae as part of phylogenetic reconstruction using combined morphological and molecular data have conducted few lineages (Albesiano and Terrazas 2012; Guerrero et al. 2011; Sánchez et al. 2018; Tapia et al. 2018; Vargas-Luna et al. 2018; Vázquez-Sánchez et al. 2019).

Scotland et al. (2003) also justify the morphological character mapping approach by the supposed idea that most of the morphological characters are ambiguous and present problems in conceptualization and codification. Conceptualization or abstraction of characters is not only dependent on organisms, but also dependent on the conceptual and operational expertise of the researcher dissecting organisms in comprehensible units (characters) and assigning similarities between taxa (character states) (Richards 2003; Winther 2009). Winther (2009) proposes some objectivity criteria for character analysis: (1) relative position (topological), (2) special similarity, (3) series of intermediate forms, (4) conjunction, (5) causal grounding, (6) interdisciplinary communication; the first four have been previously proposed (Paterson 1982; Wiley and Lieberman 2011) and well accepted in phylogenetic systematic. Following Winther (2009), the causal grounding criterion explains similarity, and interdisciplinary communication provides the means to adjust and coordinate the remaining criteria. In Cactaceae, previous studies have provided detailed descriptions regarding the morphology of stem, flower, fruit, and seed (Buxbaum 1951, 1953, 1955; Gibson and Nobel 1986), in addition to monographies based on morphological variation (e.g. Anderson 2001; Berger 1926; Britton and Rose 1919, 1920, 1922, 1923; Buxbaum 1958; Endler and Buxbaum 1974; Hunt et al. 2006; Schumann 1899). This literature represents the framework to generate a more detailed research about characters and is fairly enough to address those fifth and sixth criteria for conceptualization of morphological characters.

Combined analyses. Farris (1979) suggests the inclusion of the whole evidence to perform a phylogenetic analysis. A combined analysis (de Queiroz et al. 1995), also called total evidence analysis (Kluge 1989) or simultaneous analysis (Nixon and Carpenter 1996), produces a more supported phylogeny. This method maximizes congruence between different data, and allow the emergence of secondary phylogenetic signal (Nixon and Carpenter 1996). Morphological characters were initially discarded in probabilistic phylogenetic analyses, since evolution models were not implemented as in molecular characters. However, the incorporation of new algorithms into the software for probabilistic phylogenetic reconstruction (Lewis 2001; Nylander et al. 2004) allows to perform a combined analysis using both molecular and morphological data (e.g. Knopf et al. 2012; Sánchez et al. 2018). Additionally, Bergsten (2005) points out that the inclusion of certain morphological characters may change the topology in molecular phylogenies affected by long-branch attraction (LBA). In this sense, emerge the question is it important to include the morphological characters in phylogenetic reconstruction?

The genus Echinocereus (Cactaceae), case study. A combined analysis based on morphological and molecular evidence was performed in *Echinocereus*, which provides a good example about the importance of including morphological characters in a phylogenetic analysis. *Echinocereus* (Cactoideae, Echinocereeae) has 67 species (Hunt 2016) with short cylindrical stems, variable number of ribs, funnel-shaped flowers, fruits with spines, and black and warty seeds (Engelmann 1848). Nevertheless, *Echinocereus* includes a great variation in root, stem, spines, flower, fruit, and seed (Blum et al. 1998; Bravo-Hollis and Sánchez-Mejorada 1991; Taylor 1985, 1993). The distribution of *Echinocereus* ranges from central Mexico to the central USA, inhabiting primarily desert scrub and template woods (Taylor 1985, 1993). *Echinocereus* is a monophyletic group when *Echinocereus pensilis* (K. Brandegee) J. A. Purpus is excluded (Sánchez et al. 2014). Erumpent buds are supposed to be the synapomorphies of the genus (Sánchez et al. 2015), but these characters have not been tested in phylogenetic analyses. The molecular phylogeny of *Echinocereus*, based on chloroplast DNA sequences, recovered nine clades (Sánchez et al. 2014); however, only one clade represents the taxonomic section *Triglochidiati*, and the remaining sections (sensu Hunt et al. 2006) correspond to paraphyletic and polyphyletic groups. Particularly, species in section *Wilcoxia* (Hunt et al. 2006; Taylor 1985, 1993) were not recovered as a monophyletic group, despite its remarkable morphological and anatomical similarity (Blum et al. 2008; Loza-Cornejo and Terrazas 1996; Taylor 1985, 1993). Incongruence between morphological and molecular data may be explained due to parallel evolution of morphology (Wallace and Gibson 2002) or a method artefact such as LBA (Bergsten 2005). Therefore, in this work, authors conducted a phylogenetic analysis of *Echinocereus* that included a set of morphological and molecular characters to evaluate the possibility of an LBA artefact in the phylogenetic position of *Echinocereus poselgeri*, to obtain a series of morphological and molecular characters that supported the genus and internal clades, and to present a taxonomic treatment of *Echinocereus* and infrageneric taxa from the recovered monophyletic groups.

11.2 Materials and Methods

The analysis included 59 species of *Echinocereus* representing the morphological diversity of the genus and the eight sections (Hunt et al. 2006). Additionally, ten species were included as a sister group of the genus, according to recent phylogenies of the family (Bárcenas et al. 2011; Sánchez et al., 2014). A set of 44 morphological characters (including chromosome number) was generated in the present study by the examination of specimens collected in fieldwork and herbaria and living collections (Table 11.1). All characters were codified following the objectivity criteria proposed by Winther (2009) (see Sánchez et al. 2018). Six chloroplast DNA markers were included: the intergenic spacers *psbA-trnH* and *trnQ-rps16*; the *rpl16* intron; the region composed of the intron *trnL* and the IGS *trnL-trnF* (hereafter, *trnL-F*); the coding gene *matK*, flanked by the *trnK* intron (hereafter, *trnK/matK* marker); and the coding gene *rbcL* (see Sánchez et al. 2014 for details about primer sequences and thermal profiles in PCR amplification; see Sánchez et al. 2018 for DNA sequence accessions stored in the GenBank). DNA sequences were manually aligned and concatenated in a single matrix, and the extremes of sequences for each marker were deleted because of ambiguities. Highly variable regions that were difficult to align were not detected; only small regions of poly-A (in *rpl16* and *trnL-F*) and poly-T (in *psbA-trnH*, *rpl16*, *trnK/matK*, *trnL-F*, and *trnQ-rps16*) of different lengths were observed. Additionally, we generated a binary matrix with DNA insertion and deletion events (indels) observed on the aligned sequences (Table 11.2); these indels were coded using a simple coding method (Ochoterena 2009). Gaps generated by differences in lengths in the poly-A and poly-T regions were not coded.

Phylogenetic analyses. A first matrix using only morphological characters (morphology matrix), a second matrix using only sites of cpDNA sequences (DNA site matrix), a third matrix using sites and indels of the cpDNA sequences (molecular matrix), and finally a fourth matrix that incorporated morphological, cpDNA sequences, and indel data (combined matrix) were analysed under parsimony (MP) and Bayesian inference (BI). Also, a maximum likelihood

Table 11.1 Morphological characters and character states used in this study

1. **Habit**: (0) shrub; (1) tree
2. **Growth form**: (0) erect columnar; (1) decumbent columnar; (2) erect cylindrical; (3) decumbent cylindrical; (4) depressed globose
3. **Stem type**: (0) aerial stem; (1) caudex; (2) rhizome
4. **Stem diameter** (Tukey test (Tt), $P < 0.05$): (0) <2.2 cm; (1) 2.2–15 cm; (2) >15 cm
5. **Stem branching**: (0) branched stem; (1) mainly solitary stem
6. **Stem rib number** (Tt, $P < 0.05$): (0) <5: (1) 5–9; (2) 10–14; (3) 15–23; (4) >23
7. **Storage roots**: (0) absent; (1) thickened main root; (2) thickened lateral roots
8. **Central spine number** (Tt, $P < 0.05$): (0) 1–2; (1) 3–6; (2) >6; (3) none central spines
9. **Central spine shape**: (0) rounded; (1) angulate; (2) flattened
10. **Consistency of stem areole spines**: (0) rigid; (1) setous; (2) hairy
11. **Spine arrangement**: (0) radial; (1) pectinate with furrow; (2) pectinate without furrow
12. **Bud development**: (0) non-erumpent buds; (1) erumpent buds
13. **Stigma colour**: (0) yellow; (1) green
14. **Receptacular tube shape**: (0) regular funnel-shaped; (1) wide funnel-shaped; (2) narrow funnel-shaped
15. **Receptacular tube thickness** (Tt, $P < 0.05$): (0) <4 mm; (1) >4 mm
16. **Receptacular tube length and perianth length ratio**: (0) perianth 1.5 times longer than receptacular tube; (1) perianth more than 1.6 times longer than receptacular tube; (2) receptacular tube up to 1.5 times larger than perianth; (3) receptacular tube more than 1.6 times larger than perianth
17. **Flower length** (Tt, $P < 0.05$): (0) 4–8.5 cm; (1) >8.5 cm; (2) <4 cm
18. **Relation of the inner and outer stamen length**: (0) similar length between inner and outer stamens; (1) inner stamens longer than outer stamens
19. **Nectary length** (Tt, $P < 0.05$): (0) >9 mm; (1) 3–8 mm; (2) 1–2 mm
20. **Position of nectarial tissue**: (0) lateral; (1) basal
21. **Anther colour**: (0) yellow; (1) purple
22. **Inner tepal shape**: (0) linear; (1) oblanceolate; (2) oblanceolate–ovate
23. **Thickness of the tepal base** (Tt, $P < 0.05$): (0) <2 mm; (1) >2 mm
24. **Dominant colour of the flower** (colour chart cells): (0) white (A); (1) yellow (C9–11); (2) purple (H32–36); (3) pink (E36–39, F36–39); (4) red (E1–4, F1–4); (5) brown (I1–4)
25. **Colour tone of the flower throat**: (0) light colour; (1) dark colour; (2) without change
26. **Floral dimorphism**: (0) bisexual flowers; (1) unisexual flowers
27. **Trichome length in receptacular tube areoles** (Tt, $P < 0.05$): (0) short trichomes $\leq$ 1.5 mm; (1) long trichomes > 1.5 mm
28. **Consistency of the spines on receptacular tube areoles**: (0) rigid; (1) setous
29. **Fruit shape**: (0) round, 1:1–6:5; (1) elliptical, 2:1–3:2
30. **Fruit pulp**: (0) juicy; (1) semi-dry
31. **Chromosome number** (haploid): (0) 11; (1) 22
32. **Cell layers in hypoderm**: (0) 1–2 layers; (1) 3–5 layers; (2) >6 layers; (3) absent
33. **Silica bodies in stem epidermis**: (0) without silica bodies; (1) with silica bodies
34. **Type of secondary xylem**: (0) fibrous with rays; (1) fibrous without ray; (2) non-fibrous

(continued)

Table 11.1 (continued)

35. **Type of cortical vascular bundles**: (0) cortical bundles without phloic fibres; (1) cortical bundles with phloem fibres
36. **Tannins into epidermal cells of ovule funiculus**: (0) without tannins; (1) with tannins
37. **Tannins into epidermal cells of ovule body**: (0) without tannins; (1) with tannins
38. **Tannins into epidermal cells of stamen filaments**: (0) without tannins; (1) with tannins
39. **Tannins into epidermal cell of anther wall**: (0) without tannins; (1) with tannins
40. **Tannins into epidermal cell of tepals**: (0) epidermal cells of tepals without tannins; (1) epidermal cells of tepals with tannins
41. **Length of cotyledons**: (0) very long cotyledons > 1 mm); (1) long cotyledons > 0.5 mm; (2) short cotyledons, < 0.5 mm
42. **Seed length**: (0) very large: 3.0–3.9 mm; (1) large: 2.0–2.9 mm; (2) medium: 1.2–1.9 mm; (3) small: 0.9–1.1 mm
43. **Shape of the periclinal cell wall in the lateral region of testa seed**: (0) flattened; (1) convex; (2) hemispheric; (3) dome
44. **Ornamentation of lateral region of testa seed**: (0) smooth; (1) warty; (2) rugose

For explanation about generation and codification of characters, see Sánchez et al. (2018)

(ML) analysis using the CAT-GTR model, with the cpDNA site matrix, and the long-branch extraction test (LBE test),using the combined matrix, were made in case of incongruence between the resulting topologies from the MP and BI analyses, in order to explore LBA. It has been proposed that the use of a site heterogeneous model (e.g. CAT-GTR) in a phylogenetic analysis suppresses long-branch artefacts (Lartillot et al. 2007). LBE test assumes that a long branch is able to attract or be attracted by another long branch in a phylogenetic analysis; therefore, the exclusion of one of the long branches will allow the second long branch to be grouped in the correct clade (Pol and Siddall 2001). So, the selected taxon in phylogenetic reconstruction was excluded and an LBE test using the combined matrix and the same parameters (below) as the previous analyses was performed.

Parameters for phylogenetic analyses. All matrices were analysed under parsimony (MP) and using Bayesian inference (BI). The MP analysis was performed in TNT v. 1.1 (Goloboff et al. 2008) using parsimony-informative characters only (Table 11.3). We performed a heuristic search of 10,000 random addition sequences using ratchet, sectorial searches, drift, and tree fusing algorithms (Goloboff et al. 2008), saving 10 trees per replica. Support values were calculated from 10,000 replicas, using the same parameters as the heuristic search. The standard bootstrap support (BS) shows the absolute frequencies. The jackknife support (JK) removed 36% of the characters and shows the absolute frequencies. A strict consensus tree was computed from the most parsimonious trees. The ML analysis of the cpDNA matrix was performed using the mixture model CAT-GTR implemented in PhyloBayes 4.01 (Lartillot and Philippe 2004). The molecular and combined matrices were partitioned and analysed by BI using MrBayes 3.2.1 (Ronquist and Huelsenbeck 2003), because categorical characters can be included. For those analyses, the morphological and indel partitions were analysed under the Mkv model and coded as variable; for each cpDNA sequence, the nucleotide

Table 11.2 List of indels from cpDNA sequences included in this study

1.	Reversion of sequence in the *psbA-trnH* marker, sites 73–75
2.	Simple sequence repetition in the *psbA-trnH* marker, sites 107–110
3.	Simple sequence repetition in the *psbA-trnH* marker, sites 134–138
4.	Simple sequence repetition in the *psbA-trnH* marker, sites 167–184
5.	Insertion of sequence in the *psbA-trnH* marker, sites 266–269
6.	Deletion of sequence in the *psbA-trnH* marker, sites 222–263
7.	Simple sequence repetition in the *psbA-trnH* marker, sites 290–311
8.	Insertion of sequence in the *psbA-trnH* marker, site 378
9.	Insertion of sequence in the *psbA-trnH* marker, sites 505–510
10.	Simple sequence repetition in the *rpl16* marker, sites 202–206
11.	Simple sequence repetition in the *rpl16* marker, sites 221–224
12.	Inversion of sequence in the *rpl16* marker, sites 718–737
13.	Simple sequence repetition in the *rpl16* marker, sites 976–977
14.	Deletion of sequence in the *rpl16* marker, sites 998–1056
15.	Inversion of sequence in the *rpl16* marker, sites 1024–1038
16.	Simple sequence repetition in the *rpl16* marker, sites 1055–1056
17.	Deletion of sequence in the *trnL-F* marker, sites 301–307
18.	Deletion of sequence in the *trnL-F* marker, sites 283–496
19.	Deletion of sequence in the *trnL-F* marker, sites 309–486
20.	Deletion of sequence in the *trnL-F* marker, sites 367–371
21.	Simple sequence repetition in the *trnL-F* marker, sites 441–442
22.	Deletion of sequence in the *trnL-F* marker, sites 445–499
23.	Deletion of sequence in the *trnL-F* marker, sites 375–474
24.	Deletion of sequence in the *trnL-F* marker, sites 339–845
25.	Deletion of sequence in the *trnL-F* marker, sites 878–881
26.	Deletion of sequence in the *trnL-F* marker, sites 811–1041
27.	Simple sequence repetition in the *trnL-F* marker, sites 886–890
28.	Deletion of sequence in the *trnL-F* marker, sites 1155–1162
29.	Simple sequence repetition in the *trnL-F* marker, sites 1159–1162
30.	Deletion of sequence in the *trnQ-rps16* marker, sites 130–452
31.	Deletion of sequence in the *trnQ-rps16* marker, sites 228–232
32.	Deletion of sequence in the *trnQ-rps16* marker, sites 403–458
33.	Insertion of sequence in the *trnQ-rps16* marker, sites 472–476
34.	Simple sequence repetition in the *trnK/matK* marker, sites 168–172
35.	Deletion of sequence in the *trnK/matK* marker, sites 2117–2122

substitution was determined by the AIC using jModelTest 2 (Darriba et al. 2012; Table 11.3). The posterior probability values (BPP) were computed using two separate runs of Markov Monte Carlo chains (MCMCs), each run with four chains and 5,000,000 generations. The Markov chains were sampled every 10,000 generations, the MCMC convergence was visually examined, and 20% of the sampled trees were discarded. The remaining trees and BPP were summarized in a consensus

Table 11.3 Numerical data of aligned cpDNA sequences included in the analyses

	psbA-trnH	*rbcL*	*rpl16*	*trnK/matK*	*trnL-F*	*trnQ-rps16*	Combo
Included taxa	66	66	66	69[a]	66	66	69
Sequence length	520	578	1238	2524	1158	631	6649
Non-informative sites	485	558	1181	2471	1106	616	6404
Informative sites	35	20	57	53	53	27	245
% informative sites	6.73	3.46	4.60	2.09	4.57	4.64	3.68
Informative indels	9	0	7	2	13	4	35

[a]14 taxa include only the coding region *matK*

majority rule tree. Finally, an unequivocal character optimization was conducted using Winclada (Nixon 2002) on the strict consensus tree to understand the contribution of the characters in the phylogeny and to recognize the synapomorphies and homoplasies that defined each recovered clade. Delayed optimization (DELTRAN) and fast optimization (ACCTRAN) were explored to recognize additional characters that supported certain clades (Agnarsson and Miller 2008).

11.3 Results

Phylogenetic analyses. The MP and BI analyses of the morphological matrix resulted in a consensus tree that recovered *Echinocereus* as monophyletic group (Fig. 11.1). However, topology displayed a polytomy in which many *Echinocereus* species were collapsed at the base of the genus and only a few clades were recovered (section *Triglochidiati* and groups of species from sections *Costati*, *Erecti*, *Wilcoxia*, and *Reichenbachii*; sensu Hunt et al. 2006). Consensus tree of the MP analyses of the matrices using cpDNA sequences only, and concatenated cpDNA sequences and indels, showed a topology with many collapsed branches (Figs. 11.2a and 11.3a). BI analysis of the same molecular matrices resulted in more resolved topology (Figs. 11.2b and 11.3b); however, relationships between some clades are not resolved, and several nodes are weakly supported. Interestingly, branch support was better when indels were used in the MP analysis (Fig. 11.3a). Finally, MP and BI analyses of the combined matrix resulted in congruent topologies with minor differences in some clades (Fig. 11.4). The whole analysis recognized the genus *Echinocereus* as a monophyletic group (BS, JK, and BPP = 100, Figs. 11.1, 11.2, 11.3, 11.4 and 11.5. The last analysis combining all the available evidence shows the most resolved phylogeny, where the clade *Echinocereus* contains eight main clades with different support (Figs. 11.4a and 11.5). However, in the strict consensus tree from MP analyses (Fig. 11.5), *E. poselgeri* was grouped into clade B, which was the sister group of clade C. Together, clades B and C were the sister group of the remaining major clades (D–H; Fig. 11.5). The majority consensus tree from BI analysis included to *E. poselgeri* in

(a)

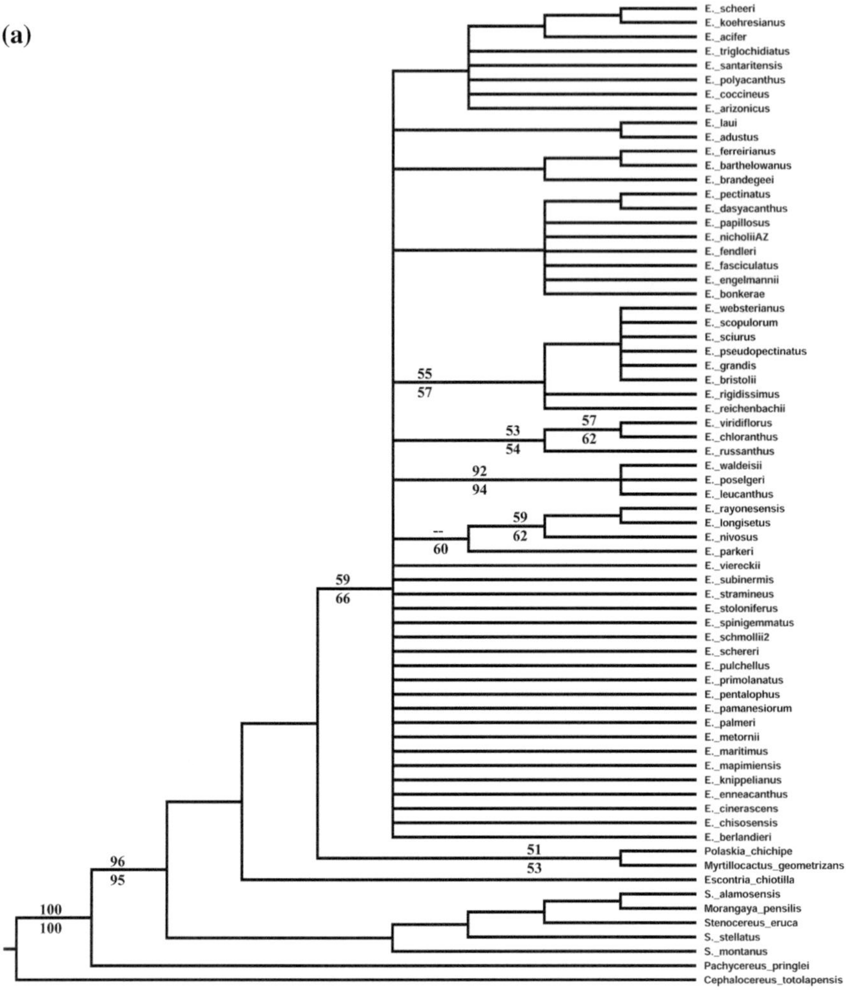

Fig. 11.1 **a** Consensus tree from the MP analysis of the morphological characters. Numbers above/below branches represent bootstrap/jackknife values. **b** Majority rule tree (shown as phylogram) from BI analysis of the morphological characters. Numbers in nodes represent posterior probabilities

clade C; clade B was recovered as a sister of the group that included clades C–H (Fig. 11.4a).

Exploration of long-branch attraction. Due to incongruence in the phylogenetic relationships of *E. poselgeri* between MP and BI analyses using the combined matrix (Fig. 11.4), a BI analysis using the model CAT-GTR and LBE test was directed to explore the sisterly of *E. poselgeri* and *E. mapimiensis* since both species with dissimilar morphology represented long branches (Fig. 11.4a). The ML analysis using the CAT-GTR model did not show significant changes in

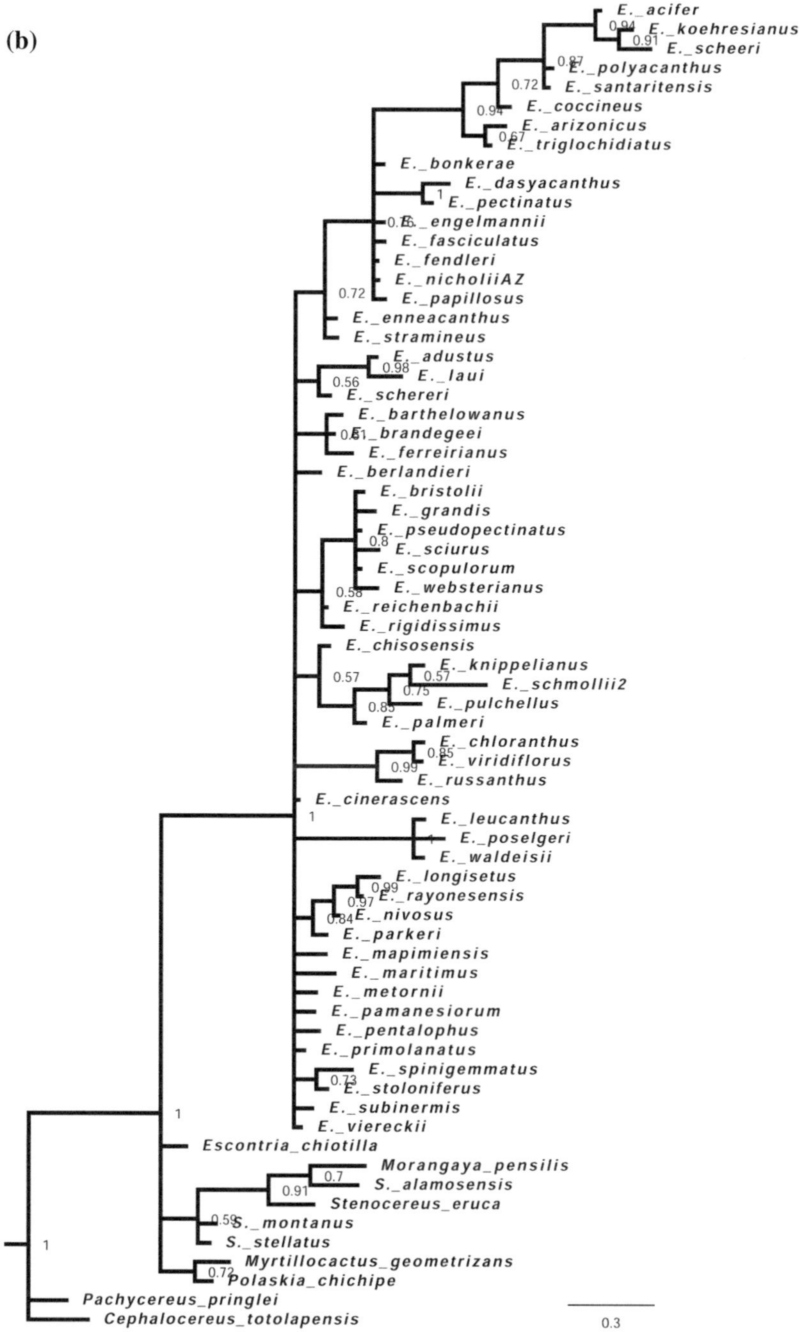

Fig. 11.1 (continued)

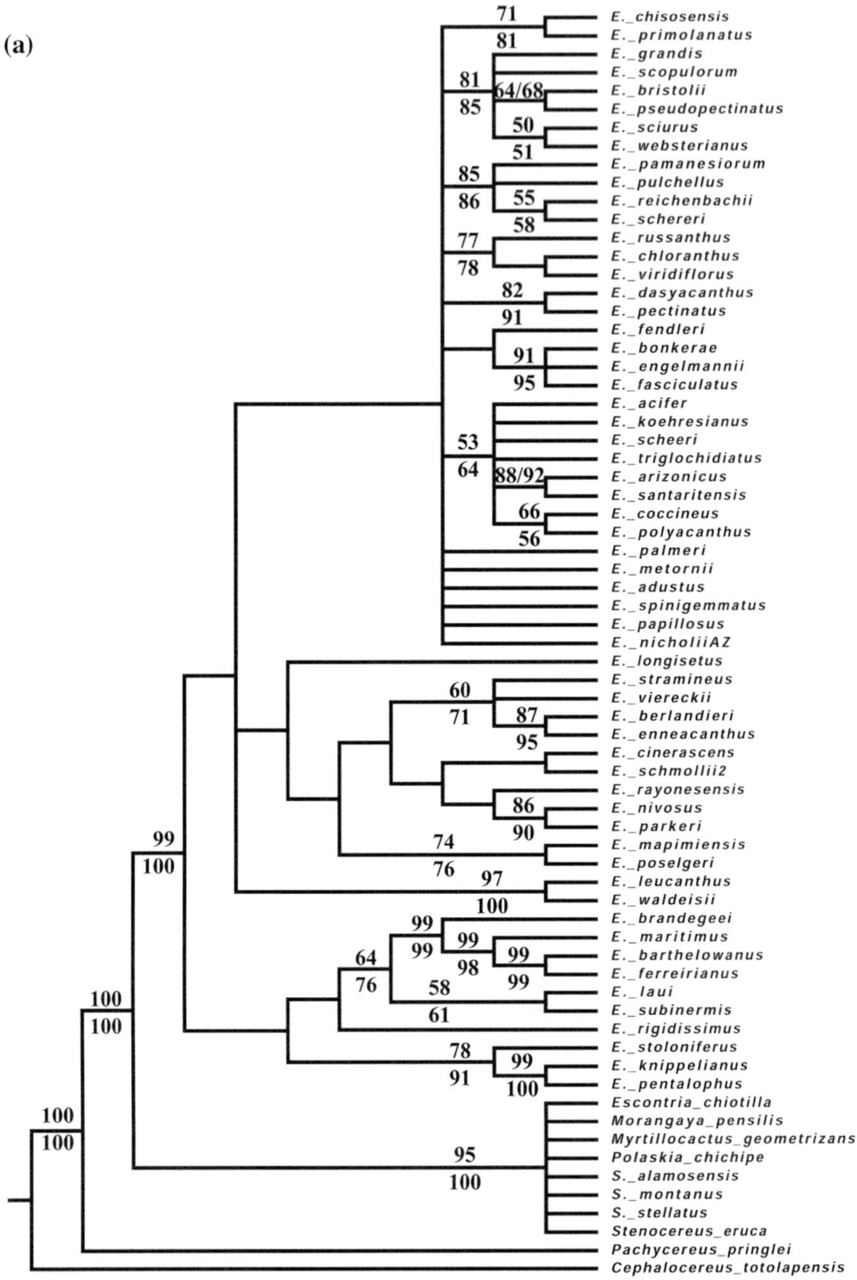

Fig. 11.2 **a** Consensus tree from the concatenated cpDNA sequences from the MP analysis. Numbers above/below branches represent bootstrap/jackknife values. **b** Majority rule tree (shown as phylogram) from BI analysis of the concatenated cpDNA sequences. Numbers in nodes represent posterior probabilities

Fig. 11.2 (continued)

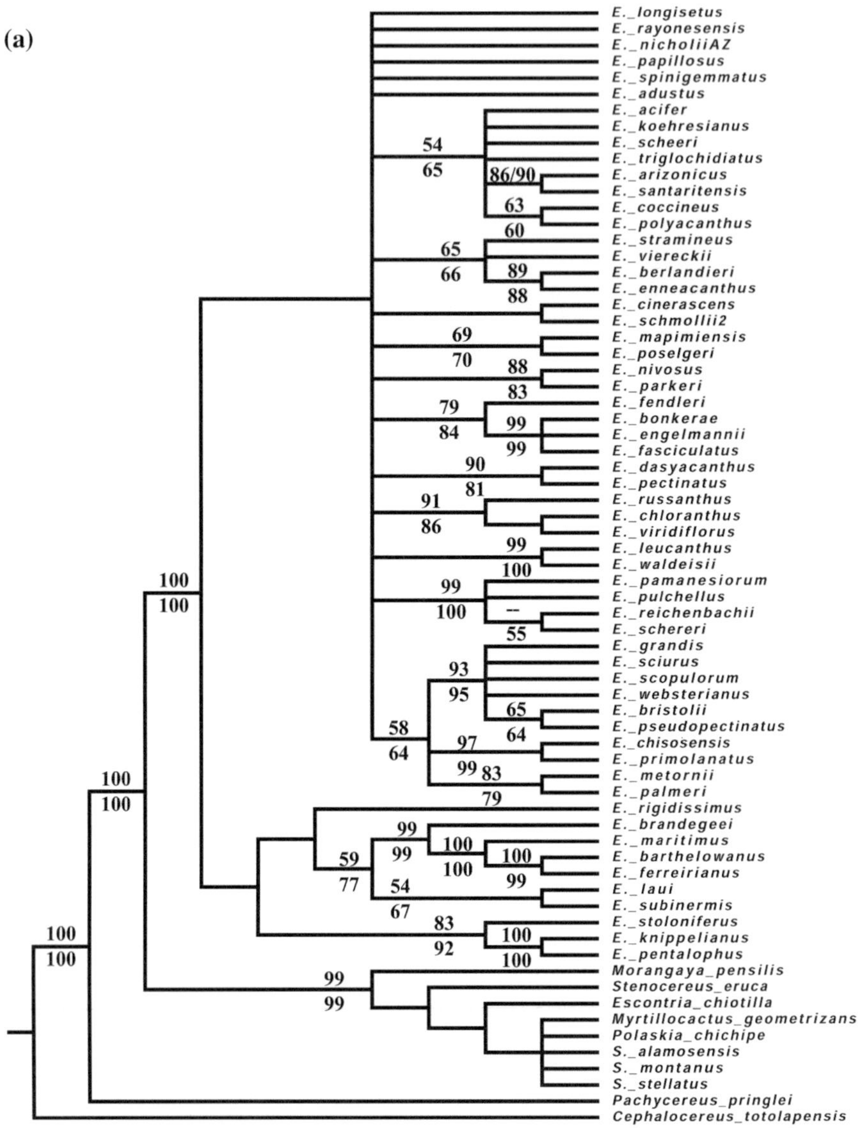

Fig. 11.3 **a** Consensus tree from the concatenated cpDNA sequences and codified indels from the MP analysis. Numbers above/below branches represent bootstrap/jackknife values. **b** Majority rule tree (shown as phylogram) from BI analysis of the concatenated cpDNA sequences and codified indels. Numbers in nodes represent posterior probabilities

topology where *E. poselgeri* was grouped with *E. mapimiensis* (Fig. 11.4c). On the other hand, the LBE test using BI method that excluded *E. mapimiensis* resulted in the grouping of *E. poselgeri* into clade B with the morphologically similar *E.*

Fig. 11.3 (continued)

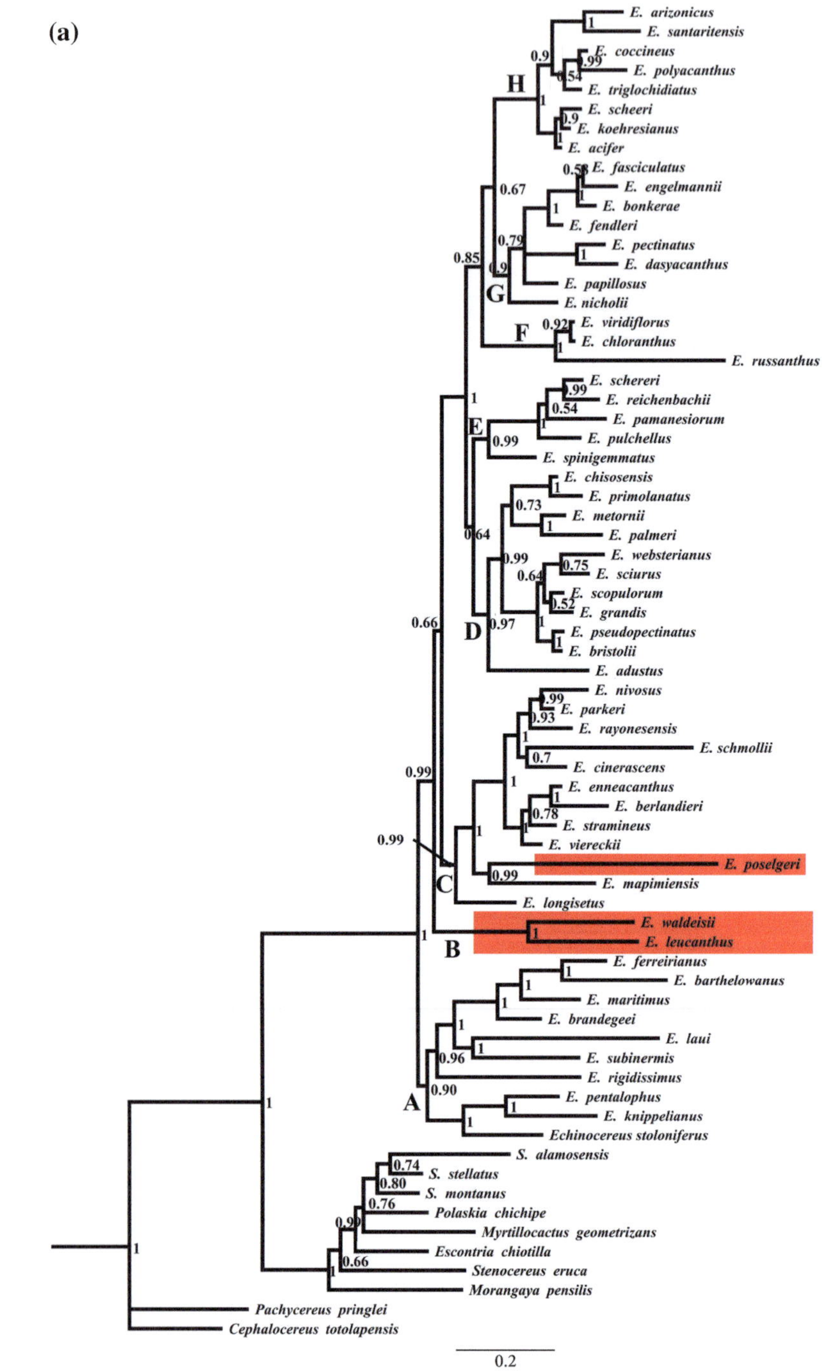

Fig. 11.4 Phylogenetic tress from probabilistic analyses of the combined matrix; the blue and red shadows show the position of *E. poselgeri* and putative sister species (Hunt et al. 2006). **a** Majority rule tree (shown as phylogram) from the BI analysis including all taxa. **b** Majority rule tree (shown as phylogram) from the LBE test excluding *E. mapimiensis*. **c** Phylogram of the best tree from the ML analysis using the CAT-GTR model

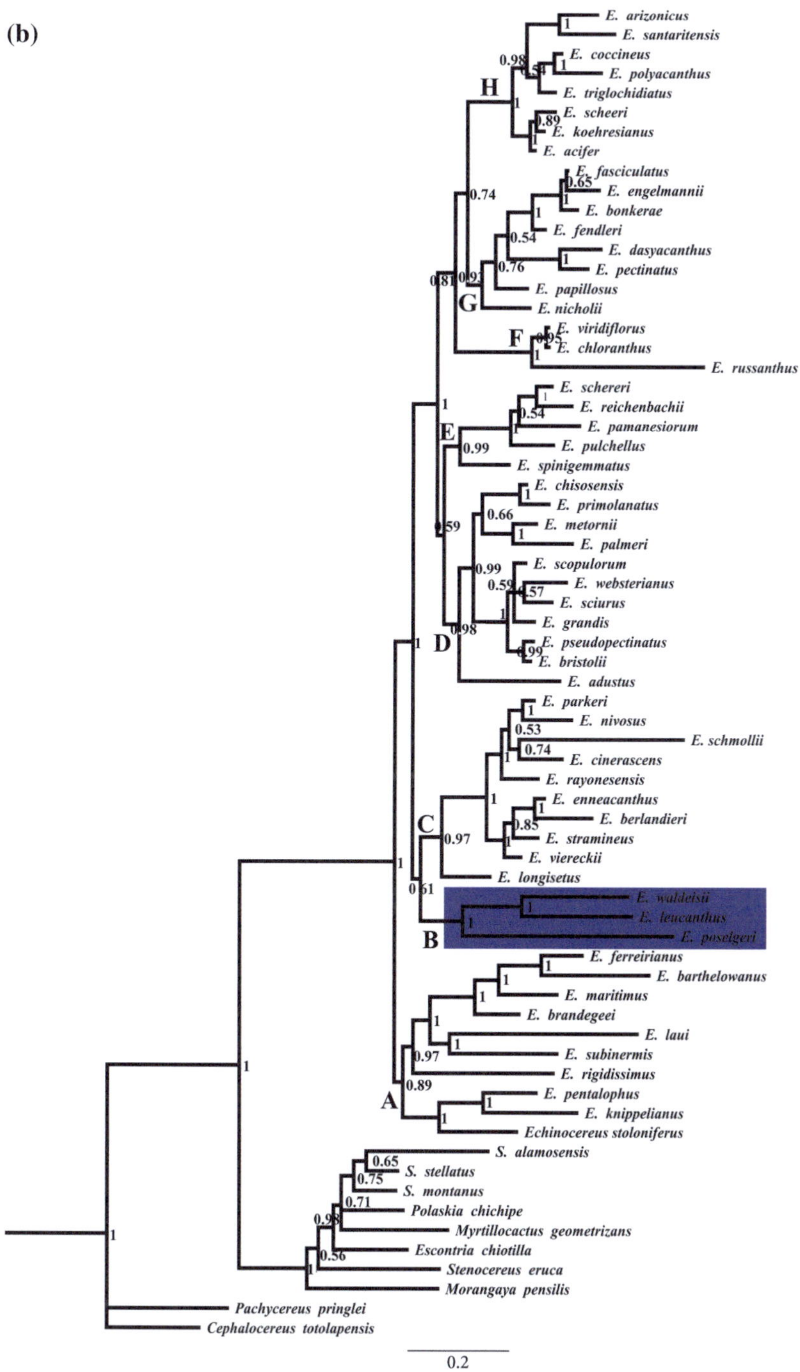

Fig. 11.4 (continued)

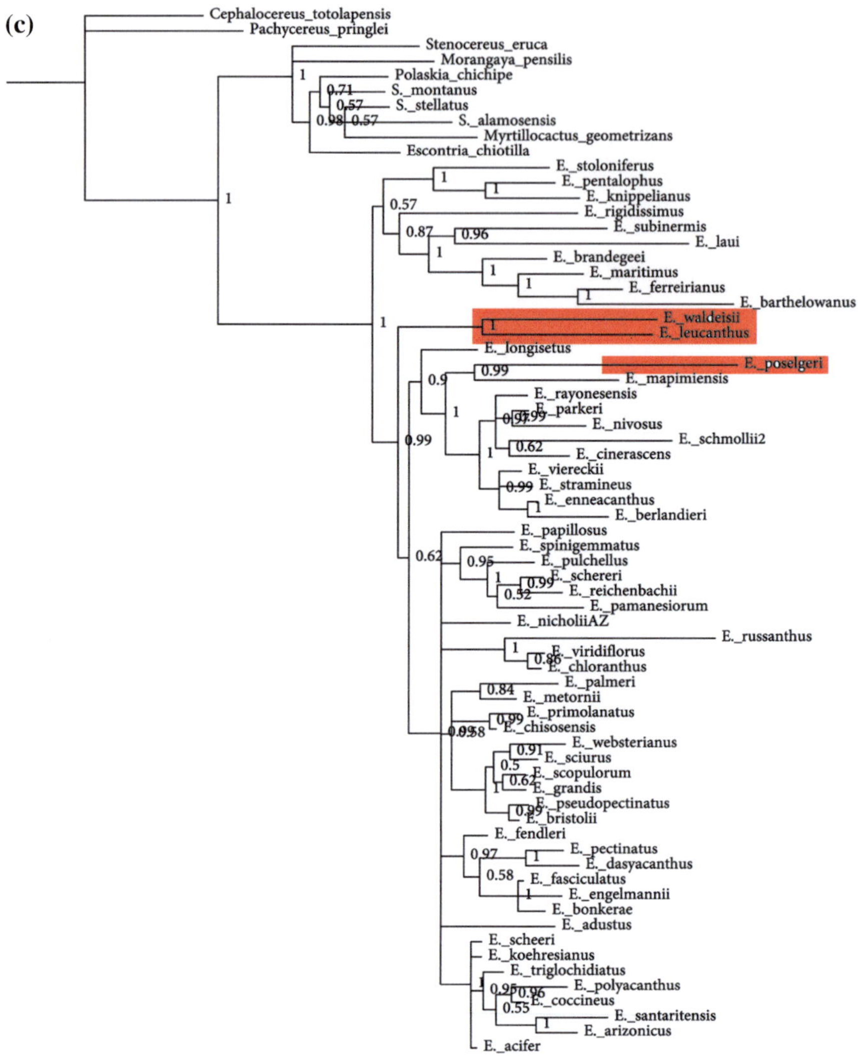

Fig. 11.4 (continued)

leucanthus and *E. waldeisii* (Fig. 11.4b). The test also showed clades B and C as sister groups; both clades B and C represented the sister group of the clade that included the remaining major clades (D–H). This result suggested an LBA bias in the original BI analysis of the combined matrix. A second LBE test that excluded *E. poselgeri* from the analysis did not show any change in the topology; *E. mapimiensis* was grouped in clade C, as in the previous analysis. LBE tests using the MP method and excluding the previous taxa also did not show any changes in topology (not shown).

11.4 Discussion

Phylogenetic analyses. Analyses showed that as we add characters (sequences, indels, and morphology), we have a positive effect improving resolution, support, and accuracy of the resulted phylogenetic hypothesis. The advantages of adding characters in phylogenetics analyses have been suggested in other analyses in several organisms (e.g. Wortley et al. 2005). The topology using only morphological characters was not resolved to support hypothesis of relationships, although it recovered some species groups traditionally recognized by Taylor (1985, 1993) and Blum et al. (1998), as well as clades recovered in the molecular phylogeny of the genus (Sánchez et al.2014) as section *Triglochidiati*, *Erecti*, and *Echinocereus* species from Baja California peninsula (Fig. 11.1). The contribution and importance of the morphological characters in the combined analysis is discussed below. Otherwise, the use of indels was significant in increasing the support of some nodes, particularly in the MP analysis. This result coincides with Simmons et al. (2001) who have demonstrated that including indel characters in sequence-based matrices often changes the topology or resolution of the strict consensus tree and including indel characters in sequence-based matrices often increases branch support values (Fig. 11.3a). This positive effect on branch support values has been recorded in other phylogenetic analyses in Cactaceae (Vázquez-Sánchez et al. 2013). Also, the inclusion of indels showed a more resolved topology in the BI analysis (Fig. 11.3b).

Long-branch attraction bias in Echinocereus. Different phylogenetic reconstruction methods (MP or BI) using the same data set commonly result in topologies with minor differences or differences in support values (Rindal and Brower 2011). The same pattern has been observed in phylogenetic analyses on Cactaceae; however, the causes of these differences in topologies have not been discussed (Vázquez-Sánchez et al. 2013). The results of the analysis of all available evidence from MP and BI analyses showed a strong inconsistency in the phylogenetic position of *E. poselgeri* (Figs. 11.4a and 11.5), a very distinctive taxon within *Echinocereus* (see discussion of section *Wilcoxia*). Although we explored the possibility of LBA effect by analysing the cpDNA sequence matrix through ML analysis using the CAT-GTR model, the topology did not show any change in the position of *E. poselgeri* (Fig. 11.4c). However, the LBE test using the same BI parameters and excluding *E. mapimiensis,* resulted in grouping *E. poselgeri* in clade B as a sister species to *E. leucanthus* and *E. waldeisii* (BPP = 1; Fig. 11.4b), as it was previously allocated in the MP analysis (Fig. 11.5). Our results showed that with the inclusion of morphological characters in the MP analysis, the LBA bias on *E. poselgeri* relationships could be avoided. Bergsten (2005) suggests adding morphological characters to the analyses as a strategy to obtain more accurate topologies and avoid LBA problems. Our results are consistent with the Kolaczkowsky and Thornton (2009) who suggest that LBA bias can affect BI analyses but that MP analysis was not the most susceptible method to improperly group taxa through the LBA bias. We surmised that the results of the MP analysis

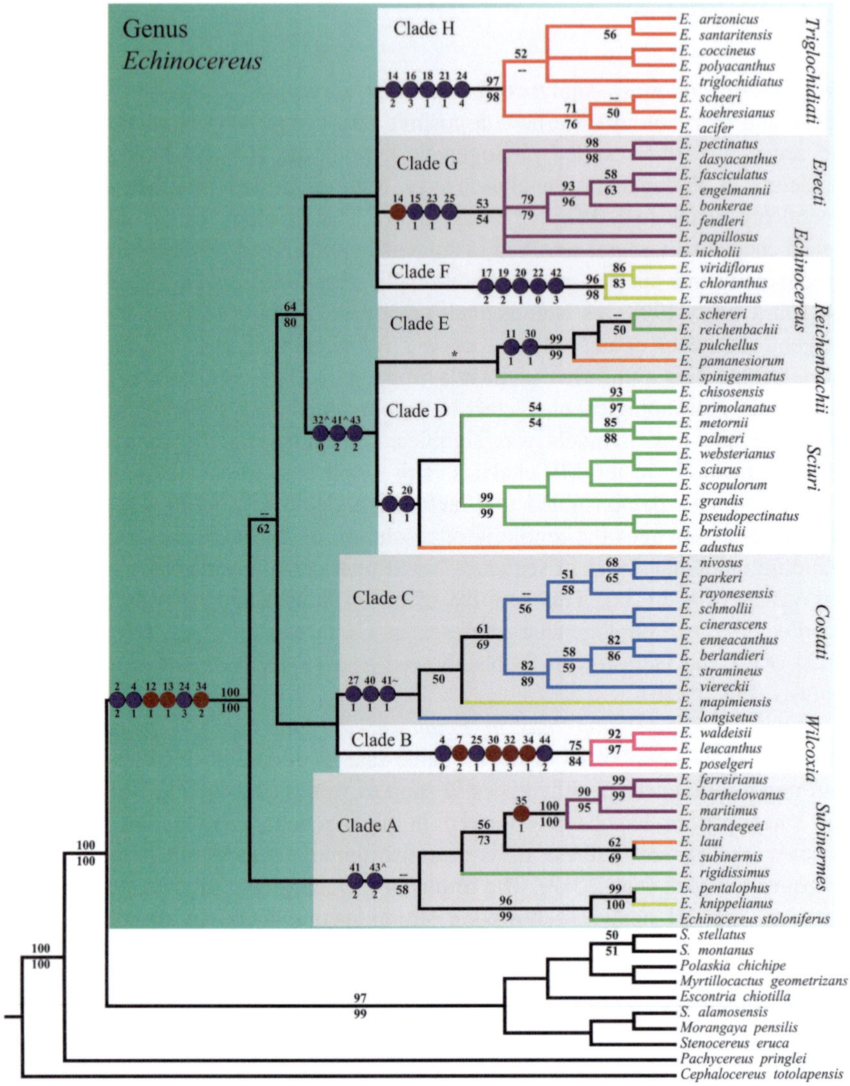

Fig. 11.5 Strict consensus tree from six most parsimonious trees from MP analysis. Numbers above/below branches represent bootstrap/jackknife values. Unambiguous character optimization is represented by circles on branches: red circle = synapomorphy and blue circle = homoplasy. Numbers above/below circles indicate character/state (see character list; Table 13.1); ^ specifies DELTRAN optimization; ~ specifies ACCTRAN optimization. * indicates that clade was supported by molecular characters rather than morphological characters. Branch colours represent the traditional sectional classification by Hunt et al. 2006, blue: *Costati*; green: *Reichenbachii*; orange: *Pulchellus*; pink: *Wilcoxia*; purple: *Erecti*; red: *Triglochidiati*; and yellow: *Echinocereus*

of the combined matrix were better because all taxa sampled were included and the analysis was not affected by LBA.

Combining molecular and morphological characters in *Echinocereus*. The combined analyses of morphological and molecular characters corroborated that the genus *Echinocereus* was a monophyletic group with high support (Figs. 11.1, 11.2, 11.3, 11.4 and 11.5), as was proposed previously using only molecular characters (Sánchez et al. 2014). Wortley and Scotland (2006) suggest that a combined analysis positively affects topology resolution but does not necessarily elevate the support values, which was partially observed in our results. The strict consensus tree showed a decrease in the support values in some clades (e.g. clade A; Fig. 11.5) and an increase in those values in some other clades (e.g. clade H; Fig. 11.5), compared with previous studies (Sánchez et al. 2014). Reduction of support values in some clades can be explained by the inclusion of several vegetative characters (i.e. stem diameter, number of ribs, and number of central spines; Sánchez et al. 2018) that were revealed as homoplasies, but have been useful in species group delimitations (Baker 2006a, b; Sánchez et al. 2013). However, our phylogenetic analyses using the combined matrix resulted in six most parsimonious trees, and according to de Carvalho (1996), an analysis resulting in few parsimonious trees is evidence of congruence among data, although several branches have low support values. Contrasting the molecular phylogenetic analyses in *Echinocereus* (Sánchez et al. 2014), the inclusion of morphological characters and the *trnK/matK* marker allowed the inclusion of *Echinocereus chloranthus*, *Echinocereus russanthus*, and *Echinocereus papillosus,* recovered a more resolved relationship of the main clades, and grouped *E. poselgeri* in a morphologically congruent clade. Therefore, based on the principles of ontological and epistemological congruency in phylogenetic analyses (Assis and Rieppel 2011), we preferred using the strict consensus tree from the combined MP analysis to describe the phylogenetic relationships in *Echinocereus* and to optimize the characters to recognize synapomorphies and homoplasies that supported the main clades. This character optimization on the MP strict consensus tree showed that morphological and molecular characters (including indels) were important in the definition of the clades recovered in *Echinocereus*. Although synapomorphies are preferred as evidence of monophyly, homoplasies are also important because they can support many of the nodes in a phylogenetic tree, and homoplasies are fundamental in a group diagnosis (Assis 2009; Assis and Rieppel 2011; de Carvalho 1996; Nixon and Ochoterena 2000). Therefore, the genus and the main clades were defined by synapomorphies (when present) and/or a combination of homoplasies, as it has been determined for other angiosperm lineages (Hughes et al. 2004; Norup et al. 2006). Dichotomic key of the sections recognized in *Echinocereus* is presented in Sánchez et al. (2018). Recently, the combined use of molecular characters and morphology has been made in other phylogenetic analyses in Cactaceae where the new circumscription of the studied genera has been clearly understood by means of the morphological characters (Vargas-Luna et al. 2018; Vázquez-Sánchez et al. 2019).

Classification of *Echinocereus* and sections. ***Echinocereus*** Engelm., Wisliz. Tour North Mexico: 91 (1848).*Cereus* subgen.
Echinocereus Engelm., Proc. Amer. Acad. Arts 3: 278 (1856). Lectotype (designed by Britton and Brown 1913): *Echinocereus viridiflorus* Engelm.

Echinocereus was defined by a combination of six morphological characters: erumpent buds, green stigmas, non-fibrous secondary xylem, cylindrical growth form, stem diameter from 3 to 15 cm, and three of them as synapomorphies (Fig. 11.5). Erumpent buds are reported for all *Echinocereus* species (Sánchez et al. 2015), and authors suggest that this trait protects buds from extremely low temperatures during winter. This trait also has favoured the lineage diversification in the temperate and semi-arid regions of Northern Mexico and the Southwest USA. Non-fibrous secondary xylem is described for several species in Cactoideae and corresponds to a specialized secondary xylem feature in small-sized cacti (Gibson 1973; Vázquez-Sánchez and Terrazas 2011). Cylindrical growth form is a distinctive character in *Echinocereus* comparing with the tribe Echinocereeae, which several lineages have a tendency to show a tree-like or scrub-like columnar growth form. A total of 15 synapomorphic sites from the cpDNA sequences defined *Echinocereus* (*psbA-trnH*: 2, *rbcL*: 3, *rpl16*: 3, *trnK/matK*: 4, *trnL-F*: 1, *trnQ-rps16*: 2) and two homoplasic sites (*psbA-trnH*: 1 and *trnQ-rps16*: 1). Additionally, the absence of two indels in *psbA-trnH* and *trnL-F* markers were diagnostic for the genus. The usefulness of those chloroplast sequences has been proven in other phylogenetic studies within Cactaceae (Arias et al. 2005; Calvente et al. 2011; Edwards et al. 2005; Hernández-Hernández et al. 2011; Hernández-Ledesma and Bárcenas 2017; Korotkova et al. 2011; Nyffeler 2002; Sánchez et al. 2014; Vázquez-Sánchez et al. 2013), resulting in a well-resolved phylogenetic hypothesis. With those results, we propose a new infrageneric classification of *Echinocereus* based on phylogenetic information because the genus was traditionally divided into eight non-monophyletic sections (Hunt et al. 2006). Consequently, the sections are presented as follows.

Section ***Subinermes*** (K. Schum.) Mich. Lange, Echinocereenfreund 8: 16 (1995). *Echinocereus* ser. *Subinermes* K. Schum., Gesamtbeschr. Kakt.: 246 (1899). Type species: *Echinocereussubinermis* Salm-Dyck ex Scheer.
Species included: *Echinocereus barthelowanus* Britton & Rose, *Echinocereus brandegeei* (J. M. Coult.) K. Schum., *Echinocereus ferreirianus* H. E. Gates, *Echinocereus knippelianus* Liebner, *Echinocereus laui* G. Frank, *E. maritimus* (M. E. Jones) K. Schum., *Echinocereus pentalophus* (DC.) Lem., *Echinocereus rigidissimus* (Engelm.) Haage, *Echinocereus stoloniferus* W. T. Marshall, *Echinocereus subinermis* Salm-Dyck ex Scheer.

Section *Subinermes* (Figs. 11.5 and 11.6a, b). Group of ten species with heterogeneous morphology included taxa previously classified in different sections (sensu Hunt et al. 2006). This section was supported by a hemispheric periclinal cell wall in the lateral region of testa seed and the cotyledon size (Figs. 11.5 and 11.6b), and one DNA site in the *rpl16* marker. Within this clade, a group formed by

Fig. 11.6 Members of the recognized sections in *Echinocereus* and their distinctive characters. **a** *E. subinermis* (*S. Arias 2007*, MEXU). **b** Periclinal wall of the lateral side of seed in *E. pentalophus* (*S. Arias 1746*, MEXU). **c** *E. leucanthus* (*S. Arias 1837*, MEXU). **d** Tuberous roots in *E. poselgeri* (*S. Arias 1492*, MEXU). **e** *E. parkeri* (JB-UNAM, s. n., MEXU). **f** Tannins into tepals epidermis of *E. berlandieri* (*S. Arias 1454*, MEXU). **g** *E. metornii* (*D. Sánchez 83*, MEXU). **h** Floral nectary of *E. palmeri* (*D. Sánchez 25*, MEXU). **i** E. *schereri* (*D. Sánchez 72*, MEXU). **j** Fruit with semi-dry pulp in *E. schereri* (*D. Sánchez 72*, MEXU). **k** *E. viridiflorus* (*D. Sánchez 80*, MEXU). **l** Flower morphology in *E. viridiflorus* (JB-UNAM, s. n., MEXU). **m** *E. engelmannii* (*D. Sánchez s. n.*). **n** Flower morphology in *E. dasyacanthus* (*D. Sánchez 47*, MEXU). **o** *E. acifer* (*D. Sánchez 50*, MEXU). **p** Flower morphology in *E. polyacanthus* (*D. Sánchez 24*, MEXU)

Echinocereus stoloniferus, *E. pentalophus*, and *E. knippelianus* was primarily supported by several DNA sites (*rbcL*: 1, *trnK/matK*: 1, *trnL-F*: 1, *trnQ-rps16*: 2). A distinctive subgroup with high support was composed of four endemic species from Baja California and the Gulf of California, which presented cortical bundles with phloem fibres and six changes in DNA sites (*rpl16*: 2, *trnK/matK*: 2, *trnL-F*: 2).

Section ***Wilcoxia*** (Britton and Rose) N. P. Taylor, Gen. Echinocereus: 134 (1985). *Wilcoxia* Britton and Rose, Contr. U.S. Natl. Herb. 12: 434 (1909). Type species: *Echinocereus poselgeri* Lem.
Species included: *Echinocereus kroenleinii* (Mich. Lange) W. Blum and Waldeis, *E. leucanthus* N. P. Taylor, *E. poselgeri* Lem., *E. tamaulipensis* (Wenderm.) Mich. Lange, *E. waldeisii* Haugg.

Section *Wilcoxia* (Figs. 11.5 and 11.6c, d). The three species that represented this section are characterized by presence of tuberous roots (Fig. 11.6d), elliptic fruits, fibrous rayless wood, and non-collenchyma hypoderm in the stem; in addition to a columnar growth form, a stem diameter less than 2.2 cm, and rugose ornamentation in the lateral region of testa seed. Also, four sites in the *trnK/matK* marker supported this clade. Section *Wilcoxia* represents a lineage with high specialization in stem and root (Taylor 1985, 1993) because the aforementioned traits allow it to clamber over surrounding bushes. The fibrous, rayless wood provides better support to the long and thin stem (Loza-Cornejo and Terrazas 1996).

Section ***Costati*** (Engelm.) N. P. Taylor, Piante Grasse 13 (4, Suppl.): 94 (1994). [1993 publ. 1994]). *Cereus* section *Costati* Engelm., Mem. Amer. Acad. Arts. ser. 2, 4: 50 (1849). Type species: *Echinocereus enneacanthus* Engelm.
Species included: *Echinocereus berlandieri* (Engelm.) Haage, *Echinocereus cinerascens* (DC.) Lem., *E. enneacanthus* Engelm., *Echinocereus longisetus* (Engelm.) Lem., *E. mapimiensis* Anderson, *Echinocereus nivosus* Glass & R. A. Foster, *Echinocereus parkeri* N. P. Taylor, *Echinocereus rayonesensis* N. P. Taylor, *Echinocereus schmollii* (Weing.) N. P. Taylor *insertae sedis*, *Echinocereus stramineus* (Engelm.) Engelm. ex F. Seitz, *Echinocereus viereckii* Werderm.

Section *Costati* (Figs. 11.5 and 11.6e, f). This clade grouped eleven species, which share short trichomes on the areolas of the receptacular tube, transparent spines on the receptacular tube in fixing solution, an embryo with large cotyledons, tannins in the epidermis of tepals supported this section, and one site in the *trnL-F* marker. Tannins in floral structures are observed in several members of the sister group of *Echinocereus* (Fuentes 2004). Tannin in tepal epidermis is easily recognized because flowers turn brown when they are fixed in formalin (Taylor 1993; Fig. 11.6f). In this study, *E. schmollii* was grouped in this clade because of molecular characters; however, its inclusion in *Costati* remains controversial because it does not share any of the diagnostic morphological characters of the section.

Section ***Sciuri*** Dan. Sánchez and S. Arias. Syst. Biodivers. 16 (1): 32 (2018). Type species: *Echinocereus sciurus* (K. Brandegee) Dams, Monatsschr. Kakteenk. 14: 130 (1904). Species included: *Echinocereus adustus* Engelm., *E. bristolii* W. T. Marshall, *Echinocereus chisosensis* W. T. Marshall, *Echinocereus grandis*

Britton and Rose*, *E. metornii* G. Frank, *Echinocereus palmeri* Britton & Rose, *Echinocereus primolanatus* Fritz Shwarz ex N. P. Taylor, *Echinocereus pseudopectinatus* (N. P. Taylor) N. P. Taylor*, *E. sciurus*(K. Brandegee) Dams*, *Echinocereus scopulorum* Britton and Rose*, *Echinocereus websterianus* G. E. Linds*. * Included in the informal species group *Sciurus* according to Blum et al. (1998).

Section *Sciuri* (Figs. 11.5 and 11.6g, h). This group included eleven species having commonly unbranched stems and a nectary with basal nectarial tissue (Fig. 11.6h). Furthermore, one site in the *rpl16* marker and one indel (simple sequence repetition of four bases) in the *psbA-trnH* marker support the clade.

Section ***Reichenbachii*** N. P. Taylor, Gen. Echinocereus: 105 (1985). Type species: *Echinocereus reichenbachii* (Terscheck ex Walp.) Haage.
Echinocereus section *Pulchellus* N. P. Taylor, Gen. Echinocereus: 140 (1985).
Species included: *Echinocereus pamanesiorum* A. B. Lau, *Echinocereus pulchellus* (Mart.) C. F. Förster ex F. Seitz, *E. reichenbachii* (Terscheck ex Walp.) Haage, *E. schereri* G. Frank, *Echinocereus spinigemmatus* A. B. Lau.

Section *Reichenbachii* (Figs. 11.5 and 11.6i, j). This clade with five species is supported by one site in the *rpl16*, and one site and one indel (deletion of 54 sites) in the *trnL-F* marker. Most of species formed a clade with strong support and shared two morphological characters: fruits with semi-dry pulp (Fig. 11.6j) and stem areoles without central spines (Fig. 11.5).

Section ***Echinocereus***. Engelm. *Cereus* sección *Sulcati* Engelm., Mem. Amer. Acad. Arts. ser. 2, 4: 50 (1849). Type species: *Echinocereus viridiflorus* Engelm. Species included*: *Echinocereus chloranthus* (Engelm.) Haage, *E. russanthus* Weniger, *E. viridiflorus* Engelm. *See discussion about other recognized species (above).

Section *Echinocereus* (Figs. 11.5 and 11.6k, l). Three species were grouped in this section, although more species have been described for this section (Blum et al. 2012). This clade was recognized by flower length less than 4 cm, nectary length 1–2 mm (Fig. 11.6l), basal nectarial tissue, and small seeds (Fig. 11.5). Also, four DNA sites in the *trnK/matK* marker supported this section. Traditionally, this section is easily identified by the combination of both flower length and flower colour (yellow and/or brown).

Section ***Erecti*** (K. Schum.) Bravo, Cact. Suc. Mex. 27: 16 (1982). *Echinocereus* serie *Erecti* K. Schum., Gesamtb. Kakt. 247 (1987). *Cereus* subsection *Erecti* (K. Schum.) Berger, Rep. (Annual) Missouri. Bot. Gard. 16: 80 (1905). Type species: *Echinocereus engelmannii* (Parry ex Engelm.) Lem.
Species included: *Echinocereus bonkerae* Thornber and Bonker, *Echinocereus dasyacanthus* Engelm., *E. engelmannii* (Parry ex Engelm.) Lem., *Echinocereus fasciculatus* (Engelm. ex S. Watson) L. D. Benson, *Echinocereus fendleri*

(Engelm.) Rümpler, *E. nicholii* (L. D. Benson) B. D. Parfitt, *E. papillosus* A. Linke ex Rümpler, *Echinocereus pectinatus* (Scheidw.) Engelm.

Section *Erecti* (Figs. 11.5 and 11.6m, n). This clade recovered eight species and is characterized by the presence of a wide funnel-shaped receptacular tube, the thickness of the receptacular tube is more than 4 mm, the thickness of the base of tepals is more than 2 mm, and flowers with dark colour tone in the throat (Fig. 11.6n). Also, one DNA site and one indel (reversion of 15 sites) in the *rpl16* marker support this clade.

Section ***Triglochidiati*** Bravo, Cact. Suc. Mex. 28: 109 (1973). *Echinocereus* subgenus *Triglochidiatus* (Bravo) W. Blum, Mich. Lange & Rutow 1998: 357. Type species: *Echinocereus triglochidiatus* Engelm.
Species included: *Echinocereus acifer* (Otto ex Salm-Dyck) Jacobi, *Echinocereus arizonicus* Rose ex Orcutt, *Echinocereus coccineus* Engelm., *Echinocereus koehresianus* (G. Frank) W. Rischer, *Echinocereus polyacanthus* Engelm., *Echinocereus santaritensis* W. Blum & Rutow, *Echinocereus scheeri* (Salm-Dyck) Scheer, *Echinocereus triglochidiatus* Engelm., *Echinocereus yavapaiensis* M. A. Baker.

Section *Triglochidiati* (Figs. 11.5 and 11.6o, p). This lineage has been largely recognized based on its distinctive floral morphology (Taylor 1985, 1993). All species of this clade shared a narrow funnel-shaped receptacular tube, a receptacular tube 1.5-fold larger than the perianth, larger inner stamens than outer ones, purple anthers, a predominantly red perianth, and an embryo with large cotyledons (Fig. 11.6p; Table 11.1). Those floral traits are cited as adaptations to hummingbird pollination syndrome (Cota 1993; Taylor 1985, 1993). Additionally, five DNA sites support this clade (*trnK/matK*: 1, *psbA-trnH*: 0, *rpl16*: 2, *trnL-F*: 2).

11.5 Perspectives and Conclusions

Recently, phylogenomics has provided a new tool to reconstruct the ancestry–descendance relationships of the organism (Delsuc et al. 2005), and some other evolutive processes have been inferred using this approach (Yang et al. 2015). However, it has been suggested that one challenge to increasing the size of phylogenomic data sets is the burden of homology identification (Walker et al. 2018), and that misidentified orthology is common in phylogenomic studies (Brown and Thomson 2017). It is a logical conclusion that the misalignment of data will cause more inferred substitutions in the gene, altering the model and likelihood of the whole data set score resulting in a wrong topology. So, homology detection, gene tree conflict, and computational complexities are problems that phylogenomic studies should address (Walker et al. 2018).

As Wipfler et al. (2016) explain, morphology will remain crucial in systematics for several reasons. Morphological characters are an independent source of data useful to critically evaluate molecular phylogenies. In contrast to cpDNA sequences, morphology provides specific and informative features as synapomorphies, homoplasies, or any combination of them, in order to characterize recovered clades. In case of fossils, older collection material or rare species, where molecular characters cannot be obtained, the only option is the analyses of morphology. The analysis of morphology is crucial in "evo-devo" research, for understanding evolutionary changes on the phenotypic level, and in the interpretation of the transformation of morphological structures in complex and meaningful evolutionary scenarios.

We concluded that the addition of morphological characters represented the rational complement to corroborate and support the cpDNA sequence-based phylogeny. Because in any dataset, conflicting signal is present, it is necessary to include several sources of character information to encourage the emerging of secondary signal. Also, conflicting signal can suggest several evolutive events in past such as ancient hybridization, horizontal gene transfer, gene duplication, and loss. This work demonstrated the importance of combining morphological and molecular evidence because morphology allowed secondary signals to arise when interacting with molecular markers and avoid long-branch attraction. Particularly, we recommend the inclusion of MP in analysis where conflicting signal is included. Because a phylogeny is typically the starting point for most subsequent analyses (e.g. divergence dating and ancestral state reconstruction), a phylogenetic analysis using some cpDNA markers and morphology characters is an ideal approach to infer the relationships on focused lineages, since phylogenomics is slowly beginning the process of filtering genes for optimal species tree inference (Walker et al. 2018). We highlight that morphology provides useful and practical information to compare and diagnose clades, and propose formal classifications.

References

Agnarsson I, Miller JA (2008) Is ACCTRAN better than DELTRAN? Cladistics 24:1032–1038

Albesiano S, Terrazas T (2012) Cladistic analysis of *Trichocereus* (Cactaceae: Cactoideae: Trichocereeae) based on morphological data and chloroplast DNA sequences: dedicated to Omar Emilio Ferrari (1936–2010). Haseltonia 17:3–23

Anderson EF (2001) The cactus family. Timber Press, Portland

Arias S, Terrazas T, Arreola-Nava HJ et al (2005) Phylogenetic relationships in *Peniocereus* (Cactaceae) inferred from plastid DNA sequence data. J Plant Res 118:317–328

Assis L (2009) Coherence, correspondence, and the renaissance of morphology in phylogenetic systematics. Cladistics 25:528–544

Assis L, Rieppel O (2011) Are monophyly and synapomorphy the same or different? Revisiting the role of morphology in phylogenetics. Cladistics 27:94–102

Alves RJV, Machado MD (2007) Is classical taxonomy obsolete? Taxon 56:287–288

Baker M (2006a) A new florally dimorphic hexaploid, *Echinocereus yavapaiensis* sp. nov. (section *Triglochidiatus*, Cactaceae) from central Arizona. Plant Syst Evol 258:63–83

Baker M (2006b) Circumscription of *Echinocereus arizonicus* subsp. *arizonicus*, Phenetic analysis of morphological characters in section *Triglochidiatus* (Cactaceae) part II.Madroño 53:388–399

Bárcenas RT, Yesson C, Hawkins JA (2011) Molecular systematics of the Cactaceae. Cladistics 27:470–489

Berger A (1926) Die entwicklungslinien der Kakteen. Fisher, Jena

Bergsten J (2005) A review of long-branch attraction. Cladistics 21:163–193

Blum W, Felix D, Bauer H (2012) *Echinocereus* Die Sektion *Echinocereus*. Der Echinocereenfreund 25:1–336

Blum W, Felix D, Waldeis D (2008) Echinocereus Die Sektion Wilcoxia. Der Echinocereenfreund 21:1–142

Blum W, Lang M, Rischer M, Rutow J (1998) Echinocereus, Monographie. Selbstverlag, Aachen

Bollback JP (2006) SIMMAP: stochastic character mapping of discrete traits on phylogenies. BMC Bioinform 7:88

Bravo-Hollis H, Sánchez-Mejorada H (1991) Las Cactáceas de México [The cacti of Mexico], vol 2. Universidad Nacional Autónoma de México, Ciudad de México

Britton NL, Rose JN (1919) The Cactaceae, vol 1. Carnegie Institution of Washington, Washington

Britton NL, Rose JN (1920) The Cactaceae, vol 2. Carnegie Institution of Washington, Washington

Britton NL, Rose JN (1922) The Cactaceae, vol 3. Carnegie Institution of Washington, Washington

Britton NL, Rose JN (1923) The Cactaceae, vol 4. Carnegie Institution of Washington, Washington

Brown JM, Thomson RC (2017) Bayes factors unmask highly variable information content, bias, and extreme influence in phylogenomic analyses. Syst Biol 66:517–530

Butterworth CA, Cota-Sanchez JH, Wallace RS (2002) Molecular systematics of tribe Cacteae (Cactaceae: Cactoideae): a phylogeny based on *rpl*16 intron sequence variation. Syst Botany 27:257–270

Buxbaum F (1951) Morphology of cacti, section I: root and stems. Abbey Garden Press, Pasadena

Buxbaum F (1953) Morphology of cacti, section II: flower. Abbey Garden Press, Pasadena

Buxbaum F (1955) Morphology of cacti, section III: fruits and seeds. Abbey Garden Press, Pasadena

Buxbaum F (1958) The phylogenetic division of the subfamily Cereoideae, Cactaceae. Madroño 14:177–206

Calvente A, Zappi DC, Forest F et al (2011) Molecular phylogeny of tribe Rhipsalideae (Cactaceae) and taxonomic implications for *Schlumbergera* and *Hatiora*. Mol Phylogenet Evo 58:456–468

Cota JH (1993) Pollination syndromes in the genus *Echinocereus*: a review. Cact Succ J (US) 65:19–26

Cruz MÁ, Arias S, Terrazas T (2016) Molecular phylogeny and taxonomy of the genus *Disocactus* (Cactaceae), based on the DNA sequences of six chloroplast markers. Willdenowia 46:145–164

Cunningham WC, Omland KE, Oakley TH (1998) Reconstructing ancestral characters states: a critical reappraisal. Trends Ecol Evol 13:361–366

Darriba D, Taboada GL, Doallo R et al (2012) jModelTest 2: more models, new heuristics and parallel computing. Nat Methods 9:772

de Carvalho MR (1996) Higher-level elasmobranch phylogeny, basal squaleans, and paraphyly. In: Stiassny MLJ, Parenti LR, Johnson GD (eds) Interrelationships of Fishes 3. Academic Press, San Diego, pp 593–660

De Pinna MG (1991) Concepts and tests of homology in the cladistic paradigm. Cladistics 7:367–394

de Queiroz A, Donoghue MJ, Kim J (1995) Separate versus combined analysis of phylogenetic evidence. Ann Rev EcolSyst 26:657–681

Delsuc F, Henner B, Hervé P (2005) Phylogenomics and the reconstruction of the tree of life. Nat Rev Genet 6:361

Demaio PH, Barfuss MH, Kiesling R et al (2011) Molecular phylogeny of *Gymnocalycium* (Cactaceae): Assessment of alternative infrageneric systems, a new subgenus, and trends in the evolution of the genus. Am J Bot 98:1841–1854
Edwards EJ, Nyffeler R, Donoghue MJ (2005) Basal cactus phylogeny, implications of *Pereskia* (Cactaceae) paraphyly for the transition to the cactus life form. Am J Bot 92:1177–1188
Endler J, Buxbaum F (1974) Die Pflanzenfamilie der Kakteen, 3rd edn. A. Philler Verlag, Miden
Engelmann G (1848) Botanical appendix. In: Wislizenus FA (ed) Memoir of a tour to Northern Mexico, connected with Col. Doniphan's Expedition, in 1846 and 1847, Tippin and Streeper, Washington, pp 87–115
Farris J (1979) The information content on the phylogenetic system. Syst Zool 28:483–519
Fuentes M (2004) Anatomía floral de algunas especies de Pachycereeae (Cactaceae) [Floral anatomy of some Pachycereeae species] (Unpublished bachelor dissertation). Facultad de Estudios Superiores Iztacala, Universidad Nacional Autónoma de México, Tlalnepantla
Gibson AC (1973) Comparative anatomy of secondary xylem in Cactoideae (Cactaceae). Biotropica 5:29–65
Gibson AC, Nobel PS (1986) The cactus primer. Harvard University Press, Cambridge
Goloboff PA, Farris J, Nixon K (2008) T.N.T. Tree analysis using new technology. Cladistics 24:774–786
Guerrero PC, Arroyo MTK, Bustamante RO et al (2011) Phylogentics and predictive distribution modeling provide insights into the divergence of *Eriosyce* subgen. *Neoporteria* (Cactaceae). Plant Syst Evol 297:113–128
Hernández-Hernández T, Hernández HM, De-Nova JA et al (2011) Phylogenetic relationships and evolution of growth form in Cactaceae (Caryophyllales, Eudicotyledoneae). Am J Bot 98:44–61
Hernández-Ledesma P, Bárcenas RT (2017) Phylogenetic utility of the *trnH–psbA* IGR and stem-loop diversity of the 3′ UTR in Cactaceae (Caryophyllales). Plant Syst Evol 1–17
Hughes CE, Lewis GP, Yomona AD et al (2004) *Maraniona*. A new dalbergioid legume genus (Leguminosae, Papilionoideae) from Peru. Syst Bot 29:366–374
Hunt DR (2012) Taxonomic implications of DNA studies relating to Cactaceae subfam. Cactoideae. Bull Int Cactaceae Syst Group 26:3–20
Hunt DR (2016) CITES Cactaceae checklist, 3rd edn. DH Books, Milborne Port
Hunt DR, Taylor NP, Charles G (2006) The new cactus lexicon. DH Books, Milborne Port
Jenner R (2004) Accepting partnership by submission? Morphological phylogenetics in a molecular millenium. Syst Biol 53:333–342
Kluge AG (1989) A concern for evidence and phylogenetic hypothesis for relationships among *Epicrates* (boide, Serpentes). Syst Zool 38:1–25
Knopf P, Schulz C, Little DP et al (2012) Relationships within Podocarpaceae based on DNA sequence, anatomical, morphological, and biogeographical data. Cladistics 28:271–299
Kolaczkowski B, Thornton JW (2009) Long-branch attraction bias and inconsistency in Bayesian phylogenetics. PLoS ONE 4:e7891
Korotkova N, Borsch T, Arias S (2017) A phylogenetic framework for the Hylocereeae (Cactaceae) and implications for the circumscription of the genera. Phytotaxa 327:1–46
Korotkova N, Borsch T, Quandt D et al (2011) What does it take to resolve relationships and to identify species with molecular markers? An example from the epiphytic Rhipsalideae (Cactaceae). Am J Bot 98:1549–1572
Larridon I, Walter HE, Guerrero PC et al (2015) An integrative approach to understanding the evolution and diversity of *Copiapoa* (Cactaceae), a threatened endemic Chilean genus from the Atacama Desert. Am J Bot 102:1506–1520
Lartillot N, Brinkmann H, Philippe H (2007) Suppression of long-branch attraction artefacts in the animal phylogeny using a site-heterogeneous model. BMC EvolBiol 7:S4
Lartillot N, Philippe H (2004) A bayesian mixture model for across-site heterogeneities in the amino-acid replacement process. Mol Biol Evol 21:1095–1109
Lewis PO (2001) A likelihood approach to estimating phylogeny from discrete morphological character data. Syst Biol 50:913–925

Loza-Cornejo S, Terrazas T (1996) Anatomía del tallo y raíz de dos especies de *Wilcoxia* Britton & Rose (Cactaceae) del noreste de México. Bol Soc Bot México 59:13–23

Luna E, Mishler BD (1996) El concepto de homología filogenética y la selección de caracteres taxonómicos. Bol Soc Bot México 59:131–146

Majure LC, Puente R, Griffith MP et al (2012) Phylogeny of *Opuntia* s.s. (Cactaceae): clade delineation, geographic origins, and reticulate evolution. Am J Bot 99:847–864

Nixon KC (2002) WinClada, version 1.00. 08. Ithaca, NY

Nixon KC, Carpenter JM (1996) On simultaneous analysis. Cladistics 12:221–241

Nixon KC, Carpenter JM (2000) On the other "phylogenetic systematics". Cladistics 16:298–318

Nixon KC, Ochoterena H (2000) Taxonomía tradicional, cladística y construcción de hipótesis filogenéticas. In: Hernández HM, García Aldrete AN, Álvarez F et al (eds) Enfoques contemporáneos para el estudio de la biodiversidad. Universidad Nacional Autónoma de México & Fondo de Cultura Económica, Ciudad de México, pp 15–37

Norup MV, Dransfield J, Chase MW et al (2006) Homoplasious character combinations and generic delimitation: a case study from the Indo-Pacific arecoid palms (Arecaceae: Areceae). Am J Bot 93:1065–1080

Nyffeler R (2002) Phylogenetic relationship in the cactus family (Cactaceae) based on evidence from *trnK/matK* and *trnL-trnF* sequences. Am J Bot 89:312–326

Nyffeler R, Eggli U (2010) A farewell to dated ideas and concepts: molecular phylogenetics and revised suprageneric classification of the family Cactaceae. Schumannia 6:1–42

Nylander JAA, Ronquist F, Huelsenbeck JP et al (2004) Bayesian phylogenetic analysis of combined data. Syst Biol 53:47–67

Ochoterena H (2009) Homology in coding and non-coding DNA sequences, a parsimony perspective. Plant Syst Evol 282:151–168

Patterson C (1982) Morphological characters and homology. Problems of phylogenetic reconstruction. Academic Press, London

Pol D, Siddall ME (2001) Biases in maximum likelihood and parsimony, a simulation approach to a 10-taxon case. Cladistics 17:266–281

Richards R (2003) Character individuation in phylogenetic inference. Philo Sci 70:264–279

Rindal E, Brower AV (2011) Do model-based phylogenetic analyses perform better than parsimony? A test with empirical data. Cladistics 27:331–334

Ritz CM, Martins L, Mecklenburg R et al (2007) The molecular phylogeny of *Rebutia* (Cactaceae) and its allies demonstrates the influence of paleogeography on the evolution of South American Mountain cacti. Am J Bot 94:1321–1332

Ronquist F, Huelsenbeck JP (2003) MRBAYES 3, Bayesian phylogenetic inference under mixed models. Bioinformatics 19:1572–1574

Sánchez D, Arias S, Terrazas T (2013) Análisis morfométrico de las especies de *Echinocereus* sección *Triglochidiati* (Cactaceae) en México. Brittonia 65:368–385

Sánchez D, Arias S, Terrazas T (2014) Phylogenetic relationships in *Echinocereus* (Cactaceae, Cactoideae). Syst Bot 39:1183–1196

Sánchez D, Grego-Valencia D, Terrazas T et al (2015) How and why does the areole meristem move in *Echinocereus* (Cactaceae)? Ann Bot 115:19–26

Sánchez D, Terrazas T, Grego-Valencia D, Arias S (2018) Phylogeny in *Echinocereus* (Cactaceae) based on morphological and molecular evidence: taxonomic implications. Syst Biodivers 1:28–44

Schlumpberger BO, Renner SS (2012) Molecular phylogenetics of *Echinopsis* (Cactaceae), polyphyly at all levels and convergent evolution of pollination modes and growth forms. Am J Bot 99:1335–1349

Schumann K (1899) Gesamtbeschreibung der Kakteen. Neudamm, Berlin

Scotland RW, Olmstead RG, Bennett JR (2003) Phylogeny reconstruction, the role of morphology. Syst Biol 52:539–548

Simmons MP, Ochoterena H, Carr TG (2001) Incorporation, relative homoplasy, and effect of gap characters in sequence-based phylogenetic analyses. Syst Biol 50:454–462

Smith ND, Turner AH (2005) Morphology's role in phylogeny reconstruction: perspectives from paleontology. Syst Biol 54:166–173

Tapia HJ, Bárcenas-Argüello ML, Terrazas T et al (2018) Phylogeny and Circumscription of *Cephalocereus* (Cactaceae) based on molecular and morphological evidence. Syst Bot 42:709–723

Taylor NP (1985) The genus Echinocereus. Kew Magazine Monograph, Middlesex

Taylor NP (1993) Ulteriori studi su Echinocereus. Piante Grasse 13:79–96

Vargas-Luna MD, Hernández-Ledesma P, Majure LC et al (2018) Splitting *Echinocactus*: morphological and molecular evidence support the recognition of *Homalocephala* as a distinct genus in the Cacteae. PhytoKeys 111:31–59

Vázquez-Sánchez M, Sánchez D, Terrazas T et al (2019) Polyphyly at the iconic cactus genus *Turbinicarpus* (Cactaceae) and its generic circumscription. Bot J Linean Soc 190:405–420

Vázquez-Sánchez M, Terrazas T (2011) Stem and wood allometric relationships in Cacteae (Cactaceae). Trees 25:755–767

Vázquez-Sánchez M, Terrazas T, Arias S et al (2013) Molecular phylogeny, origin and taxonomic implications of the tribe Cacteae (Cactaceae). Syst Biodivers 11:103–116

Walker JF, Yang Y, Feng T et al (2018) From cacti to carnivores: improved phylotranscriptomic sampling and hierarchical homology inference provide further insight into the evolution of Caryophyllales. Am J Bot 105:446–462

Wallace RS (1995) Molecular systematic study of the Cactaceae: using chloroplast DNA variation to elucidate cactus phylogeny. Bradleya 13:1–12

Wallace RS, Gibson AC (2002) Evolution and systematics. In: Nobel PS (ed) Cacti biology and uses. University of California Press, Berkeley, pp 1–21

Wheeler QD (2004) Taxonomic triage and the poverty of phylogeny. Philos Trans Royal Soc B: Biol Sci 359:571–583

Wiley EO, Lieberman BS (2011) Phylogenetics: theory and practice of phylogenetic systematics. Wiley, Hoboken

Winther RG (2009) Character analysis in cladistics: abstraction, reification and the search for objectivity. Acta Biotheor 57:129–162

Wipfler B, Pohl H, Yavorskaya MI et al (2016) A review of methods for analysing insect structures—the role of morphology in the age of phylogenomics. Curr Opin Insect Sci 18:60–68

Wortley AH, Rudall PJ, Harris DJ et al (2005) How much data are needed to resolve a difficult phylogeny? Case study in Lamiales. Syst Biol 54:697–709

Wortley AH, Scotland RW (2006) The effect of combining molecular and morphological data in published phylogenetic analyses. Syst Biol 55:677–685

Yang Y, Moore MJ, Brockington SF et al (2015) Dissecting molecular evolution in the highly diverse plant clade Caryophyllales using transcriptome sequencing. Mol Biol Evol 32:2001–2014

Chapter 12
Chemotaxonomic Survey on the Genus *Sedum* L. (Crassulaceae) Based on Distribution and Variability of the Epicuticular Wax Constituents

Snežana Č. Jovanović, Bojan K. Zlatković and Gordana S. Stojanović

Abstract In this chapter are given general considerations on biosynthesis, chemical composition, function and importance of epicuticular wax constituents, specially their utilization and usefulness in chemotaxonomy. A special attention is given to the components of epicuticular waxes isolated from different representatives of the genus *Sedum* L. from the territory of Europe, emphasizing the importance of certain classes of compounds for chemotaxonomic purposes and at certain taxonomic levels of classification. Also, the chapter brings out survey of literature data on distribution, classification and phylogenetic relationships of the genus *Sedum*, as well as consideration of usefulness of alkanes and triterpenoids from epicuticular wax in taxonomical investigations of the family Crassulaceae, with special attention to the genus *Sedum*.

Keywords *Sedum* L. · Epicuticular wax · Variability · Chemotaxonomy

12.1 Introduction

Each living organism possesses a kind of coating as interface with its environment. It protects the organism as a physical barrier and presents an active participant in various processes that are essential for the given organism. Therefore, this layer is

S. Č. Jovanović (✉) · G. S. Stojanović
Department of Chemistry, Faculty of Science and Mathematics, University of Niš, Višegradska Street no 33, Niš, Serbia
e-mail: coka.sslagalica@gmail.com

G. S. Stojanović
e-mail: gocast@pmf.ni.ac.rs

B. K. Zlatković
Department of Biology and Ecology, Faculty of Science and Mathematics, University of Niš, Višegradska Street no 33, Niš, Serbia
e-mail: bojanzlat@pmf.ni.ac.rs

K. G. Ramawat (ed.), *Biodiversity and Chemotaxonomy*, Sustainable Development and Biodiversity 24, https://doi.org/10.1007/978-3-030-30746-2_12

characterized by a certain role, appearance and composition. The development and utilization of microscopy and methods of chemical analysis enabled the detailed study of the surfaces of living organisms, and thus, characteristic features and dynamic processes intrigued scientists from various fields to explore it from different aspects and purposes.

The plants interact with their environment via cuticle, a thin hydrophobic layer which covers epidermis. Its presence primarily characterizes evolutionarily more developed plants—vascular plants that usually inhabit more or less dry types of land habitats, but also could be seen in some ferns and mosses (Heredia 2003). Above-ground parts of the plant, such as leaves, fruits, flowers, seeds and non-woody stems, are covered with a cuticle. It has been recognized as heterogeneous formation in plants due to protective and regulatory role: protection from water loss, regulating gas exchange, UV radiation, as physical barrier protecting against mechanical injury caused by microorganisms, pests or the environment (Kunst and Samuels 2003; Domínguez et al. 2011; Buschhaus and Jetter 2012). The structure of plant cuticle is complex and varies depending on plant species, genotypes, plant organ, development phase, physiological status and environmental conditions (Heredia 2003; Borisjuk et al. 2014; Fernández et al. 2016). As far as the chemical composition is concerned, the cuticle is made up of different chemical compounds. In general, all chemical constituents could be classified as follows: waxes as deposits on the surface or embedded in the cutin matrix, biopolyesters cutin and cutan made of cross-linked C_{16} and C_{18} fatty acids and glycerol molecules by ester bonds, polysaccharides, phenolics (primarily cinnamic acid derivatives and flavonoids as minor compounds) and mineral elements (Kunst and Samuels 2009; Domínguez et al. 2011; Fernández et al. 2016).

12.1.1 Plant Epicuticular Wax: General Consideration on Biosynthesis, Chemical Composition, Function and Importance

Many studies have been dealt with the morphological characteristics and chemical composition of the uppermost part of the cuticle called cuticular wax or plant wax (Jeffree et al. 1976; Barthlott et al. 1998; Kunst and Samuels 2003, 2009; Zlatković et al. 2017). Interestingly, the appearance of the plant surfaces could vary notably between different species. For example, leaves of vine are coated with amorphous wax, or very fine epicuticular wax crystals in lotus, and whitish or bluish appearance in succulents from the genera *Dudleya* Britton and Rose and *Sedum* L. (Stevens et al. 1994a, b; Taylor 2011). This phenomenon shows a variation in the chemical content of the wax from green to glaucous variants, as well as correlation between micromorphological characters and composition of surface waxes in plants (Herbin and Robins 1969).

As already mentioned, plant wax components are arranged into the surface layer in two ways: epicuticular wax exists as deposits on the surface which, in some cases, could be perceived without microscope; or intracuticular wax embedded in the cutin matrix (Domínguez et al. 2011). Some studies differentiated the entire wax coating on epicuticular and intracuticular wax due to observed physical and chemical differences. Thus, epicuticular wax is a waxy layer on the surface of cuticle and appears as crystalloid or smooth structure, while intracuticular wax, a lower layer closer to the cuticle, refers as an amorphous mixture of lipids. Detail analysis of these two layers means the subsequently physical separation which results in comparative data on both type of waxes (Buschhaus and Jetter 2012; Kunst and Samuels 2003; van Maarseveen and Jetter 2009; Taylor 2011). However, many researches follow procedure of plant wax isolation which implies lipid removal from plant surface (or from a specific plant organ) by brief immersion in a nonpolar organic solvent, such as hexane or dichloromethane (Stevens et al. 1994a, b; Jovanović et al. 2015a, 2016; Mitić et al. 2016; Zlatković et al. 2017). A scanning electron microscopy (SEM) is a method of choice for the characterization of wax micromorphology, while gas chromatography coupled with mass spectrometry or flame ionization detector (GC-MS and GC-FID) for the chemical composition analysis of lipid mixture (Stevens et al. 1994a, b; Barthlott et al. 1998; Bush and McInerney 2013; Jovanović et al. 2015a, 2016; Zlatković et al. 2017; Mitić et al. 2016, 2017). Since that the recent studies gain a lot of data and heterogeneous characteristics (variables), very often, the studies include statistical analysis as final step which allows observation and proper interpretation on all data (Stevens et al. 1994a, b; Maffei 1994, 1996a, b; Maffei et al. 2004, 1997; Medina et al. 2006; Bojović et al. 2012; Jovanović et al. 2015a, 2016; Mitić et al. 2016, 2017; Zlatković et al. 2017).

The biosynthesis of components of epicuticular wax is performed in epidermal cells. Subsequently, they migrate through cuticle membrane (pores) to the surface of the plant forming a fine crystalline structure (Jeffree et al. 1976). The biosynthesis comprises several biochemical steps which include complex mechanism of chemical compounds transformation and intermediation of multienzyme complexes. Generally speaking, it includes elongation of fatty acids, reduction in fatty acids to aldehydes and primary alcohols, decarbonylation of the resulting aldehyde to alkane, hydroxylation of alkanes to secondary alcohols and oxidation of secondary alcohol to ketone. Therefore, whole process leads to the production of very long-chain fatty acid derivatives (Lemieux 1996; Seigler 1998; Kunst and Samuels 2003). This lipid mixture primarily contains long-chain fatty acid derivatives, terpenoids and minor compounds—sterols and flavonoids (Stevens et al. 1994a, b; Kunst and Samuels 2003; Jovanović et al. 2016; Mitić et al. 2018). Among fatty acid derivatives, normal-chain alkanes (C_{19}–C_{37}) are characterized by domination in content, particularly with an odd number of C atoms (C_{27}–C_{31}). Beside normal-chain alkanes, although limited distribution, epicuticular wax could contain branched alkanes, unsaturated hydrocarbons, alcohols (primary alcohols C_{12}–C_{40} and secondary alcohols C_{21}–C_{35}), fatty acids (C_{12}–C_{36}), aldehydes (C_{14}–C_{36}), ketones (C_{21}–C_{35}), *n*-alkyl-esters (C_{30}–C_{60}) and other related compounds. Cyclic compounds are possible to be found in mixture of wax, but they are usually limited in distribution (Manners and Davis 1984; Barthlott et al. 1998; Kunst and Samuels 2003; Marin 2003; Müller and Riederer 2005; Jovanović et al. 2015a, 2016). Although the morphological markers of the wax are formally separated from the chemical characters, there is a correlation between these two. It was confirmed that the morphological features of wax layer depend on its chemical composition. Humans perceive whitish or bluish coloration of aerial plant parts such as leaves and fruits which actually present experience of the reflection of light by characteristic three-dimensional wax structures. In other case, the coloration is not obvious due to lower amount of wax, or because of different type of surface—smooth layer (Müller and Riederer 2005). Thus, studies of Barthlott et al. (1998) and Jeffree (2006) distinguished several morphological wax types. Jeffree (2006) pointed out main morphological wax types: films, crusts, tubules, platelets and rodlets (Koch and Bartholtt 2009). For example, platelet type in case of high content of primary alcohols, crusts when alkanes and aldehydes dominate, β-diketones and asymmetric secondary alcohols indicate tubular type, glaucous appearance was observed in wax with significant content of triterpenes and secondary alcohols, etc. (Jefree et al. 1976; Stevens et al. 1994b; Lemieux 1996; Osborne and Stevens 1996; Koch and Barthlott 2009).

The significance and function of wax have been discussed from functional and ecological aspects. The unique chemical composition of the wax with consequential hydrophobicity provides barrier or defence mechanisms against abiotic and biotic stress: water loss, radiation, variable temperature (extremely high or low), bacterial and fungal pathogens, insects, high salinity, etc. (Xue et al. 2017).

12.1.2 Epicuticular Wax as a Valued Criteria in Taxonomic Studies

Many chemotaxonomic studies have emphasized the importance of metabolites (or markers) from the epicuticular wax in the differentiation and systematics at different taxonomic levels. The variability in the composition and structural diversity of epicuticular wax components, most often *n*-alkanes, are important parameters in the plant taxonomy, since they provide insights in taxonomic grouping within the various taxonomic ranks. It is possible to distinguish between higher taxa (families and genera), differentiate groups within sections, or even identify species, thus understanding evolutionary and phylogenetic relationships in the plant world (Stevens et al. 1994a, b; Maffei 1994, 1996a, b; Marin 2003; Maffei et al. 1997, 2004; Medina et al. 2006; Bojović et al. 2012; Jovanović et al. 2015a).

A plant wax represents a relatively constant characteristic of a specific plant species, therefore can be used as a very reliable parameter in plant taxonomy. In other words, individuals of the same species, that growing under the influence of the same factors, are characterized by a similar morphological structure and chemical composition of the wax. Variability of wax morphological and chemical characteristics usually happens due to variation regarding various ecological factors (Jeffree et al. 1976). Also, the differences occur depending on the plant species, plant organ, position (e.g. leaves with different position on the plant may have different wax composition), or certain ontogenetic phase (Koch et al. 2008). According to a recent survey, Tomaszewski and Zielinski (2014) claimed indisputable systematic significance of epicuticular wax markers, but also, they emphasized the fact that most of the studies were focused on the leaf epicuticular wax. They studied wax structures of about 340 taxa from 80 plant families in order to examine wax structures of both stems and leaves and determine relationships between these two plant organs. In general, they concluded that there were quite differences regarding wax micromorphology between leaves and stems. Therefore, all the above-mentioned observations and characteristics support the application of wax markers for chemotaxonomic purposes. Furthermore, if the study considers other types of plant markers, it could open the possibility of clarifying more complex phylogenetic relationships within the given taxonomic rank. For example, Denton (1994) studied the wax pattern of twelve cultivated taxa and hybrids at the population level of *Sedum* sect. *Gormania* by the SEM analysis. Although the taxa were morphologically similar, Denton observed differences regarding some morphological characteristics between cultivated taxa, wild-growing taxa and hybrids. Despite the fact that a uniform pattern regarding the type of wax deposits (plate-like appearance) was established, certain deviations were also identified. The size of the platelets did not vary significantly within the population of the same taxon, but it varied among the different taxa. Interestingly, hybrid taxa showed tendency to form "intermediate" pattern on wax structures in relation to the corresponding parent species.

Although wax is a mixture of various compounds as potential biomarkers, most chemotaxonomic studies chose to monitor the distribution and variability of *n*-alkanes. Why? The answer lies in the many advantages of alkanes in relation to other classes of compounds that have been detected in the epicuticular wax. From a functional point of view, alkanes are significant components of the wax because they contribute to the hydrophobicity. *n*-Alkanes (C_{19}–C_{37}) are ubiquitous in plant waxes; they can be relatively easily isolated, not only from plant material but also from different deposits, soil and fossil materials; chemically stable compounds due to the absence of any functional group in structure, and thus, in the function of the time, alkanes are highly resistant to changes which qualify them as valuable markers from the environment. The content of alkanes in waxes varies depending on the phases of development. During the spring and early summer, when the intensive development of leaves on the plant occurs, the amount of *n*-alkane is constantly increasing, while the content of these hydrocarbons is relatively constant during further development. As other secondary metabolites, variability of *n*-alkane is affected by various ecological factors (Bush and McInerney 2013; Ardenghi et al. 2017). Finally, the content of *n*-alkanes varies depending on the plant species, genus and family, so identification, quantification and consideration of their distribution in waxes are very often used in chemotaxonomic studies. In general, alkanes can be characterized as good markers at lower levels of classification (level of the species and lower levels of classification), except in cases of their high variability in samples. Sometimes, in the case when taxa at the species or subspecies level morphologically do not differ each other significantly, the characteristic pattern on alkane distribution would be useful in their differentiation. Alkanes can be used as reliable markers at higher levels of classification, such as tribes, subfamilies and families (Marin 2003). Beside these ubiquitous markers of wax, many studies valued other markers of epicuticular wax. Opposite to alkanes, these are often considered as limited distribution markers that allow better differentiation within certain taxonomic category (Manheim et al. 1979; Manners and Davis 1984; Stevens et al. 1994a, b; Matas et al. 2003; Medina et al. 2006; Jovanović et al. 2015a, 2016; Mitić et al 2018).

12.2 Genus *Sedum* L. (Crassulaceae)

The genus *Sedum* L. (subfam. Sempervivoideae Arn., tribe Sedeae Fr.) represents one of the most numerous and widely distributed genera within the family of succulent plants—Crassulaceae. The genus comprises about 420 species (Thiede and Eggli 2007) mainly distributed in temperate and subtropical parts of Europe, North Africa, Asia with Near East and North America, a few species in South America and Central to East Africa. Most of the species of the genus *Sedum* have been found to be used in horticulture because of their extremely attractive appearance. Figure 12.1 displays fifteen of the many species of the genus *Sedum* that grows on the territory of the Balkan Peninsula.

Additionally, many species are known in traditional medicine due to their healing properties. Numerous studies tested and proved biological activities of extracts or pure compounds of *Sedum* species: *S. sediforme* (Jacq.) Pau, *S. acre* L., *S. sarmentosum* Bunge, *S. takesimense* Nakai, *S. dendroideum*, *S. album* L. and many others (He et al. 1998; Kang et al. 2000; Thuong et al. 2007; Jung et al. 2008; De Melo et al. 2009; Stanković et al. 2012; Ertaş et al. 2014). Some species are traditionally used as food, such as *Sedum sarmentosum* Bunge in Korea (Lim and Choi 2017). Also, they are favoured candidates for green roofs or "living roofs" installation that are increasingly popular and applicable in worldwide, especially in Europe and North America. Some of the main reasons of their large-scale use in green roofs architecture are the following: specific morphology of root (shallow roots), prominent ability to storage water (succulent organs), characteristic Crassulacean Acid Metabolism (CAM) that reduces water loss and allows survival in extreme conditions of the habitat (Li and Yeung 2014; Lim and Choi 2017). Therefore, multiple economic benefits arise from cultivation of the genus representatives, and any further scientific research could improve their status and application.

Representatives of the genus *Sedum*, commonly known as stonecrops, inhabit moderate areas of the northern hemisphere ('t Hart 1991), but also, some species

Fig. 12.1 Selected wild-growing *Sedum* species from the territory of the Balkan Peninsula (**a** *S. acre*, **b** *S. album*, **c** *S. amplexicaule*, **d** *S. atratum*, **e** *S. aetnense*, **f** *S. caespitosum*, **g** *S dasyphyllum*, **h** *S. hispanicum*, **i** *S. magellense*, **j** *S. rubens*, **k** *S. ochroleucum*, **l** *S. sexangulare*, **m** *S. stefco*, **n** *S. urvileii*, **o** *S. tuberiferum*)

could grow in the subtropical regions of the Earth. A lot of members are distributed in Europe, where southern parts of the continent have been described as one of the most important areas of distribution the genus diversity, including area of Balkan Peninsula, with relatively large number of the taxa (Stojanović et al. 2015). Regarding distribution on the territory of Europe, the genus *Sedum* is represented by 30–50 species ('t Hart 1991). The Mediterranean areas, North Africa and the Middle East, are, in a broader sense, distinguished as one of the centres of taxonomic diversity of the aforementioned genus ('t Hart 1995).

Due to their specific morphological characteristics, most species of this genus show great infraspecific variability, which classifies genus *Sedum* into a group of more complex genera, with insufficiently explained and clarified taxonomic relations. In general, the genus comprises succulent, glabrous to pubescent, annual to perennial, herbaceous plants or rarely chamaephytes. Underground part is represented by fibrous root, sometimes subterranean rootstock, woody in the base. Steams succulent, sometimes creeping, in upper part usually covered with succulent, alternate, rarely decussate or in whorls of 3 or 4, sessile or rarely (semi-) petiolate, usually (semi-)terete, rarely flat leaves. Inflorescence usually terminal, many-flowered pleiochasia with single or double cincinni, or corymboid, with bracts is usually present. Flowers (3–)5(–12)-merous, sessile to pedicellate. Sepal sessile, equal or strongly unequal, is usually smaller than petals. Petals are usually free, spreading or erect, yellow, white, pink, purple or reddish. Filaments are usually free or connate at base with petals. Nectary scales are usually present, variable in shape. Carpels are usually sessile, broad in the base, slightly connate, or completely free. Fruits are follicles, containing ovoid to ellipsoid, costate-bipapillate or reticulate-papillate to reticulate seeds (Thiede and Eggli 2007).

12.2.1 Systematics of the Genus **Sedum** *L.*

From the taxonomic aspect, the genus *Sedum* is usually mentioned in the context of controversial genera of the Crassulaceae family, especially in relation to the group of species named as Acre evolutionary line (Gontcharova et al. 2006; Carrillo-Reyes et al. 2009). According to Schönland, the genus *Sedum* represents one of the most primitive and oldest genera of the Crassulaceae family and it is often distinguished as the "core" from which many other genera of the paraphyletic family Crassulaceae have been potentially developed. Generally, there are numerous series of taxa of the Crassulaceae family, phylogenetic undefined, whose study on classification presents a challenge ('t Hart 1991; Gontcharova and Gontcharov 2007).

The first official classification of the genus *Sedum*, based on the studies on of the morphological characteristics of the species, was proposed by Koch and Boisser. Over time, the theories on the classification of the genus *Sedum* have changed, depending on the application of new methodologies in research work. According to Berger, none of the methods applied were good enough for the classification scheme of the genus. Nevertheless, it can be said that Berger's classification system or the concept suggested a split of the Crassulaceae family and former genus *Sedum* into several close, smaller taxa of higher taxonomical rank, in order to simplify their recognition. That principle to divide the heterogeneous genus *Sedum* was accepted in the practice during the twentieth century ('t Hart 1991). 't Hart (1991) proposed classification based on the evolutionary relationships of European representatives of the genus *Sedum*, i.e. *Sedum* L. subgen. *Sedum*, sect. *Sedum*. It includes morphological, cytological and molecular characters, i.e. their variability within the genus, hybridization pattern and taxon distribution. Also, the relationship between the 27 derived series of that genus was determined on the basis of selected characters that were cladistically analysed. The most recent and accepted taxonomic concept of the genus *Sedum* divides it into two subgenera *Sedum* subgen. *Gormania* which representatives are mostly distributed in North America and *Sedum* subgen. *Sedum* with distribution relates to territory of Europe, Africa and Asia (Thiede and Eggli 2007).

Several studies consider infrageneric taxonomy of the genus *Sedum* as quite complex ('t Hart 1991, 1995; 't Hart et al. 1999). Observations come from the fact that the genus comprises a large number of taxa and many of them are not clearly defined even in morphological terms. Cytological studies and hybridization experiments have shown that the number of chromosomes in the species of the mentioned genus is characterized by a wide range, and therefore, the cytological criterion, due to its pronounced diversity, is less favourable in the systematics. On the other hand, the hybridization pattern of the European species of the genus *Sedum* is not in agreement with the morphological characteristics traditionally used in the infrageneric classification. Based on specific morphological characteristics, certain genera are separated from the genus *Sedum* (e.g. *Rhodiola* L., *Hylotelephium* H. Ohba, *Phedimus* Raf.). It is required to clarify the origin of the groups that are separated to define the phylogenetic relations of the separated taxa in relation to the genus *Sedum*. Molecular studies have confirmed the high presence of paraphyletic groups in the given genus ('t Hart 1991; Mayuzumi and Ohba 2004). According to Mayuzumi and Ohba (2004), the genera *Rhodiola*, *Hylotelephium*, *Phedimus* and *Umbilicus* are derived from the genus *Sedum*, but the monophyletic origin of the *Telephium* clade has not been proven, which raises new questions in the phylogeny of this evolutionary line.

12.2.2 *Phytochemical and Chemotaxonomic Studies on the Genus* Sedum *L.*

The most commonly investigated phytoconstituents of the genus *Sedum* are: alkaloids, phenolic acids and their derivatives, flavonoids and corresponding heterozids, tannins and coumarins and compounds of epicuticular wax (Nahrstedt et al. 1982; Wolbis and Królikowska 1988; Wolbis 1989; Stevens et al. 1992; Sakar et al. 1993; Stevens et al. 1993, 1994a; b, 1995; Kim et al. 1996; Stevens et al. 1996; He et al. 1998; Korul'kin 2001, Thuong et al. 2007; Niu et al. 2011; Carrascoa et al. 2014; Xu et al. 2015; Jovanović et al. 2015a, b; Stojanović et al. 2015; Jovanović et al. 2016). Modern phytochemical studies, in service to chemotaxonomy, have been dedicated to the research of secondary metabolites of the genus *Sedum* with intention to recognize chemotaxonomic markers that can be used for clarification of complex taxonomic relations, at the level of the genus, and furthermore within the entire family of Crassulaceae. Many infrageneric relationships were discussed on the basis of examined patterns of different classes of compounds: alkaloids, tannins, flavonoids or epicuticular wax compounds. Apart from the correlation between chemical characters, relations with other types of taxonomic markers were also considered. For example, by examining the distribution of piperidine and pyrrolidine alkaloids within the Crassulaceae family, it was found that their occurrence was limited to the Acre clade, comprising *Sedum* and several smaller genera, whose species are characterized by specific ornamentation of seed coat and often large presence of alkaloids instead of tannins (Thiede and Eggli 2007). Distribution of alkaloids in this case is in agreement with the phylogeny of a family based on molecular data. In the phylogenetic sense, the presence of alkaloids is characteristic of primitive family members (which is supported by the existence of primitive morphological characters in the same species), while the complete absence of alkaloids is found in the phylogenetically advanced species of the family. Also, the structural complexity and diversity of these alkaloids are associated with the evolution of the taxon. Namely, the presence of more complex piperidine and pyrrolidine alkaloids in species indicates to a phylogenetic more advanced taxon. This is one of the well examples when different types of taxonomic markers are correlated into meaningful phylogenetic representation. Furthermore, many other useful scientific facts about the genus originated from chemotaxonomic studies and improved its phylogenetic status (Stevens et al. 1992, 1994a, b, 1995; Kim et al. 1996; 't Hart et al. 1999; Stevens et al. 1996; Korul'kin 2001; Jovanović et al. 2015a; Stojanović et al. 2015; Jovanović et al. 2016).

12.3 Chemical Markers of the Epicuticular Wax in the Service of Clarification on Infrageneric Relationships: Case of the Genus *Sedum* L.

The determination of the wax chemical profiles and consideration of distribution patterns of wax constituents (potential taxonomic markers) are valued steps and reliable criterion in chemotaxonomy, which necessarily leads to the classification of taxa at different taxonomic ranks. Almost all representatives of the genus *Sedum* possess a well-developed layer of epicuticular wax, often visible to the human eye. This fact has encouraged many studies to test the wax compounds as markers for taxonomic purposes. In general, epicuticular waxes of the *Sedum* species are characterized by the high content of the fatty acid derivates, mainly alkanes, followed by triterpenoids (Stevens et al. 1994a).

Stevens and co-workers (1994a) studied a large number of European taxa of the genus *Sedum,* in order to determine chemical profiles of their epicuticular waxes and implement distribution patterns of selected wax constituents in cladistics analysis of data. Within the group of long-chain fatty acid derivatives—alcohols, aldehydes, ketones, fatty acids and esters were reported. Beside normal-chain alkane fraction (C_{29}, C_{31}, C_{33} were dominant), there were identified branched (*iso-*) alkanes (*i*-C_{31} and *i*-C_{33} as most presented). Alcohols and fatty acids (C_{16}–C_{34}) were characterized as minor compounds, while, among all identified (C_{42}–C_{50}) esters of palmitic or stearic acid, C_{46} and C_{48} were dominant. Regarding long-chain aldehydes, frequently distributed were those with C_{30}, C_{32} and C_{34}. In about half of study samples, triterpenoids were detected. Analysis of isolated compounds determined structure of most triterpenoids and classified them into several groups (in accordance with polycyclic structure and characteristic substitution): ursane (α-amyrenyl acetate), oleanane (β-amyrenyl acetate, germanicyl acetate, and germanicyl formate), multiflorane (multiflorenyl acetate), taraxerane (taraxeryl acetate and taraxeryl formate), fernane (fernenyl formate) and lupane. In chemotaxonomic study on *Sedum* representatives belonging to the ser. *Rupestria*, characteristic distribution and content of triterpene series very well support phylogeny of studied species (Stevens et al. 1994b). Recent studies on epicuticular waxes isolated from Balkan representatives of the genus *Sedum* display similar profiles: normal-chain alkanes (C_{20}–C_{35}) as frequently distributed and most variable compounds and triterpenoid series (oleanane, lupane and taraxerane) which showed limited distribution among examined waxes (Jovanović et al. 2015b, 2016). Also, in case of epicuticular waxes of *Sedum rupestre* ssp. *rupestre,* only two alkanes and one triterpene were identified: hentriacontane, tritriacontane and germanicyl formate. The influence of habitat (natural vs. horticultural) was considered, and thus, the results indicated that habitat did not affect the qualitative but only the quantitative composition (Jovanović et al. 2015b). Generally, beside alkanes, which are generally considered as taxonomic markers, with more or less variable distribution, triterpenes from epicuticular waxes have proven to be good taxonomic character due to limited distribution within the genus. Their presence in epicuticular waxes is associated with some qualitative characteristics of the surface of the leaf and other plant organs related to wax micromorphology (Stevens et al. 1994a, b; Jovanović et al. 2015a, 2016; Jovanović 2016) (Fig. 12.2).

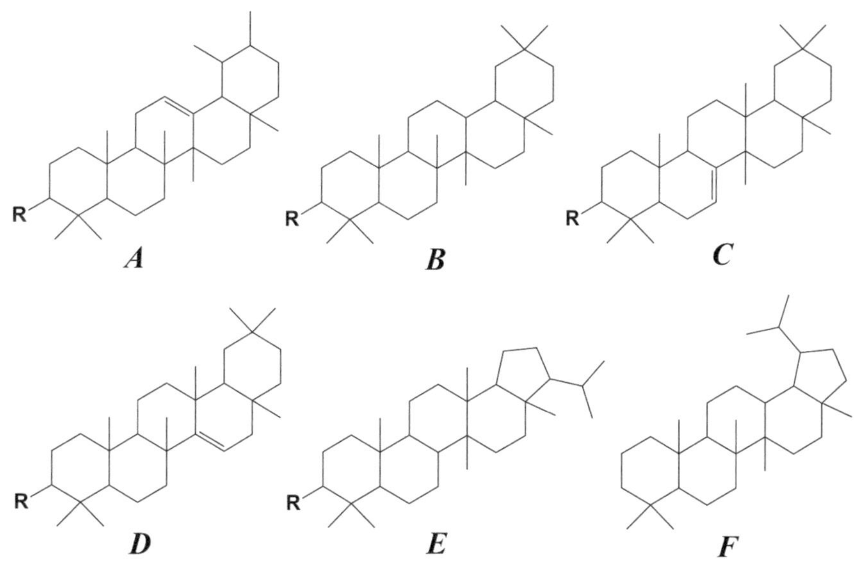

Fig. 12.2 Triterpene series identified in the epicuticular waxes of the studied *Sedum* taxa (Stevens et al. 1994a, b; Jovanović et al. 2015b, 2016). **a**. *Ursane series*: α-amyrenyl acetate (R–OAc); **b**. *Oleanane series*: β-amyrenyl acetate (R–OAc, Δ^{12}); germanicyl acetate (R–OAc, Δ^{18}); germanicyl formate (R–OOCH, Δ^{18}); **c**. *Multiflorane series*: multiflorenyl acetate (R–OAc); **d**. *Taraxerane series*: taraxeryl acetate (R–OAc); taraxeryl formate (R–OOCH); taraxerone (R=O); taraxerol (R–OH); **e**. *Fernane series*: Δ^{7}, Δ^{8}, $\Delta^{9(11)}$ fernenyl formate (R–OOCH), fernenone (R=O); fernenyl acetate (R–OOCH, Δ^{8}); **f**. *Lupane series*

12.3.1 *Chemotaxonomic Considerations on* Sedum L. *Based on Variability of Chemical Markers*

It has been observed that *n*-alkane and triterpene distribution pattern is distinguished from the distribution of other waxes by relevance to the systematics (Stevens et al. 1994a). Statistical analyses of characteristic distribution pattern of both alkanes and triterpenoids brought out relations among examined taxa which were discussed in relation to phylogeny and accordance with existing classification systems. Table 12.1 displays literature data on the examined taxa of the genus *Sedum* whose epicuticular waxes have been investigated for chemotaxonomic purposes.

In order to test usefulness of chemical markers and consequently figure out the relationships between taxa of *Sedum* sect. *Sedum*, Stevens et al. (1994a) investigated leaf waxes of 30 *Sedum* species, mostly collected at the territory of Europe. The characteristic alkane brought out separation of investigated samples of *Sedum* taxa into three groups, mainly due to difference in amount of alkane with odd number of C atoms: C_{29}, C_{31} and C_{33}. The obtained clustering was discussed in accordance with morphology of examined taxa and their current status in the taxonomy, taking into account status of evolutionary "primitive" as well as evolutionary "advanced" species. The primitive species, from series *Acria*, *Macaronesia*, *Anglica* and *Melananthera*, were distributed over all compared units in analyses since these series did not show distinct alkane(s) distribution pattern. The interesting situation happened in case of representatives from ser. *Macaronesia*. Distribution of alkanes replaced one of representatives, *Sedum fusiforme* Lowe from other studied taxa from the ser. *Macaronesia*. This evidence was one more which indicated that the species *S. fusiforme* should be excluded from the assigned series since previously recognized differences regarding other types of employed markers and characters (morphological traits, petals colour, ploidy level, hybridization pattern and alkaloids). Also, the content of triterpenoids confirmed that this taxon has more similarities with *Sedum anglicum* and *Sedum melanantherum* from other two series. Members of the advanced series of the genus *Sedum* —*Alpestria*, *Alba*, *Litorea*, *Lydia*, *Pedicellata* and *Rupestria* conglomerate together and formed a special cluster together with representatives of other genera of Crassulaceae, *Aeonium* Webb and Berth and *Sempervivum* L., which had been considered as taxa that evolved from Sedum-like ancestor. The characteristic distribution to a satisfactory extent supported existing classification of studied taxa. Other compounds of epicuticular waxes, listed under long-chain fatty acid derivatives, showed minor taxonomic significance, except non-characteristic high content of fatty acid methyl ester in single sample of *S. acre* from Turkey (Stevens et al. 1994a). In the studies of Jovanović et al. (2015a, 2016), a set of studied taxa had been supplemented regarding additional series and species of the genus (Table 12.1). These studies visualized chemotaxonomic tendencies between *Sedum* taxa from the Balkan Peninsula in Europe. The studies included representatives from less or more related genera from the Crassulaceae family—*Hylotelephium*,

Table 12.1 Taxa of the genus *Sedum* from the territory of Europe whose chemical markers from epicuticular waxes were investigated

Section	Series	No.	Species and subspecies	Country of origin	Literature sources
Sedum	*Acria* Berger	1.	*S. acre* L.	Turkey, France, Germany, Serbia, Macedonia	Stevens et al. (1994a), Jovanović et al. (2015a, 2016)
	Macaronesia Fröd	2.	*S. brissemoreti* Raym.-Hamet	Madeira	Stevens et al. (1994a)
		3.	*S. farinosum* Rose	Madeira	Stevens et al. (1994a)
		4.	*S. fusiforme* Lowe	Madeira	Stevens et al. (1994a)
		5.	*S. nudum* Aiton	Madeira	Stevens et al. (1994a)
	Anglica 't Hart	6.	*S. anglicum* Huds.	France	Stevens et al. (1994a)
	Melananthera 't Hart	7.	*S. melanantherum* DC.	Spain	Stevens et al. (1994a)
	Alpestria Berger	8.	*S. alpestre* Vill.	Turkey, France, Serbia	Stevens et al. (1994a), Jovanović et al. (2015a, 2016)
		9.	*S. annuum* L.	Italy, Turkey	Stevens et al. (1994a)
		10.	*S. annuum* L. subsp. *annuum*	Serbia, Macedonia	Jovanović et al. (2015a, 2016)
		11.	*S. apoleipon* 't Hart	Greece	Stevens et al. (1994a)
		12.	*S. borissovae* Balk.	Ukraine	Stevens et al. (1994a)
		13.	*S. euxinum* 't Hart and Alpinar	Turkey	Stevens et al. (1994a)
		14.	*S. grisebachii* Boiss. and Heldr.	Turkey, Bulgaria, Serbia	Stevens et al. (1994a), Jovanović et al. (2015a, 2016)
		15.	*S. laconicum* Boiss. and Heldr. subsp. *pallidum* 't Hart and Van Ham	Israel	Stevens et al. (1994a)
		16.	*S. laconicum* Boiss. and Heldr.subsp. *laconicum*	Greece	Stevens et al. (1994a)
		17.	*S. multiceps* Coss. and Durieu	North Africa	
		18.	*S. sexangulare* L.	Netherlands, Italy, Poland, Serbia	Stevens et al. (1994a), Jovanović et al. (2015a, 2016)
		19.	S. *tuberiferum* Stoj. and Stef.	Bulgaria, Serbia, Macedonia	Stevens et al. (1994a), Jovanović et al. (2015a, 2016)
		20.	*S. tuberosum* Coss. and Letell.	Tunisia	Stevens et al. (1994a)
		21.	*S. ursi* 't Hart	Turkey	Stevens et al. (1994a)
		22.	*S. urvillei* DC.	Bulgaria, Turkey, Serbia, Macedonia	Stevens et al. (1994a), Jovanović et al. (2015a, 2016)

(continued)

Table 12.1 (continued)

Section	Series	No.	Species and subspecies	Country of origin	Literature sources
	Litorea 't Hart	23.	*S. litoreum* Guss.	Israel, Greece	Stevens et al. (1994a), Jovanović et al. 2015a), 2016
	Alba Berger	24.	*S. album* L.	Spain, Albania, Bulgaria, Greece, Macedonia, Montenegro, Serbia	Stevens et al. (1994a), Zlatković et al. (2017)
		25.	*S. album* L. subsp. *album*	Macedonia, Serbia	Jovanović et al. (2015a, 2016)
		26.	*S. micranthum* Bastard ex DC.	Serbia	Jovanović et al. (2015a, 2016)
	Pedicellata 't Hart	27.	*S. brevifolium* DC.	Portugal	Stevens et al. (1994a)
	Lydia 't Hart	28.	*S. lydium* Boiss.	Turkey	Stevens et al. (1994a)
	Rupestria Berger	29.	*S. forsterianum* Sm.	Spain, Morocco, Portugal, England, France	Stevens et al. (1994a, b)
		30.	*S. montanum* Songeon and E. P. Perrier subsp. *montanum*	France, Italy	Stevens et al. (1994a, b)
		31.	*S. montanum* Songeon and E.P. Perrier subsp. *orientale* 't Hart	Italy, Yugoslavia	Stevens et al. (1994a, b)
		32.	*S. ochroleucum* Chaix	France, Macedonia, Serbia, Montenegro, Turkey, Greece	Stevens et al. (1994a, b), Jovanović et al. (2015a, 2016)
		33.	*S. rupestre* L. subsp. *rupestre*	France, Netherlands, Sweden, Germany, Serbia	Stevens et al. (1994b), Jovanović et al. (2015a, b, 2016)
		34.	*S. rupestre* L. subsp. *erectum* 't Hart	Italy	Stevens et al. (1994a, b)
		35.	*S. amplexicaule* DC.	Spain, Greece, Morocco, Turkey	Stevens et al. (1994b)
		36.	*S. amplexicaule* DC. subsp. *tenuifolium* (Sm.) Greuter	Macedonia	Jovanović et al. (2015a, 2016)
		37.	*S. sediforme* (Jacq.) Pau	Portugal, Israel, Italy, Turkey, France, Morocco	Stevens et al. (1994a, b)
		38.	*S. pruinatum* Link ex Brot.	Spain, Portugal	Stevens et al. (1994b)
	Aithales (Webb and Berthel.) 't Hart	39.	*S. rubens* L.	Macedonia	Jovanović et al. (2015a, 2016)
	Cepaea (Koch) Fröd.	40.	*S. cepaea* L.	Macedonia	Jovanovic et al. (2015a, 2016)
	Dasyphylla 't Hart	41.	*S. dasyphyllum* L. subsp. *dasyphyllum*	Macedonia, Serbia	Jovanović et al. (2015a, 2016)

(continued)

Table 12.1 (continued)

Section	Series	No.	Species and subspecies	Country of origin	Literature sources
	Glauco-rubens Fröd.	42.	*S. hispanicum L.*	Macedonia, Serbia	Jovanović et al. (2015a, 2016)
		43.	*S. hispanicum L. var. minus*	Serbia	Jovanović et al. (2015a, 2016)
	Magellensia 't Hart	44.	*S. magellense* Ten. subsp. *olympicum* (Boiss.) Greuter and Burdet	Montenegro	Jovanović et al. (2015a, 2016)
	Rubra Boriss.	45.	*S. caespitosum* (Cav.) DC.	Macedonia, Serbia	Jovanović et al. (2015a, 2016)
	Sedella (Fourr.) 't Hart	46.	*S. atratum* L. subsp. *carinthiacum* Hoppe ex Pacher	Montenegro	Jovanović et al. (2015a, 2016)
	Stefco 't Hart	47.	*S. stefco* Stef.	Macedonia	Jovanović et al. (2015a, 2016)

Echeveria DC., *Crassula* L. and *Kalanchoe* Adans. According to phylogenetic tree of the Crassulaceae family (Thiede and Eggli 2007), showing the eight major clades, *Crassula* and *Kalanchoe* are more distanced than *Hylotelephium* and *Echeveria* in regard to *Sedum* taxa. Even more, American genus *Echeveria* is traditionally nested in the same group with majority of *Sedum* taxa and a few more related genera (Acre clade). The position of employed outgroup elements (*Hylotelephium*, *Echeveria*, *Crassula*, *Kalanchoe*) in chemotaxonomical studies of Jovanović et al. (2015a, 2016), derived on alkane pattern distribution, showed good agreement with the phylogenetic study of Thiede and Eggli (2007). Exception is *Crassula* which was shown to be more similar with representatives of S. ser. *Rubra*, *Dasyphylla* and *Rupestria*. The segregation of *Hylotelephium* taxa was more obvious onto a PCA (Principal component analysis) plot, as well as closeness of other outgroup genera (*Echeveria*, *Crassula*, *Kalanchoe*) with examined *Sedum* taxa. Generally, among all identified normal alkanes, C_{27}, C_{29}, C_{31}, C_{33}, and C_{35} were dominant in majority of analysed *Sedum* samples, thus their dominance or deficit in epicuticular waxes reflected on the outcome of the cluster (agglomerative hierarchical analysis, AHC) and the (PCA) analyses. Since several examined *Sedum* taxa had pretty equal *n*-alkane composition (therefore, poor variability), their positions in AHC and PCA diagram were close each other. That was samples of species *Sedum stefco* Stef., *Sedum annuum* L., *S. album* L., *Sedum dasyphyllum* L., *Sedum rubens* L., *Sedum urvillei* DC. and *Sedum sexangulare* L. Regardless this fact, many of Balkan Peninsula representatives of the genus *Sedum* showed a certain level of absence of taxon grouping within the assigned series. This situation arises from considerable variation in *n*-alkane content in taxa of the genus *Sedum* collected in this region, which consequently disputes applicability of alkanes as markers in the clarification of the infrageneric relations of the genus *Sedum*, while their application at higher taxonomic ranks is not disputed (Jovanović et al. 2015a). In addition, the same set of samples was chemotaxonomically considered based on triterpenoid distribution. Since in some samples the triterpenoid fraction was absent, the AHC was conducted on matrices which comprise samples of wax in which whether oleanane, taraxerane or lupane series was detected. Therefore, the aforementioned absence of triterpenoids segregated a particular group of taxa, like "homogenous" group consisted of *S. album*, *Sedum alpestre* Vill., *Sedum caespitosum* (Cav.) DC., *S. stefco*, also separately clustered in case of alkane pattern. Other samples were clustered in accordance with their affiliation to the assigned series, but, in taxa of ser. *Rupestria*, *S. rupestre* L., *Sedum amplexicaule* DC. and *Sedum ochroleucum* Chaix although placed under the same cluster, these taxa were distanced from each other. Except unequal distribution—from totally absence of triterpenoids to a certain amount of them in wax, the high variability was also observed. However, the relations among other taxa of *Sedum* were in accordance with their phylogeny and systematics. Segregation of included outgroup taxa is different compared with those in case when alkanes were tested. At first, distribution of triterpenoids detached *Hylotelephium* and *Crassula* due to the absence of triterpenoids series in their waxes, which is in accordance with mentioned phylogenetic study of Thiede and Eggli (2007). But, reverse situation was observed in case of *Echeveria* the most detached in cluster in relation to *Sedum* taxa, while *Kalanchoe* was situated in the same group with samples of *S. ochroleucum* which contained taraxerane series in waxes. The position of these two outgroups is contradictory with the existing classification system (Mort et al. 2001; Mayuzumi and Ohba 2004; Thiede and Eggli 2007).

Furthermore, Stevens and co-workers (1994b) dedicated a study to the waxes of the taxonomically complicated *Rupestria* series and allied artificial hybrids (55 samples of seven species). As in the study of Jovanović et al. (2016), some species showed the absence of triterpenoids of wax, part of them were characterized due to low content and some of samples were abundant in triterpenoids. In order to clarify infraserial situation among taxa of ser. *Rupestria*, authors included one more, non-numerical morphological character—colour or appearance of the leaf surface. Since is known that the surfaces of leaf depend on micromorphology of epicuticular wax, they correlated glossy and/or glaucous appearance with the content of triterpenoids. Based on numerical data on content it had been already observed that glossy leaves did not possess triterpenoid or the content was very low. On the opposite side, there were glaucous leaves with identified triterpenoid in very high percentage. These statements were confirmed by conducting multivariate PCA analysis (Stevens et al. 1994b). Also, the mentioned correlation between content of triterpenoids and micromorphology was confirmed by scanning electron microscopy. Glossy leaves without triterpenoids or with very small amount possessed amorphous wax layer. As the content of triterpenoids was higher, the crystalline deposits were detected onto amorphous layer. The alkane distribution pattern of ser. *Rupestria* taxa was similar, except *Sedum forsterianum* Sm. which formed separate branch in the cluster due to high content of n-C_{35} alkane. Overall, alkanes are not valuable markers in case of ser. *Rupestria* despite standardization of all factors which could influence production and variation of them in waxes, which was not the case with triterpenoid pattern where quantitative variation is related to the environmental conditions (also shown in recent study of Jovanović et al. 2015b); while in qualitative sense this series is unique due to presence of taraxerane type of triterpenoid in wax, different than other European taxa of Crassulaceae subfam. Sempervivoideae (formerly Sedoideae) studied by the same author. The studied distribution and variation of triterpenoids were estimated as compatible to the previous classification based on variation of other markers (morphological, molecular, and chemical).

Except for taxa from the ser. *Rupestria*, going to the lower taxonomic rank, complex of taxa under *S. album* group has been recognized as diverse in taxonomic recognition, which consequently intrigues taxonomist. The survey of Zlatković et al. (2017) implemented different types of chemotaxonomic markers in order to analyse *S. album* species complex from Balkan Peninsula more detailed, starting from the assumption that high number of features such as leaf epidermal characteristics together with variability pattern of alkanes and triterpenes from leaf waxes could suggest logical relationships within the investigated population of traditionally reported species: *S. album*, *Sedum micranthum* Bastard ex DC., *Sedum athoum* DC. and *Sedum serpentini* Janch. The standardization of plant material was done by cultivating samples in the greenhouse prior to analyses, and thus, the influence of the environmental factors was possibly prevented. Taking into account all detailed observation and discussion on examined markers processed via canonical discriminant (CDA) and AHC analysis of epidermal characters, on the one hand, and wax constituents, on the other hand, similar but not always the same output could be derived. Both types of characters, epidermal as well as phytochemical, confirmed the unique position of *S. serpentini* in relation to the rest of the taxa from the complex. However, the differentiation between *S. athoum* and typical *S. album* is primarily based on differences in the chemical composition, less than in the case of epidermal characters. That evidence suggested *S. athoum* as specific chemotype, possibly affected by environmental factor or by some of the species genetic properties. Finally, conducted patterns pointed to somewhat insufficient differentiation between the *S. album* and *S. micranthum*, and therefore, their delimitation should be reconsidered in further analyses that will include reliable taxonomic markers of possibly different nature.

Taking into account the questionable applicability of wax character at the lower levels of classification, the research of Jovanović (2016) tested the existing patterns of alkanes and triterpenes at the higher, serial level. *n*-Alkanes with an odd number of C atoms are predominantly present in the waxes of all investigated representatives of the genus *Sedum*, specially *n*-C_{29}, C_{31} and C_{33} are the most common. In terms of the distribution of triterpenes series, the oleanane series was most frequently distributed triterpene series present in waxes, and often in significant percentage. In general, several patterns of distribution of the alkanes and triterpenes were recognized within studied samples from the central part of the Balkan Peninsula:

a. The average content of *n*-C_{31} is significantly lower than that of *n*-C_{33}:
    i. The average content of triterpenes is low (series *Acria*, *Alpestria*, *Litorea* and *Sedella*).
    ii. The average content of triterpenes is significant (series *Aithales*, *Rupestria*, *Cepaea*, *Glauco-rubens* and *Magellensia*).
b. The average content of *n*-C_{31} is significantly higher than that of *n*-C_{33}:
    i. The absence of triterpenes, while the average content of *n*-C_{29} is greater than the average content of *n*-C_{33} (series *Stefco* and *Rubra*).
    ii. Absence of the triterpenoids, while the average content of *n*-C_{29} is close to the content of *n*-C_{33} (ser. *Alba*).
    iii. The absence of a triterpenoid fraction, while the average content of *n*-C_{29} is lower than the content of *n*-C_{33} (ser. *Dasyphylla*).

The variability and the way of grouping of the investigated taxa at the level of the series or at a higher taxonomic level (genus level, when the representatives of the other genera are included) depend on the type of applied character and type of analysis. Statistical significance testing through the ANOVA highlighted a large number of wax characters. However, the oleanane series among triterpenes and

n-C_{33} among n-alkanes were emphasized. The results of conducted AHC, PCA and CDA showed that the representativeness of the results depends on the applied analysis and wax chemical character, whether alkane and triterpene distribution is observed separately, mutually, or with other types of markers being considered. On the other hand, by testing qualitative character of wax (glaucous, pruinose and green appearance of leaf surface), the results of the correspondence analysis (CA) indicate the existence of a clear model of variability of qualitative characters. The variation pattern is conditioned by the presence, or absence of the triterpene fraction in the epicuticular wax, which reflects the clear separation of the groups. The presence of certain triterpene series in wax not only affects the coloration/appearance of leaf surface but also plays a significant role in the differentiation of the taxa within the genus *Sedum*.

12.4 Conclusion

Taking into account all considered scientific data of the chemical characters (markers) in chemotaxonomy of the genus *Sedum* L., it could be noticed that epicuticular wax was considered in taxonomic delamination at several ranks. Among all identified constituents of epicuticular waxes isolated from numerous *Sedum* taxa, n-alkane and triterpenes distribution pattern were proved to be the most adequate in clarification of infrageneric relationships. The basic difference between these two groups of chemical markers is that alkanes are ubiquitous in waxes, while the distribution of triterpenes is limited. Structural diversity within the mentioned chemical markers plays an important role in chemotaxonomy. In alkane fraction odd number n-C_{29}, C_{31} and C_{33} are the most common alkanes, while triterpene distribution pattern was considered through several series: ursane, oleanane, lupane, taraxerane, fernane and multiflorane. Each series could be characterized by its own chemical diversity, which has not been investigated for chemotaxonomic purposes. Therefore, general discussions suggest that alkanes as chemotaxonomic markers are more applicable to higher levels (series) of classification of the investigated genus. Further, it is necessary to correlate different markers in the clarification of relationships within the genus *Sedum* and other genera of the family Crassulaceae. Finally, the genus itself requires great scientific attention: on the one hand there is a need for the systematization of existing data, and on the other hand, innovativeness in the scientific approach.

Acknowledgements The authors are grateful to the Ministry of Education, Science and Technological Development for the financial support through the grant No 172047.

References

Ardenghi N, Mulch A, Pross J, Niedermeyer EM (2017) Leaf wax n-alkane extraction: an optimised procedure. Org Geochem 113:283–292

Barthlot W, Neinhuis C, Cutler D et al (1998) Classification and terminology of plant epicuticular waxes. Bot J Linn Soc 126:237–260

Bojović S, Šarac Z, Nikolić B et al (2012) Composition of n-alkanes in natural populations of *Pinus nigra* from Serbia—chemotaxonomic implications. Chem Biodivers 9:2761–2774

Borisjuk N, Hrmova M, Lopato S (2014) Transcriptional regulation of cuticle biosynthesis. Biotechnol Adv 32:526–540

Buschhaus C, Jetter R (2012) Composition and physiological function of the wax layers coating arabidopsis leaves: β-Amyrin negatively affects the intracuticular water barrier. Plant Physiol 160:1120–1129

Bush RT, McInerney FA (2013) Leaf wax n-alkane distributions in and across modern plants: implications for paleoecology and chemotaxonomy. Geochim Cosmochim Acta 117:161–179

Carrascoa V, Pintoa LA, Wolff Cordeirob K et al (2014) Antiulcer activities of the hydroethanolic extract of *Sedum dendroideum* Moc et Sessé ex DC. (balsam). J Ethnopharmacol 158:345–351

Carrillo-Reyes P, Sosa V, Mort ME (2009) Molecular phylogeny of the Acre clade (Crassulaceae): dealing with the lack of definitions for *Echeveria* and *Sedum*. Mol Phylogenet Evol 53: 267–276

De Melo GO, Malvar Ddo C, Vanderlinde FA et al (2009) Antinociceptive and anti-inflammatory kaempferol glycosides from *Sedum dendroideum*. J Ethnopharmacol 124:228–232

Denton MF (1994) SEM analysis of leaf epicuticular waxes of *Sedum* section *Gormania* (Crassulaceae). Brittonia 46:296–308

Domínguez E, Cuartero J, Heredia A (2011) An overview on plant cuticle biomechanics. Plant Sci 181:77–84

Ertaş A, Boğa M, Yılmaz MA, Yeşil Y et al (2014) Chemical compositions by using LC-MS/MS and GC-MS and biological activities of *Sedum sediforme* (Jacq.) Pau. J Agric Food Chem 62:4601–4609

Fernández V, Guzmán-Delgado P, José Graça et al (2016) Cuticle structure in relation to chemical composition: re-assessing the prevailing model. Front Plant Sci 7:427

Gontcharova SB, Gontcharov AA (2007) Molecular phylogenetics of Crassulaceae. Genes, Genomes Genomics 1:40–46

Gontcharova SB, Artyukova EV, Gontcharov AA (2006) Phylogenetic relationships among members of the subfamily Sedoideae (Crassulaceae) inferred from the ITS region sequences of nuclear rDNA. Russ J Genet 42:654–661

He A, Wang M, Hao H et al (1998) Hepatoprotective triterpenes from *Sedum sarmentosum*. Phytochemistry 49:2607–2610

Herbin GA, Robins PA (1969) Patterns of variation and development in leaf wax alkanes. Phytochemistry 8:1985–1998

Heredia A (2003) Biophysical and biochemical characteristics of cutin a plant barrier biopolymer. Biochim Biophys Acta 1620:1–7

Jeffree CE (2006) The fine structure of the plant cuticle. In: Riederer M, Müller C (eds) Biology of the plant cuticle. Annual Plant Reviews, vol 23. Blackwell Publishing, Oxford, pp 11–125

Jeffree CE, Baker EA, Holloway PJ (1976) Origins of the fine structure of plant epicuticular waxes. In: Dickinson CH, Preece TF (eds) Microbiology of aerial plant surfaces. Academic Press, London, pp 119–158

Jovanović S (2016) Secondary metabolites of the *Sedum* L. (Crassulaceae) representatives from the central Balkan Peninsula and their chemotaxonomic significance. Dissertation, University of Niš

Jovanović S, Zlatković B, Stojanović G (2015a) Distribution and variability of n-Alkanes in Epicuticular waxes of *Sedum* Species from the Central Balkan Peninsula: chemotaxonomic importance. Chem Biodivers 12:767–780

Jovanović S, Zlatković B, Stojanović G (2015b) The chemical composition of *Sedum rupestre* L. ssp. *rupestre* epicuticular waxes: horticultural versus the natural plant habitat. FU Phys Chem Tech 13:77–82

Jovanović S, Zlatković B, Stojanović G (2016) Chemotaxonomic approach to the central balkan *Sedum* species based on distribution of triterpenoids in their epicuticular waxes. Chem Biodivers 13:459–465

Jung HJ, Kang HJ, Song YS et al (2008) Anti-inflammatory, anti-angiogenic and anti-nociceptive activities of *Sedum sarmentosum* extract. J Ethnopharmacol 116:138–143

Kang TH, Pae HO, Yoo JC et al (2000) Antiproliferative effects of alkaloids from *Sedum sarmentosum* on murine and human hepatoma cell lines. J Ethnopharmacol 70:177–182

Kim JH, 't Hart H, Stevens JF (1996) Alkaloids of some Asian *Sedum* species. Phytochemistry 41:1319–1324

Koch K, Barthlott W (2009) Superhydrophobic and superhydrophilic plant surfaces: an inspiration for biomimetic materials. Phil Trans R Soc A 367:1487–1509

Koch K, Bhushan B, Barthlott W (2008) Diversity of structure, morphology, and wetting of plant surfaces. Soft Matter 4:1943–1963

Korul'kin DY (2001) Chemical composition of certain *Sedum* species of Kazakhstan. Chem Nat Compd 37:219–223

Kunst L, Samuels AL (2003) Biosynthesis and secretion of plant cuticular wax. Prog Lipid Res 42:51–80

Kunst L, Samuels L (2009) Plant cuticles shine: advances in wax biosynthesis and export. Curr Opin Plant Biol 12:721–727

Lemieux B (1996) Molecular genetics of epicuticular wax biosynthesis. Trends in Plant Sci 1: 312–318

Li WC, Yeung KKA (2014) A comprehensive study of green roof performance from environmental perspective. IJSBE 3:127–134

Lim MS, Choi SH (2017) Estimation of phylogeny of nineteen Sedoideae Species cultivated in Korea inferred from Chloroplast DNA analysis. J Jpn Soc Hortic Sci. https://doi.org/10.2503/hortj.okd-087

Maffei M (1994) Discriminant analysis of leaf wax alkanes in the Lamiaceae and four other plant families. Biochem Syst Ecol 22:711–728

Maffei M (1996a) Chemotaxonomic significance of leaf wax n-Alkanes in the Umbelliferae, Cruciferae and Leguminosae (Subf. Papilionoideae). Biochem Syst Ecol 24:531–545

Maffei M (1996b) Chemotaxonomic significance of leaf wax alkanes in the Gramineae. Biochem Syst Ecol 24:53–64

Maffei M, Meregalli M, Scannerini S (1997) Chemotaxonomic significance of surface wax n-Alkanes in the Cactaceae. Biochem Syst Ecol 25:241–253

Maffei M, Badino S, Rossi S (2004) Chemotaxonomic significance of leaf wax n-alkanes in the Pinales (Coniferales). J Biol Res 1:3–19

Manheim BS Jr, Mulroy TW, Hogness DK, Kerwin JL (1979) Interspecific variation in leaf wax of *Dudleya*. Biochem Syst Ecol 7:17–19

Manners GD, Davis DG (1984) Epicuticular wax constituents of North American and European *Euphorbia esula* biotypes. Phytochemistry 23:1059–1062

Marin P (2003) Biohemijska i molekularna sistematika. Miroslav Damjanović, Beograd, Srbija

Matas AJ, José Sanz M, Heredia A (2003) Studies on the structure of the plant wax nonacosan-10-ol, the main component of epicuticular wax conifers. Int J Biol Macromol 33:31–35

Mayuzumi S, Ohba H (2004) The phylogenetic position of Eastern Asian Sedoideae (Crassulaceae) inferred from chloroplast and nuclear DNA sequences. Syst Bot 29:587–598

Medina E, Aguiar G, Gómez M et al (2006) Taxonomic significance of the epicuticular wax composition in species of the genus *Clusia* from Panama. Biochem Syst Ecol 34:319–326

Mitić Z, Zlatković B, Jovanović S et al (2017) Geographically related variation in epicuticular wax traits of *Pinus nigra* populations from Southern Carpathians and Central Balkans—taxonomic considerations. Chem Biodivers 13:931–942

Mitić Z, Zlatković B, Miljković M et al (2017) First insights into micromorphology of needle epicuticular waxes of south-eastern european *Pinus nigra* J. F. Arnold populations. Iheringia Ser Bot 72:373–379

Mitić Z, Zlatković B, Jovanović S et al (2018) Diversity of needle n-alkanes, primary alcohols and diterpenes in Balkan and Carpathian native populations of *Pinus nigra* J. F. Arnold. Biochem Syst Ecol 80:46–54

Mort ME, Soltis DE, Pamela S. Soltis PE et al (2001) Phylogenetic relationship and evolution of Crassulaceae inferred from matK sequence data. Am J Bot 88:76–91

Müller C, Riederer M (2005) Plant surface properties in chemical ecology. J Chem Ecol 31: 2621–2651

Nahrstedt A, Walther A, Wray V (1982) Sarmentosin epoxide, a new cyanogenic compound from *Sedum cepaea*. Phytochemistry 21:107–110

Niu XF, Liu X, Pan L, Qi L (2011) Oleanene triterpenes from *Sedum lineare* Thunb. Fitoterapia 82:960–963

Osborne R, Stevens JF (1996) Epicuticular waxes and Glaucousness of *Encephalartos* leaves. Phytochemistry 42:1335–1339

Sakar MK, Petereit F, Nahrstedt A (1993) Two phloroglucinol glucosides, flavan gallates and flavonol glycosides from *Sedum sediforme* flowers. Phytochemistry 33:171–174

Seigler DS (1998) Plant secondary metabolism. Springer, Boston, MA

Stanković M, Radojević I, Ćurčić M et al (2012) Evaluation of biological activities of goldmoss stonecrop (*Sedum acre* L.) Turk J Biol 36:580–588

Stevens JF, 't Hart H, Hendriks H, Malingré TM (1992) Alkaloids of some European and Macaronesian Sedoideae and Sempervivoideae (Crassulaceae). Phytochemistry 31:3917–3924

Stevens JF, 't Hart H, Hendriks H, Malingré TM (1993) Alkaloids of the Sedum acre-group (Crassulaceae). Plant Syst Evol 185:207–217

Stevens JF, 't Hart H, Bolck A et al (1994a) Epicuticular wax composition of some European *Sedum* Species. Phytochemistry 35:389–399

Stevens JF, 't Hart H, Pouw AJA et al (1994b) Epicuticular waxes of *Sedum* series *Rupestria*. Phytochemistry 36:341–348

Stevens JF, 't Hart H, Vanham CHJ et al (1995) Distribution of alkaloids and tannins in the Crassulaceae. Biochem Syst Ecol 23:157–165

Stevens JF, 't Hart H, Elema ET, Bolck A (1996) Flavonoid variation in Eurasian *Sedum* and *Sempervivum*. Phytochemistry 41:503–512

Stojanović G, Jovanović S, Zlatković B (2015) Distribution and taxonomic significance of secondary metabolites occurring in the methanol extracts of the Stonecrops (*Sedum* L., Crassulaceae) from the Central Balkan Peninsula. Nat Prod Commun 10:941–944

't Hart H (1991) Evolution and classification of European *Sedum* species. Flora Mediterranea 1:31–61

't Hart H (1995) The evolution of the *Sedum acre* group (Crassulaceae). Bocconea 5:119–128

't Hart H (1997) Diversity within Mediterranean Crassulaceae. Lagascalia 19:93–100

't Hart H, van Ham RDHJ, Stevens JF et al (1999) Biosystematic, molecular and phytochemical evidence for the multiple origin of sympetaly in Eurasian Sedoideae (Crassulaceae). Biochem Syst Ecol 27:407–426

Taylor P (2011) The wetting of leaf surfaces. Curr Opin Colloid In 16:326–334

Thiede J, Eggli U (2007) Crassualaceae. In: Kubitzki K (ed) Flowering plants eudicots. Springer, Berlin, pp 83–118

Thuong PT, Kang HJ, Na M et al (2007) Anti-oxidant constituents from *Sedum takesimense*. Phytochemistry 68:2432–2438

Tomaszewski D, Zielinski J (2014) Epicuticular wax structures on stems and comparison between stemsand leaves—a survey. Flora 209:215–232

van Maarseveen C, Jetter R (2009) Composition of the epicuticular and intracuticular wax layers on *Kalanchoe daigremontiana* (Hamet et Perr. de la Bathie) leaves. Phytochemistry 70:899–906

Wolbis M (1989) Flavonol glycosides from *Sedum album*. Phytochemistry 28(8):2187–2189

Wolbis M, Królikowska M (1988) Flavonol glycosides from *Sedum acre*. Phytochemistry 21:3941–3943

Xue D, Zhang X, Lu X et al (2017) Molecular and evolutionary mechanisms of cuticular wax for plant drought tolerance. Front Plant Sci 621

Zlatković B, Mitić Z, Jovanović S et al (2017) Epidermal structures and composition of epicuticular waxes of *Sedum album* sensu lato (Crassulaceae) in Balkan Peninsula. Plant Biosyst 151:1–11

Chapter 13
Chemical Ecology of *Ruta* sp.: VOC, Chemotaxonomy and Allelochemistry

Zineb Bennaoum and Hachemi Benhassaini

Abstract Among the Rutaceae, the genus Ruta is an ideal model for understanding the structure and the origin of the diversity of volatile organic compounds (VOCs), since they have been for several years the subject of a thorough investigation of their interest in chemotaxonomy and pharmacology. The characteristics of these species are the presence of schizogenic secretory cavities that represent the original site of the synthesis of aromatic compounds. The power of these smells is also effective between a plant and its environment. In this chapter, we describe how, where, and by which pathways these species produce the volatile essence. We define the specific chemotype of the genus Ruta within the Rutaceae family. Moreover, we evoke the interest of VOCs in the foundations of chemical ecology to understand the role and the importance of chemical molecules in plant–insect–environment interactions.

Keywords Chemical ecology · VOC · Ruta · Schizogenic · Chemotaxonomy

13.1 Introduction

Ecology and chemistry are a priori not propitious to be combined. However, a science dares to associate the two terms: chemical ecology. She is interested in the chemical messages that govern the relationships between individuals of one or more species, animal, or vegetal. The power of odors is also valid between a plant and an insect. This latter can not colonize a plot without being attracted by an odor, especially if it is monophagous. This is observed with flowers that diffuse certain compounds to attract pollinating insects. These are not pheromones, which serve as intermediates between individuals of the same species, but inter-species mediators.

Z. Bennaoum (✉) · H. Benhassaini
Laboratory of Plant Biodiversity, Conservation and Enhancement, Faculty of Natural and Life Science, University of Djillali Liabes Sidi Bel-Abbes, Sidi Bel-Abbes, Algeria
e-mail: zinnaoum@yahoo.fr

K. G. Ramawat (ed.), *Biodiversity and Chemotaxonomy*, Sustainable Development and Biodiversity 24, https://doi.org/10.1007/978-3-030-30746-2_13

According to Amsler and Fairhead (2006), chemical ecology is described as the study of the chemical–mediated interactions that exist between organisms and/or between organisms and their chemical environment. These interactions are classified in three categories: the chemical communication between the organisms, the detection, and the adapted response of the organisms to the variations of their environment and finally, the chemical defense; that is, the synthesis by organisms of chemical compounds to deal with predators, grazers, pathogens, and other competitors. Generally, the molecules involved in chemical ecology, particularly in chemical defense belong to the secondary metabolism. They are called secondary metabolites or natural substances. This distinguishes them from compounds that play a role in primary metabolism in the cells of many taxa (Amsler and Fairhead 2006).

Vascular plants produce, store, and emit volatile terpene compounds (CTV) (isoprene (C5), monoterpenes (C10), sesquiterpenes (C15), diterpenes (C20)) and oxygenates (alcohols, aldehydes, ketones, esters), especially from the foliar organs. Owing to their low molecular weight, these substances are classified as volatile organic compounds of biogenic origin (BVOC).They are very reactive and have a very short lifespan (Lafuente 2006). Volatile organic compounds (VOCs) produced by plants, represent 1% of the secondary metabolites known nowadays (Guitton et al. 2010) and are mainly terpenes, phenylpropanoids/benzenoids and amino acid and fatty acid derivatives. They can easily cross cell membranes and be released into the atmosphere or soil (Dudareva and Negre 2006) due to their lipophilic features and high vapor pressures. The diversity and quantity of VOCs formed varies among species under the influence of biotic and abiotic factors (Pichersky and Gershenzon 2002), in the whole plant depending on the season (Angioni and Barra 2006; Atti-Santos and Pansera 2004; Johnson and Kazantzis 2004) and at the plant or some of its organs over their development (Dudareva and Negre 2006; Sangwan et al. 2001). In addition, the fruits emit VOCs essentially when they are mature (Aharoni et al. 2004), the roots produce them during attacks by herbivores (Chen et al. 2004; Rasmann and Turlings 2008) and in the leaves, and their composition varies with age and leaf position (Singh and Luthra 1989; Holm and Laakso 1997).

As secondary metabolites, they have a very important chemical diversity (Lafuente 2006) used as a means of communication between plants to interact with their environment. They do not have a universal function (Levin et al. 2003). Thereby, these VOCs allow plant to fight off against biotic and abiotic stress and against herbivores and pathogen (Wink 1988; Pichersky and Gershenzon 2002; Wink 2003), to emit chemical signals through which the plant communicates with its environment (Jeaun et al. 2005), to promote reproduction by attracting pollinators (Pichersky and Gershenzon 2002) and allow colonization of environment by limiting the growth of other plants by allelopathic mechanisms (Muller 1966). In addition, these volatile substances not only allow the plant to communicate with neighboring plants (Baldwin et al. 2006) but also between its own organs (Heil and Silva Bueno 2007). These volatile compounds are differently represented in several family, gender, and species, so that they are used as chemotaxonomic markers (Lafuente 2006).

Among these VOCs, the essential oils molecules, which are volatile and odoriferous principles, secreted and then excreted by aromatic plants (Duraffourd and Lapraz 2002). A real quintessence of nature, these rare products are more discreet. Of course, aromatic plants, but which ones, synthesize them? How and why the plants produce these hundreds of volatile substances.

13.2 Families and Genus Producing Essential Oils in the Vegetal Kingdom

In the plant kingdom, essential oils are usually found in higher plants (Richter 1993). Their elaboration is very dependent on factors of intrinsic origin (genetics, vegetative stage) and of extrinsic origin (environmental), in particular, solar radiation whose absence affects the yield and even the quality in aromatic principles. Chemically, essential oils are an assembly of different molecules (Franchomme and Penoel 1990), dissolved one in the other and combined to create homogeneous solutions. Consequently, they have multiple and multifaceted properties (Samate 2002). The genus able to elaborate these natural substances is distributed across fifty families, many of which belong to the orders of the Lamiales, Asterales, Rutales, Laurales, and Magnoliales (Mann 1987). Currently, there are about 800,000 plant species and among them, only 10% are able to synthesize essence, which are aromatic plants (Balz 1986). They can be stored inside one or more organs where the chemical composition can vary between the organs. Secretory glands, almost found on all parts of plant (Bruneton 1999), develop these aromatic essences. Quantitatively, the levels of essential oils of plants that can contain them are very low, often less than 1%. Greater percentage, as clove flower bud (15%) is rare and exceptional.

Among the main families of aromatic plants, we quote the following Bachelot et al. (2005):

- Abietaceae (Pinaceae): 200 species divided into 10 genera, represented by conifers (***Abies***, cedrus, Larix, Picea, and pinus).
- Cupressaceae: including trees and shrubs usually with shucked leaves. There are Cupressus, Thuja, and Juniperus.
- Lamiaceae (Labiatae): This is an important family of dicotyledonous plants with about 6000 species distributed in 9 sub-families themselves divided into 210 genus (Lavandula, Mentha, Rosmarinus …). This family is a great source of essential oils.
- Myrtaceae: Represented by trees and shrubs. There are 3000 species distributed in 130 genera (Eucalyptus, Syzygium, Myrtus …).
- Lauraceae: 2000–2500 species. These are green and wild trees, found mostly in Brazil. There are Cinnamomum (from China), Laurus, Sassafras … etc.

- Ericaceae: Woody plants, found in temperate and tropical regions. There are 3500 species. These are very precious plants in therapeutics (Gaultheria, Ledum).
- Asteraceae: They form the largest family of the plant kingdom. There are more than 20,000 species, especially in dry and arid areas such as Chamaemelum, Artemisia, and Santolina.
- Rutaceae: There are 900 species distributed in 50 genera and belonging to the order of sapindales, which are in tropical and subtropical zones. These are trees, shrubs or more rarely herbaceous plants, great producers of essential oils (lemon, lime, mandarin). Their glandular punctuated leaves, the production of limonoids, and the ordinary presence in the parenchyma and pericarp of secretory cavities containing volatile organic compounds distinguish them.

The most recognized genera by the phytochemical variability of essential oils in Rutaceae family are the genus Citrus and the genus Ruta. This latter, native to the Mediterranean Basin (Salvo et al. 2010) is widely replicated in southern Europe, western Asia and northern Africa (Fig. 13.1). Its habitat is characterized by the heliophilic behavior of its species, with light shading on dry and rocky substratum, with often low water reserves, thus characterizing a Chaméphytaie/Hémicryptophytaie type plant formation (Bennaoum and Benhassaini 2017). The preferred type of reproduction is entomogamy and the mode of seed dispersal is barochorous (Monterde 1986). These species are not lacking originality, but they have the disadvantage of giving off a strong and very unpleasant smell like the common rue (*Ruta graveolens*). They are considered repulsive for insects, especially aphids (Lamnaour 2006).

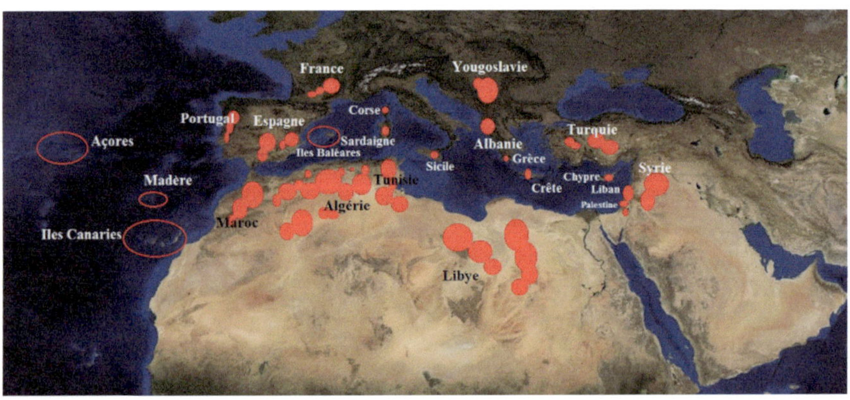

Fig. 13.1 Distribution of the genus Ruta in the world

13.3 VOC Shaping and Accumulation Sites in *Ruta* sp.

Essential oils are produced in the cellular protoplasm of aromatic plants and represent the products of cellular metabolism called "secondary" (Mann 1987). The synthesis and accumulation of these metabolites in an organ is associated with the presence of specialized histological structures (Deysson 1978). According to Raymond (2005), aromatic plants have three main categories of secretory apparatus: the epidermal glandular hairs, pockets and glandular canals schizogènes and schizolysigènes. These are cavities located in the parenchyma of leaves, stems, and fruits, in some species. Glandular structures can be found in all plant organs, vegetative, and reproductive. Few plants in which these structures are present in one organ, most are provided in all their parts.

The excretory system in the genus Ruta is mainly represented by secretory cavities and rarely by hairs (Bennaoum et al. 2015). According to the literature, these are absent from the leaves of species of the genus Ruta. However, ciliates and capitates trichomes (Fig. 13.2) may appear on the leaflets of *Ruta angustifolia* Pers. even if they do not develop trichomes over the year except during the period of spring season (Bennaoum et al. 2015). This is due to the ecological conditions of the environment where these taxa grow. Moisture and temperature can affect the trichome yield of plants. Changing temperatures and humidity through seasons or between environments can stimulate a more stressful environment, in which the plant thinks it should increase trichome production (Tirse et al. 2017).

According to the study on micro-phytodermic seasonal characters of the genus Ruta (Bennaoum et al. 2015), the secretory cavities are equidistant often arranged in a quincuncial; they are composed of cells with different forms according to their ontogeny. The numbers of secretory cells vary between 2 and 8 cells (rarely two cells) surrounded by a polyhedral cells' number ranging from 6 to 9 in rosettes form (Fig. 13.2). Their density increases during the autumn and winter following the degeneration of the central cells, which results in the formation of the gland lumen. These leaf glands are first unicellular, then they dispose in a continuous set of 2–4 specialized epidermal cells (Haberlandt 1914), where the central cell walls tear and disappear leaving in their place a vacuum filled by oleoresin (Chatin 1875). It was a characteristic of the Rutaceae family.

The secretory cavities in Rutaceae develop schizogenous and schizolysigenous processes (Tschirch and Stock 1933; Zeybek and Zeybek 1994). On the leaves of *R. graveolens* L., *Ruta Chalepensis* L. and *Ruta montana* clus. L., schizolysigenic and lysigenic glands can be observed (Peterson et al. 1978; Bennaoum et al. 2015). The mother cell divides into four (Fig. 13.3), and between these new cells that the pocket is formed, which are called schizogen. Then, after cell division, several grouped cells develop a pocket by lysis of their wall, where the secretory products accumulate in the central space by syneresis. Lysis of the most inner cells accompanies the lysis of the epidermis. This development is called lysigenes (Kienast 1885). However, there are two types of glands in Ruta genus mainly at the level of *R. chalepensis* leaves; the first is found in the leaf parenchyma forming a

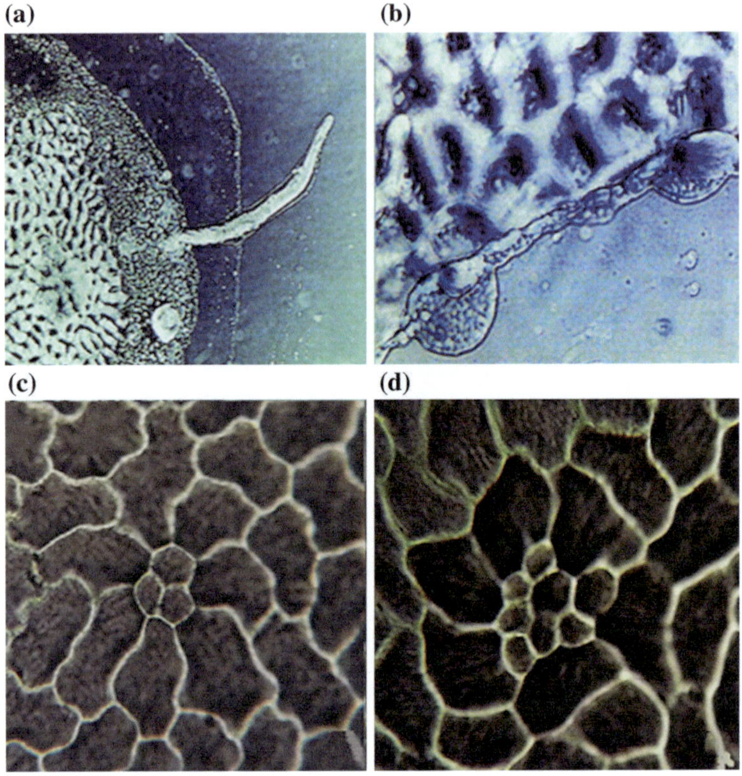

Fig. 13.2 Secretory cavities and trichomes in leaves of Ruta species **a** Ciliate trichomes; **b** Capitate trichomes (glandular hair); **c** Secretory cavities with 3 cells; **d** secretory cavities with 8 cells (Bennaoum et al. 2015)

schizolysigenous gland sac. The second type of gland is adjacent to epidermis forming lysigenous pockets (Antunes 1982). These pockets, always superficial, have an epidermal origin. This explains why it suffices to slightly crush a soft part of a Ruta leaf so that a strong smell of essence emerges.

13.4 Chemotaxonomy Within Ruta Genus

Chemotaxonomy is science, mainly applied to plants, which attempts to classify living things according to their chemical composition (Gibbs 1974; Harborne and Turner1984) Chemotaxonomy studies complete botanical studies and had a considerable impact on plant systematics (Dahlgren 1980; Hegnauer 1986; Wink 2003). However, the chemical criterion does not pretend to replace the other morphological, anatomical, cytological, genetic, and embryonic characters, but it brings its own contribution and it intervenes, especially when there are doubts in the botanical

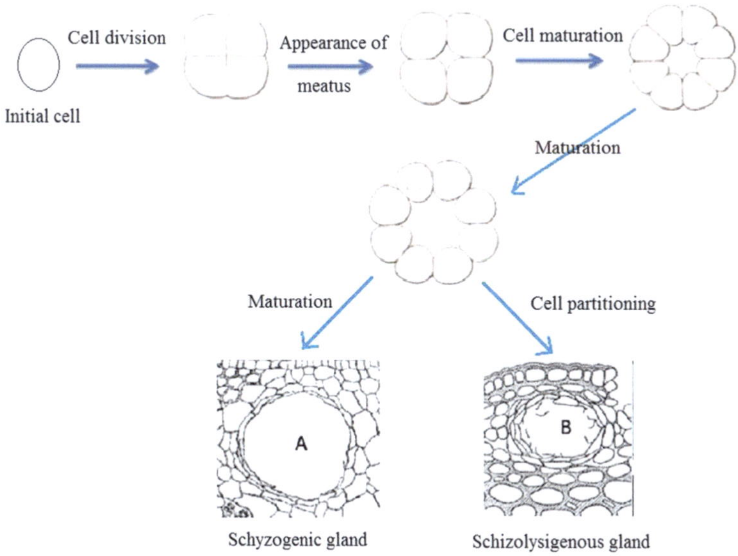

Fig. 13.3 Stages of development of secretory cells

classification. There are often correlations between chemical components of vegetal and the organization of the plant (Paris and Delaveau 1965). A real natural classification must be based on the analysis and harmonization of all organs (Singh 2016). They still hit problems related to the high intraspecific variability of secondary metabolites, from a qualitative and quantitative viewpoint. Thus, factors such as the age of the individual, the geographical area, or the substrate on which it develops, influence the chemical composition of the plants (Hegnauer 1986). Phenolic compounds, alkaloids, terpenoids, and non-protein amino acids are the four important groups widely used for chemotaxonomic classification (Smith 1976). These groups of compounds have a wide variety in distribution, function, and chemical diversity (Smith 1976; Hegnauer 1986). The chemotaxonomic classification system is based on the chemical similarity of the taxon (Atal 1982; Rasool et al. 2010).

Because of their phytochemical content, Rutaceae has a priori a big interest in a pharmacological and biological control area. Most Rutaceae produce essential oil (HE) in the schizolysigenous pockets on both the lower and upper sides of leaves, as previously mentioned. These HEs generally contain volatile mono-, sesqui- and diterpenes (Edris 2007). This family also contains often alkaloids, coumarins, flavonoids, lignans, and limonoids (Bruneton 1999; Silva et al. 1988). The presence of oxygenated terpene derivatives within Rutaceae family is often correlated in plants with a biological activity, including antiviral action (Champagne et al. 1992). This is the case for many groups of alkaloids, flavonoids and coumarins that characterize this family. Therefore, the essential oil of *Clausena excavates* Burm. has a strong larvicidal activity on the larvae of the mosquito *Aedes aegypti* and *Aedes albopictus* (Cheng et al. 2003) and some coumarins are not neglected for their activity against Flaviviridae (Mazzei et al. 2007; Giampieri et al. 2009).

In view of families, chemotaxonomy allows sometimes to include families of uncertain systematic position in well-defined orders (Paris and Delaveau 1965). In genus, the chemotaxonomy brings arguments in favor, either of the homogeneity of the various species, or of the heterogeneity and the fragmentation in subgenera. Rutaceae can be divided into two subfamilies: Rutoïdées whose genera are represented by Ruta, Dictamnus, Pilocarpus, Xantoxylum, Phellodendron, and Aurantioidea, represented by the genus Citrus (Chase et al. 1999). The five genera belonging to the Ruteae tribe have the same origin in the molecular tree (Salvo et al. 2008). They share a number of morphological and phytochemical traits, including the presence of actinomorphic flowers, cream white to bright yellow (Engler 1931), high levels of lignans (Waterman 1983; Da Silva et al. 1988), a specific class of coumarins and acridones (Waterman 1975). Therefore, these genera appear to form a coherent taxonomic group. However, does the genus Ruta have a well-defined chemical specificity within Rutaceae? What are the interspecific and intraspecific variations that it can present? Is it possible to develop a biochemical classification within the genus?

13.4.1 Chemotype of Ruta Genus

Although slight morphological differences in species of the genus Ruta, each one has unique genetic and biochemical traits that help identify it. However, among the sixty species of genus, only ten have been the subject of volatile molecules analysis nowadays. Therefore, only the phytochemical compositions of *R. graveolens* L., *R. chalepensis* L., *R. angustifolia* Pers., *R. montana* Clus. L., *Ruta corsica* Dc., *Ruta pinnata* L., *Ruta lamarmorea* L., *Ruta microcarpa* Svent., *Ruta oreojasm* Webb and Berthel and *Ruta tuberculata* Forsk have been reported in the literature (Inigo et al. 1981; Baser et al. 1996; Bagchi et al. 2003; Bertrand et al. 2003; Dob et al. 2008; Merghache et al. 2009; Zellagui et al. 2012). Ethnobotanical data teaches us that some of these species whose VOC composition is unknown have been used for centuries as natural remedies (Hammiche and Azzouz 2013). The main characteristic of the VOCs secreted by the different species of the genus Ruta is their richness in ketones and terpenes (Bennaoum et al. 2017). Thus, more than 200 different molecules are characterized in all species analyzed (Ulubelen and Öztürk 2006; Ferhat et al. 2014; Khadhri et al. 2014). 2-undecanone, 2-nonanol, and 2-dodecanone are the main constituents of Ruta species (Başer et al. 1996; Verzera et al. 2000; Rustaiyan et al. 2002; Merghache et al. 2009; Mejri et al. 2010; Khadhri et al. 2014) (Table 13.1) alongside furanocoumarins (Milesi et al. 2001; Ulubelen and Öztürk 2006). However, there is significant interspecific variation in the concentration of these compounds and/or the presence of other constituents (Bagchi et al. 2003; Gibka et al. 2009; Mejri et al. 2012). Secondary metabolites are omnipresent in plants. The presence and/or amount of each compound may vary and the composition of the chemical mixture can vary too, so that the chemical diversity can be distributed within and between individuals (Moore et al. 2013).

Table 13.1 Summary of the chemotyped classification of *Ruta* sp

Genus	Species	Chemotype	Population
Ruta	*R. graveolens*	Undecan-2-one/2-Nonanone	China/Italy/Egypt/Malaysia
		Undecan-2-one/2-Heptanolacetate	North of Iran
		n-Hex-4-en-3-one/*n*-Pent-3-one	India
		2-Nonanone/Undecan-2-one	Colombia
		2-Undecanone/Chalepensin	Colombia
	R. chalepensis	2-Undecanone/2-Nonanone	Northeastern Algeria/Iran/Morocco
		Pregeijerene/2-Decanone	Northwestern Algeria
		2-Nonanone/2-Methyl octyl acetate	Algeria
		2-Undecanone/1-Nonene	Tunisia
		2-Acetoxytridecane/2-Acetoxytetradecane	Northeastern Algeria
		2-Methyl octyl acetate/β-Phellandrene	Greece
		2-Undecanone/2-Nonylacetate	Turkey
		2-Undecanone/Chalepensin	Northeastern Algeria
		Methyl nonyl Ketone	Portugal
	R. montana	2-Undecanone/(E)-Caryophyllene	Algeria
		2-Undecanone/Z-8-3,5-Dimethyl-4-Hydroxyphenyl-2-octene	Tunisia
		2-Undecanone/1-Nonene	Turkey/Tunisia
		Ketone C6-C13	North-central Algerian
		2-Undecanone/Resorcinol	Northeastern Algeria
	R. angustifolia	2-undecanone/2-nonanone	Northeastern Algeria
		2-undecanone/2-Decanone	North-central Algerian/Portugal
	R. corsica	2-nonyl acétate/β Caryophyllene	Corsica
Haployphyllum	*H. tuberculatum* (*R. tuberculata*)	germacrene D/bicyclogermacrene	Spain

In the taxon *R. graveolens*, 2-Undecanone, 2-Nonanone, *n*-Hex-4-en-3-one, and the sesquerepenoids are the most responsive constituents (Aboutabl et al. 1988; Yaacob et al. 1989; De Feo et al. 2002; Soleimani et al. 2009; Tang et al 2011; Malik et al. 2013); 2-nonyl acetate, 2-Undecyl acetate, and β Caryophyllene in *R. corsica* (Bertrand et al. 2003); 2-Undecanone, (E)-Caryophyllene, Resorcinol in *R. montana* (Bagchi et al. 2003; Boutoumi et al. 2009; Djarri et al. 2013; Bennaoum et al. 2017); 2-Undecanone, 2-Nonanone, 2-Methyl octyl acetate, Methyl nonyl Ketone in *R. chalepensis,* and *R. angustifolia* (Dob et al. 2008; Chibani et al. 2013; Bennaoum et al. 2017). Otherwise, some species as *R. chalepensis*, *R. montana*, *R. graveolens*, and *R. corsica* synthesize more than 100 different terpenes (Bertrand et al. 2003; Ferhat et al. 2014), whereas, other species such as *R. angustifolia* express a lower or even null terpenic diversity (Dob et al. 2008; Bennaoum et al. 2017). This phytochemical diversity formed by the species of Ruta shows that exists within species a population expressing different chemotypes.

Therefore, intraspecific variations in VOC composition have been reported in *R. angustifolia* species from Algeria (Dob et al. 2008; Bennaoum et al. 2017); on *R. chalepensis* and *R. montana* from Tunisia (Khadhri et al 2014; Majdoub et al. 2014); on *R. Corsica* from Corsica (Bertrand et al. 2003), or even on *R. graveolens* from Italy (De Feo et al. 2002; Mancuso et al. 2015). These variations correspond to the synthesis of different compounds or to different proportions of some compounds. These variations are generally, whether of individual order (spatial and temporal variations) or of population order. For example, within *R. montana* from Algeria, populations have 2-undecanone as the first constituent of essential oil and second (E)-Caryophyllene (chemotype 2-Undecanone/(E)-Caryophyllene) (Bennaoum et al. 2017), other populations with first 2-Undecanone and second Resorcinol (chemotype 2-Undecanone/Resorcinol) (Djarri et al. 2013), and others with a chemotype Ketone C6–C13 (Boutoumi et al. al. 2009). This latter chemotype is very particular for its rarity in the other populations, of Tunis (Khadhri et al. 2014) of Turkey (Gibka et al. 2009). The chemistry of these plants makes it possible to discriminate between species, like *R. corsica* which has as major constituents two irregular esters and a sesquiterpene (Bertrand et al. 2003; Oussalah et al. 2007).

Based on chemotypes, a chemotaxonomic classification succeeds between species of Ruta genus (Chase et al. 1999; Salvo et al. 2008). That said the nine species currently attributed to the Ruta genus (Salvo et al. 2010) form a strongly supported clade that is a brother to Haplophyllum. The latter characterized by the taxon *R. tuberculata* Forsk is currently better known chemically by a chemotype (germacrene D/bicyclogermacrene), which separates it from the genus Ruta to confirm it in the genus Haplophyllum so taking the name *Haplophyllum tuberculatum* (Salvo et al. 2008) (Fig. 13.4). This group (Haplophyllum/Ruta) shares chemotaxonomic affinities with the Boenninghausenia clade. *R. chalepensis* and *R. angustifolia* form a brother clade to *R. graveolens*; similarly, these three species form a close group to *R. corsica* (clade I). The three species from the Canary Islands (*R. pinnata*, *R. microcarpa*, and *R. oreojasm*) form a highly supported clade (Cade III). However, the relationship between *R. montana* (clade II) and the other two clades (clade I and clade III) remains unresolved. These taxa have a chemical

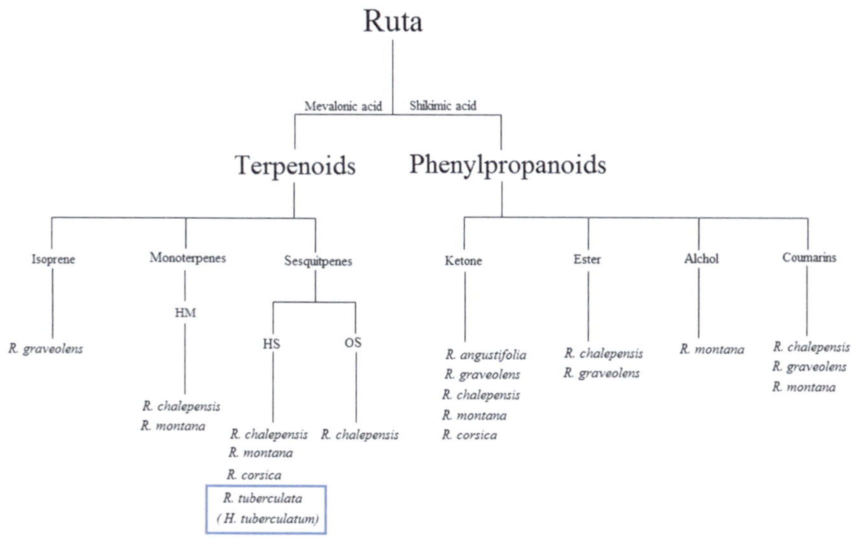

Fig. 13.4 General classification of principal Ruta species by chemical filiation

specificity that distinguishes the genus within Rutaceae. This specificity comes down to the chemotype 2-Undecanone, a ketone that characterizes most species of the genus Ruta and differentiates them from other genus; Haplophyllum (germacrene D chemotype), Citrus (limonene chemotype), and Myrtopsis (lupeol chemotype).With *R. chalepensis*, *R. graveolens* and *R. montana* (Table 13.1), different chemotypes exist between populations living in different bioclimatic zones (Ferhat et al. 2014). Specific environmental conditions could, therefore, induce a differential selection pressure on the production of some compounds in populations of different bioclimatic zones. These chemotypic variations observed in nature are kept when individuals from different populations are transplanted and cultivated on the same site (Gotsiou et al. 2002). Genetic bases are responsible for these chemotypic mutations (Iijima et al. 2004a). Consequently, small variations in the gene sequence responsible for terpene synthesis may cause changes in chemotypes (Iijima et al. 2004b). However, the molecular mechanisms responsible for chemotypic variability in aromatic species cannot be limited at punctual mutations on genes.

13.4.2 Spatial and Temporal Variations in VOC Production and Emissions Within the Genus

According to the law of nature, plants react to the surrounding environment and throughout their lifetime, the chemical composition of their metabolites may evolve (Johnson and Kazantzis 2004). Extracted from plants, essential oils composition varied in quality and quantity, depending on the season (Angioni and Barra 2006; Atti-Santos and Pansera 2004), depending on the plant or organ (Dudareva and Negre 2006; Sangwan et al. 2001), within the same species (phenomenon of chemical polymorphism) (Yamaura et al. 1989; Figueiredo et al. 1995) and according to geographical, climatic, and soil conditions (Samate 2002). Therefore, in *R. graveolens* of Colombia, the stem gives an essence rich in chalepensin, the leaves provide a volatile extract where 2-nonanone is predominant, the flowers give 2-undecanone as the major compound, whereas in the essential oil obtained from the roots Geijérene predominates (Stashenko et al. 2000). The composition of essential oils varies with age and mode of development. Thus, in fringed rue (*R. chalepensis* L.), the rate of chalepensin is 50% higher in old leaves than in young leaves (Merghache et al. 2009). In *R. montana*, the decrease in 2-Undecanone content is observed during the flowering period (Bennaoum et al. 2017). The phenology of plant at the time of harvest is difficult to verify and control (Derbesy 1997). Rue harvested during the autumn period perceive its rate of Pregeijerene (E)-Caryophyllene and Chalepensin greatly evolve compared to its content of 2-Undecanone (Merghache et al 2009; Bennaoum et al. 2017).

The position of the organs on the plant presents the same phenomenon. The leaves or the flowers on the same stem do not appear simultaneously and according to their age, do not have the same composition (Touche 1997). This phenomenon is observed, especially in *Cymbopogon flexuosus* (Singh and Luthra 1989), *Artemisia* sp. (Holm and Laakso 1997), *Salvia officinalis* (Croteau and Felton 1981), *Mentha piperita* (Gershenzon et al. 2000; Voirin and Bayet 1996), and *Lavandula latifolia* (Muñoz-Bertomeu et al. 2008). The emission of volatile terpene compounds by flower coincides with the pollinator presence period and the sexual maturity of plant. This is the case of rose, where a peak emission of terpene compounds is observed during the opening of bud flower (Bergougnoux et al. 2007), as well as variations in the day with a maximum in diurnal periods (Hendel-Rahmanim et al. 2007). Some Ruta taxa show the same variations, where VOC production is favorable during the spring period (Dob et al. 2008; Bennaoum et al. 2017).

The phytochemical variability that exists between Ruta species is due to the impact of environmental factors on the nature of volatile compounds, although the genetic factors favored this diversity (Duarte et al. 2010). This influence of environmental factors is mainly due to climatic conditions (temperature, humidity, and precipitation), altitude, and soil type (physico-chemical properties) (Guignard 1983). These can influence the content of plants in active substances (fertilization with nitrogen usually favors the synthesis of alkaloids). Livestock does not graze these species. This is explained by the concentration of alkaloids considered as toxic compounds. However, chemotypic variations at the vegetative stage may also contribute to chemotaxonomic or phylogenetic relationships within the genus.

The diversity of secondary metabolic pathways (Fig. 13.4) of the genus Ruta is reflected in the chemistry of its species. 2-undecanone is the chemotype in most taxa. We can conclude that it is a chemotaxonomic marker of Ruta species. For some taxa of different populations, the essential oil has other major constituents, so we can define an intermediate chemotype. We can cited, the chemotype 2-Nonanone, chemotype 2-Acetoxytridecane, chemotype 2-Methyl octyl acetate, chemotype (E)-Caryophyllene, chemotype Pregeijerene, chemotype β-Phellandrene, chemotype

2-nonyl acetate, chemotype Chalepensin, chemotype Resorcinol (Inigo et al. 1981; Baser et al. 1996; Bagchi et al. 2003; Khadhri et al. 2014; Bennaoum et al. 2017).

The types of chemical compositions determined allowing to the major constituents may be associated with the biosynthetic pathways thus defined: the terpenoids pathway, leading to isoprene, monoterpenes and sesquiterpenes and the phenylpropanoid pathway leading to ketones, ester, and alcohol. This made it possible to consider the general classification of some Ruta species by chemical filiation as indicated in Fig. 13.4. This variability confirms a chemical polymorphism in a plant possibly related to environmental factors and phenological influences. The research for a specific chemotype for each taxa is inconclusive and confirms the variability of the chemical constituents of volatile organic compounds.

13.5 From Phytochemistry to Chemical Ecology

The majority of volatile organic compounds (VOCs) is classified as secondary metabolites and is particularly involved in plant–environment interactions in the context of reproduction, defense, or symbiosis (Dudareva et al. 2004; Paschold et al. 2006; Gershenzon and Dudareva 2007). Furthermore, they play an important role in plant growth and development, since they lead to the synthesis of certain phytohormones such as gibberellins (Guitton et al. 2010). Figure 13.5 summarizes the different VOCs synthesized by *Ruta* sp. cells and their roles in species–environment interactions.

Among these volatile compounds, terpenoids form an essential part of direct and indirect defenses against herbivores and pathogens. The low molecular weight terpenoids that are often released as volatile plant substances before or during the attack by a predatory organism may act a direct defense against bacteria, fungi, insects, or herbivores such as toxins, antibiotics, or repellents (Kessler and Baldwin 2002; Bohlmann and Keeling 2008; Unsicker and Kunert 2009). These molecules called "*elicitors*" are molecules that can induce plant resistance against different pathogens by activating their defense responses. They may also constitute chemical and physical barriers to insect feeding or oviposition (Keeling and Bohlmann 2006; Heiling et al. 2010), or act as analogues of insect hormones. Several studies have shown that insect feeding affects the emission and synthesis of terpenes, notably in maize (Turlings et al. 1990), cotton (Rose et al. 1996), tobacco (Kessler and Baldwin 2002), and conifers (Martin et al. 2003; Miller et al. 2005).

Various compounds as ester, methyl salicylate, monoterpenes, myrcene, β-ocimene, and sesquiterpene are emitted a few hours after infestation (Dudareva and Negre 2006). Systemic release of VOC is one of the specific responses to herbivores best-studied (War et al. 2011). These molecules defend the plants directly either by repelling, dissuading, and infecting the herbivores, or indirectly by attracting the attackers' natural enemies, and thus protect plants from further damage (Dudareva and Negre 2006; Maffei 2010). Metabolites of lipoxygenase pathway, shikimic acid, and products of the terpenoid pathway play an important

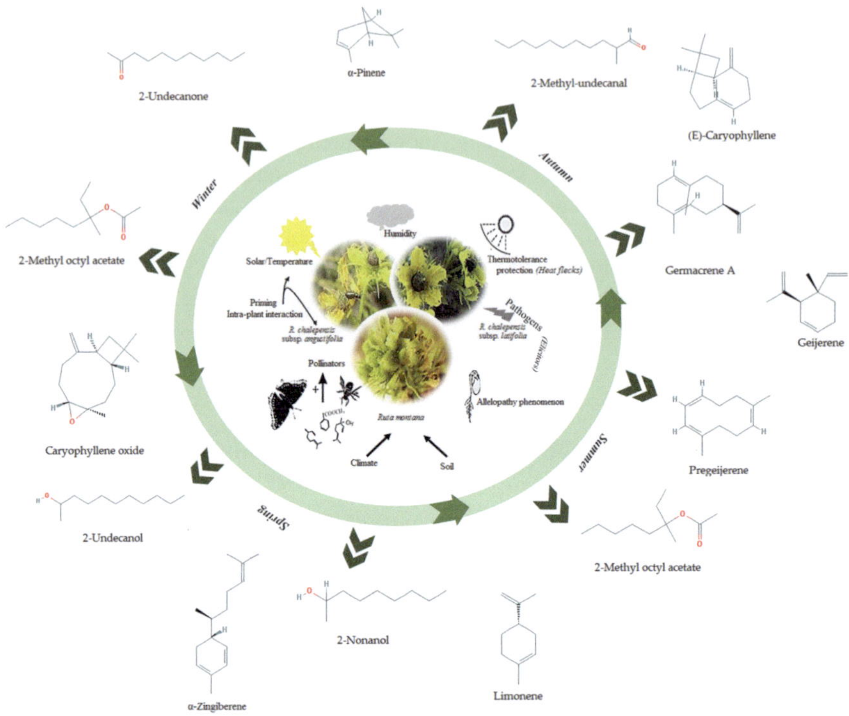

Fig. 13.5 Schema of interactions of Ruta species with their environment via VOCs

role in the defense of plants, directly and indirectly (Gill et al. 2010). In addition to the volatile compounds released from the aerial parts of the plant, several studies also show that the roots release various volatile compounds. Among these molecules, 1, 8-cineole, a volatile monoterpene, which is toxic and repellent to some insects and (E)-β-caryophyllene that protect plants from underground pests by acting as antimicrobials and antiherbivores and by attracting the natural enemies of insect pests (Rasmann et al. 2005; Dudareva and Negre 2006; Ehlting et al. 2006).

Despite the recognized toxicity in the genus Ruta (Majdoub et al. 2014; Bouabidi et al. 2015), the flowers of these species are visited by some insects such as Hymenoptera including *Apis mellifera* L., Coleoptera including *Oxythyrea funesta* Poda. and, especially Lepidoptera whose the Swallowtail caterpillar or *Papilio machaon* L. butterfly that is related to these species (Bennaoum et al. 2015). Methyl ketone, 2-tridecanone, and 2-undecanone play a key role in the biological anti-xenotic interactions by causing the mortality of some insects mainly larvae of *Helicoverpa zea* (L.), *Manduca sexta* (L.) *Keiferia lycopersicella* (Walshingham), *Spodoptera exigua* (Hubner) (Lin et al. 1987) and adults of *Tribolium castaneum* (Herbst) (Majdoub et al. 2014).In most lepidopteran insects, resistance is expressed as increased mortality during secondary larval phases (Michael Smith 1985).

Although their ability to intoxicate or repel herbivores is a central point of toxic metabolites production, nature accomplished frequently to the opposite effect; it is the semantic of nature. Often in some species, the production of such deterrent compounds can harm the plant. VOC emissions by plants or an organ at the same time of an herbivore attack do not necessarily have an immediate defense effect, but prepare the plant for future attacks (Guitton et al. 2010). This phenomenon is known as "*priming*" (Dudareva and Negre 2006; Ton and D'Alessandro 2007; Unsicker and Kunert 2009). Indeed, insects that have acquired resistance against one of them are able to perceive it as a sign of recognition of the plant, which allows them to locate the host whose can feed themselves without damage (Hopkins 2003, Horiuchi et al. 2003).

These volatile molecules can also play other ecological roles, as attracting pollinators (Pichersky and Gershenzon 2002). Floral fragrances are complex and diversified mixtures of volatile compounds with low molecular weight (Knudsen et al. 2006). Many plants to attract pollinating insects use these floral scents. Therefore, they represent a key element in plant reproduction (Pellmyr 1986). This is observed especially in general plants that favor general pollinators that pollinate a large proportion of angiosperms (Fortuna and Bascompte 2006). The volatility of floral fragrance compounds allows them to act as long-distance chemical messengers to provide information on the location and identity of flower. However, with some species, cheaters use terpenoids of attraction. For example, parasites may recognize their host plant by its odor; an odor used to attract the pollinator (Mattiacci et al. 2000). Olfactory signals may vary over time (e.g., with the different stages of maturation of a flower), and in space (between several populations) (Chartier et al. 2014). The nature of these variations can be quantitative (number and relative proportion of compounds emitted) and qualitative (nature of compounds).

The flowering of most Ruta species is very nectariferous and appreciated by pollinators (*A. mellifera* L., *O. funesta* Poda., *P. machaon* L.) due to the presence of an intrastaminal nectariferous disk in their flowers, thus ensuring an entomogam pollination (Spichiger et al. 2004). Pollinating insects produce several pheromones or chemical signals through several glands in order to find a partner via plant odors. Therefore, the bee uses the production of chemical substances such as terpenoid including geraniol, nerolic acid, citric and geranic acid. These pheromones contain more than 40 different compounds, including pentyl acetate, butyl acetate, 1-hexanol, *n*-butanol, 1-octanol, hexylacetate, octylacetate, and 2-nonanol (Kelemu et al. 2015).

Abiotic factors such as light, temperature, and water deficit influence the production and emission of volatile compounds (Bertin and Staudt 1996; Loreto and Schnitzler 2010; Peñuelas and Staudt 2010). Thus, volatile oils have a role of mobilizing light energy and thermal regulator in favor of the plant (Croteau 1986). They regulate diurnal perspiration by absorbing ultraviolet rays by their unsaturated constituents. The presence of essential oils in plants as well as their content is, therefore, related to photochemistry (Croteau 1986). The authors have demonstrated a very strong action of light since no emission is detected without light (Singsaas et al. 1997; Delfine et al. 2000; Velikova et al. 2011). In addition, the total quantity of volatile organic compound emissions is inversely proportional to the increase in soil moisture (Despinasse 2015). However, in some species, VOCs do not react in the same way. The amounts of linalool (monoterpene), β-caryophyllene (sesquiterpene), and (E)-α bergamotene (sesquiterpene) increase with humidity, while geranyl acetate (monoterpene) and (E)-β-farnesene (sesquiterpene) are scarce when soil moisture is high (Despinasse 2015). Isoprenes and monoterpenes, through their thermo-protection ability, could play a different role protecting plants against heat stress, due to a sudden increase temperature in leaf caused by the sun; a phenomenon called "*heat flecks*" (Delfine et al. 2000; Penuelas and Llusia 2002; Copolovici et al. 2005; Sharkey et al. 2008). They can still have a role in fertilization (Gouinguene and Turlings 2002) and in the protection against photo-oxidative stress and thermo-tolerance (Peñuelas and Munné-Bosch 2005).

Volatile biogenic compounds are thought to participate in the inhibition of seed germination (Fischer et al. 1994; Tarayre et al. 1995) and/or the growth of some neighboring species (Stevens 1984), phenomenon known "*allelopathy,*" grouping all processes involving secondary metabolites in the inhibition of growth and/or development of a biological organism. It is a complex phenomenon, because it involves, in addition to the two plants, respectively, "producer" and "target" molecules, an intermediate, the soil whose abiotic and biotic characteristics (in particular the microfauna) are fundamental for the expression of this allelopathic potential (Gallet and Pellissier 2002). This complexity explains the many controversies that still exist regarding the ecological importance of these interactions, as well as the difficulty to demonstrate them.

These allelochemical compounds are very abundant in *Ruta* sp. cells. Thus, *R. graveolens* from Italy possess significant allelopathic activity (De Feo et al. 2002). 2-ketones, which represent the chemotype, are inactive vis-a-vis to the germination of plants, while terpenoids, organic acids and aliphatic alcohol 2-nonanol express an appreciable inhibitory activity in dependent dose. However, studies have reported that aliphatic ketones generally have good inhibitory activity (Rice 1984). Monoterpene ketones are also very potent allelochemicals, correlating the toxicity of other monoterpenoids to their ease to being metabolically altered at ketones (Asplund 1968, 1969).

Other monoterpene compounds (1,8-cineole and a-pinene) active in the genus Ruta constitute potent phytotoxins (De Feo et al. 2002). Other monoterpene compounds (1,8-cineole and a-pinene) active in the genus Ruta constitute potent phytotoxins (De Feo et al. 2002). They are considered as potent inhibitors of oxygen uptake by mitochondrial suspensions, probably resulting from the gradual penetration of terpene across the mitochondrial membrane to the site of action (Muller et al. 1969). Consequently, it has been proposed that oxygenated monoterpenes may be used as bio-herbicides (Kordali et al. 2007). Terpenes can also influence plant–plant interactions (Singh et al. 2006), plant-microbe/fungi associations (Bednarek and Osbourn 2009), and tree species phenology (Becerra 2007) by underground allelopathic effects via the translocation of essential oils where they remain absorbed in order to exert a long-term inhibitory activity (Halligan 1975;

Friedman 1987; Kholi 1994). These effects have the potential to affect biodiversity and the evolution of terrestrial ecosystems.

Moreover, the presence of xanthotoxin and psoralen in the oils of *R. angustifolia* and *R. montana,* Rutarin, bergapten, coumarins umbelliferone in *R. graveolens* (Reyes and Gonzales 1970; Milesi et al. 2001; Ulubelen and Öztürka 2006), pinnarin and furopinnarin in *R. pinnata* (Reyes and Gonzales 1970)and Chalepensin in *R. Chalepensis* (Merghache et al. 2009) give them considerable allelopathic properties (Aliotta et al. 1994, 1996). The role of these furanocoumarins may be important to explain the allelopathic activity of essential oils, whose components can probably be synergistic.

13.6 Conclusion

The schyzogenic pockets present in the family Rutaceae are not only their structure and common feature, but also the primary sites of synthesis, secretion, and emission of volatile organic compounds. Thus, the genus Ruta is distinguished by a very rich and varied chemistry. This chemical specificity is reflected in the diversity of the secondary metabolic pathways of its species, from which organic constitutes including triterpenoids, aromatic compounds, and coumarins have been isolated. These phytochemical molecules contribute to improving taxonomic studies in the genus/family. The presence or absence of a particular phytochemical in taxon within the same population or between different populations, as well as knowledge of its biochemical synthetic pathways, can be used to define chemotaxonomic markers in the species complex Ruta. This chemotypic variability within the same taxon is strongly influenced by genetic and environmental factors. These VOCs are considered as chemical mediators to create complex and varied links between individuals of the same species or different species. The potential of chemical ecology in the genus Ruta is distinct by a role of protection, defense, pollinators, and allelopathy.

References

Aboutabl EA, Elazzouny AA, Hammerschmidt F (1988) The essential oil of *Ruta graveolens* L. growing in Egypt. Sci Pharm 56:121–124

Aharoni A, Giri AP et al (2004) Gain and loss of fruit flavor compounds produced by wild and cultivated strawberry species. Plant Cell 16(11):3110–3131

Aliotta G, Cafiero G, De Feo V, Sacch R (1994) Potential allelochemicals from *Ruta graveolens* L. and their action on radish seeds. J Chem Ecol 20:2761–2775

Aliotta G, Cafiero G, De Feo V, Palumbo AD, Strumia S (1996) Infusion of rue for control of purslane weed: biological and chemical aspects. Allelopathy J 3:207–216

Amsler CD, Fairhead VA (2006) Defensive and sensory chemical ecology of brown algae, Ed. Elsevier Academic Press, Amsterdam, Boston

Angioni A, Barra A (2006) Chemical composition, seasonal variability, and antifungal activity of *Lavandula stoechas* L. ssp. stoechas essential oils from stem/leaves and flowers. J Agric Food Chem 54(12):4364–4370

Antunes T (1982) Ultrastructure of secretory leaf glands of *Ruta chalepensis* L. Bull de la Société botanique de France 129(1):79–82

Asplund RO (1968) Monoterpenes: relationship between structure and inhibition of germination. Phytochemistry 7:1995–1997

Asplund RO (1969) Some quantitative and qualitative aspects of the phytotoxicity of monoterpenes. Weed Sci 17:454–455

Atal CK (1982) Cultivation and utilization of aromatic plants, 1st edn. Council of Scientific and Industrial Research, New Delhi

Atti-Santos AC, Pansera MR (2004) Seasonal variation of essential oil yield and composition of *Thymus vulgaris* L. (Lamiaceae) from South Brazil. J Essent Oil Res 16(4):294–295

Bachelot C, Blaise A, Corbel T, Le G (2005) Les huiles essentielles. 27 p

Bagchi GD, Dwivedi PD, Singh A, Haider F, Naqvi A (2003) Variations in essential oil constituents at different growth stages of *Ruta chalepensis* on cultivation at north Indian plains. J Essent Oil Res 15:263–264

Baldwin IT, Halitschke R et al (2006) Volatile signaling in plant-plant interactions: "talking trees" in the genomics era. Science 311(5762):812–815

Balz R (1986) Les huiles essentielles et comment les utiliser, Ed. Rodolphe BALZ, 152 p

Baser KH, Ozek CT, Beis SH (1996) Medicinal constituents of the essential oil of *Ruta chalepensis* L. from Turkey. J Essent Oil Res 8:413–414

Becerra JX (2007) The impact of herbivore-plant coevolution on plant community structure. Proc Natl Acad Sci USA 104:7483–7488

Bednarek P, Osbourn A (2009) Plant–microbe interactions: chemical diversity in plant defense. Science 324:746–748

Bennaoum Z, Benhassaini H (2017) Autoecology and place of species of the Ruta genus in vegetation of north-western Algeria. Ecologia Mediterr 43(1):19–35

Bennaoum Z, Benhassaini H, Larabi F, Tirse M (2015) Macro and micro-phytodermic seasonal characters of wild Ruta species (*Ruta montana*, *Ruta chalepensis* subsp. latifolia and *Ruta chalepensis* subsp. angustifolia) in northwestern Algeria. J G A E 3:56–68

Bennaoum Z, Benhassaini H, Falconieri D, Piras A, Porcedda S (2017) Chemical variability in essential oils from Ruta species among seasons, and its taxonomic and ecological significance. Nat Prod Res 31(19):2329–2334

Bergougnoux V, Caissard J-C, Jullien F, Magnard J-L, Scalliet G, Mark Cock J, Hugueney P, Baudino S (2007) Both the adaxial and abaxial epidermal layers of the rose petal emit volatile scent compounds. Planta 226(4):853–866

Bertin N, Staudt M (1996) Effect of water stress on monoterpene emissions from young potted holm oak (*Quercus ilex* L.) trees. Oecologia 107:456–462

Bertrand C, Fabre N, Moulis C (2003) Composition of the essentials oil of *Ruta Corsica* DC. J Essent Oil Res 15(2):98–99

Bohlmann J, Keeling CI (2008) Terpenoid biomaterials. Plant J. 54(4):656–669

Bouabidi W, Hanana M, Gargouri S, Amri I, Fezzani T, Ksontini M, Jamoussi B, Hamrouni L (2015) Chemical composition, phytotoxic and antifungal properties of *Ruta chalepensis* L. essential oils. Nat Prod Res 29(9):864–868

Boutoumi H, Saad M, Khodja M (2009) Essential oil from *Ruta montana* L. (Rutaceae) chemical composition, insecticidal and larvicidal activities. J Essent Oil Bear Pl 12:714–721

Bruneton J (1999) Pharmacogonosie, phytochimie, plantes médicinales, 3rd edn. Lavoisier, Paris

Champagne DE, Koul O, Isman MB, Scudder GGE, Neil Towers GH (1992) Biological activity of limonoids from the rutales. Phytochemistry 31:377–394

Chartier M, Gibernau M, Renner SS (2014) The evolution of pollinator/plant interaction types in the Araceae. Evolution 68:1533–1543

Chase MW, Morton CM, Kallunki JA (1999) Phylogenetic relationships of Rutaceae: a cladistic analysis of the subfamilies using evidence from rbcL and atpB sequence variation. Am J Bot 86:1191–1199

Chatin J (1875) Etudes histologiques et histogéniques sur les glandes foliaires intérieures et quelques productions analogues: annales des sciences naturelle. Ed Imprimerie Martinet, Paris

Chen F, Ro DK et al (2004) Characterization of a root-specific arabidopsis terpene synthase responsible for the formation of the volatile monoterpene 1,8-Cineole. Plant Physiol 135 (4):1956–1966

Cheng SS, Chang HT, Chang ST, Tsai KH, Chen WJ (2003) Bioactivity of selected plant essential oils against the yellow fever mosquito *Aedes aegypti* larvae. Bioresour Technol 89:99–102

Chibani S, Bouratoua A, Kabouche A, Laggoune S, Semra Z, Smati F, Kabouche Z (2013) Composition and antibacterial activity of the essential oil of *Ruta chalepensis* subsp. *angustifolia* from Algeria Der Pharm Lett 5(5):252–255

Copolovici LO, Filella I et al (2005) The capacity for thermal protection of photosynthetic electron transport varies for different monoterpenes in *Quercus ilex*. Plant Physiol 139:485–496

Croteau R (1986) Biochemistry of monoterpenes and sesquiterpenes of the essentials oils herbs, spices and médicinal plants. Recent Adv Botany Hortic Pharmacol 1:81–133

Croteau R, Felton M (1981) Relationship of camphor biosynthesis to leaf development in Sage (*Salvia officinalis*). Plant Physiol 67(4):820–824

Da Silva MF, das GF, Gottlieb OR, Ehrendorfer F (1988) Chemosystematics of the Rutaceae: suggestions for a more natural taxonomy and evolutionary interpretation of the family. Plant Syst Evol 161:97–134

Dahlgren RMT (1980) A revised system of classification of the angiosperms. Bot J Linean Soc 80:91–124

De Feo V, Desimone F, Senatore F (2002) Potential allele chemicals from the essential oil of *Ruta graveolens*. Phytochemistry 61:573–578

Delfine S, Csiky O, Seufert G, Loreto F (2000) Fumigation with exogenous monoterpenes of a nonisoprenoid-emitting oak (*Quercus suber*): monoterpene acquisition, translocation, and effect on the photosynthetic properties at high temperatures. New Phytol 146:27–36

Derbesy M (1997) Reproductibilité des extraits naturels industriels. Ed Palums Actualités Cosmétiques, 132 p

Despinasse Y (2015) Diversité chimique et caractérisation de l'impact du stress hydrique chez les lavandes, 181 pp

Deysson G (1978) Organisation et classification des plantes vasculaires. ed SEDES, Paris

Djarri L, Ferhat M, Merabet G, Chelghoum A, Laggoune S, Semra Z, Smati F, Kabouche Z (2013) Composition and antibacterial activity of the essential oil of *Ruta montana* from Constantine (Algeria). Der Pharmacia Lett 5(4):70–73

Dob T, Dahmane D, Gauriat-Desrdy B, Daligault V (2008) Volatile constituents of the essential oil of *Ruta chalepensis* L. subsp *angustifolia* (Pers.) P. Cout J Essent Oil Res 20:306–309

Duarte LP, Figueiredo RC, Sousa GF, Soares DBS, Rodrigues SBV, Silva FC, Silva GDF (2010) Chemical constituents of *Salacia elliptica* (Celastraceae). Quim Nova 33:900–903

Dudareva N, Negre F (2006) Plant volatiles: recent advances and future perspectives. Crit Rev Plant Sci 25(5):417–440

Dudareva N, Pichersky E, Gershenzon J (2004) Biochemistry of plant volatiles. Plant Physiol 135 (4):1893–1902

Durаffourd C, Lapraz JC (2002) Traité de phytothérapie clinique. Ed. Maloine, Paris

Edris AE (2007) Pharmaceutical and therapeutic potentials of essential oils and their individual volatile constituent. Phytother Res 21:308–323

Ehlting J, Hamberger B, Million-Rousseau R, WerckReichhart D (2006) Cytochromes P450 in the phenolic metabolism. Phytochem Rev 5:293–308

Engler A (1931) Rutaceae. The natural plant families. Ed Die naturlichen Pflanzen familien. 2 Aufl 19 a. Leipzig (Germany): Engelmann

Ferhat M, Kabouche A, Kabouche Z (2014) Comparative compositions of essential oils of three Ruta species growing in different soils. J Mater Environ Sci 5:735–738

Figueiredo AC, Barroso JG et al (1995) Composition of the essential oil of *Lavandula pinnata* L. fil. var. *pinnata* grown on madeira. Flavour Fragrance J 10(2):93–96

Fischer NH, Williamson GB et al (1994) In search of allelopathy in the florida scrub-the role of terpenoids. J Chem Ecol 20:1355–1380

Franchomme P, Penoel D (1990) L'aromathérapie exactement, encyclopédie de l'utilisation thérapeutique des huiles essentielles. Ed. Jollois, Limoges

Friedman J (1987) Allelopathy in desert ecosystems. In: Waller GR (ed) Allelochemicals: role in agriculture and forestry. A. C. S. Symposium

Gallet C, Pellissier F (2002) Interactions allélopathiques en milieu forestier. Rev For Fr LIV 6:567–576

Gershenzon J, Dudareva N (2007) The function of terpene natural products in the natural world. Nat Chem Biol 3(7):408–414

Gershenzon J, McConkey ME et al (2000) Regulation of monoterpene accumulation in leaves of peppermint. Plant Physiol 122(1):205–214

Giampieri M, Balbi A, Mazzei M, La Colla P, Ibba C, Lodd R (2009) Antiviral activity of indole derivatives. Antiviral Res 83:179–185

Gibbs RD (1974) Chemotaxonomy of flowering plants. Ed McGill, Queen's University Press

Gibka J, Kunicka-Styczynska A, Glinski M (2009) Antimicrobial activity of undecan- 2-one, Undecan-2-ol and their Derivatives. J Essent Oil Bearing Plants 12:605–614

Gill RS, Gupta K, Taggar GK, Taggar MS (2010) Role of oxidative enzymes in plant defenses against herbivory. Acta Phytopathol Entomol Hung 45:277–290

Gotsiou P, Naxakis G et al (2002) Diversity in the composition of monoterpenoids of *Origanum microphyllum* (Labiatae). Biochem Syst Ecol 30(9):865–879

Gouinguene SP, Turlings TCJ (2002) The effects of abiotic factors on induced volatile emissions in corn plants. Plant Physiol 129(3):1296–1307

Guignard JL (1983) Abrégé de botanique. Ed Masson, Paris, 259 p

Guitton Y, Nicolè F et al (2010) Differential accumulation of volatile terpene and terpene synthase mRNAs during lavender (*Lavandula angustifolia* and *L. x intermedia*) inflorescence development. Physiol Plant 138:150–163

Haberlandt G (1914) Physiological plant anatomy. Ed Macmillan, London. 777 p

Halligan JP (1975) Toxic terpenes from *Artemisia californica*. Ecology 56:999–1003

Hammiche V, Azzouz M (2013) Les rues: ethnobotanique, phytopharmacologie et toxicité. Phytothérapie 11:22–30

Harborne JB, Turner BL (1984) Plant chemosystematics. Ed Academic Press

Hegnauer R (1986) Phytochemistry and plant taxonomy-an essay on the chemotaxonomy of higher plants. Phytochemistry 25:1519–1535

Heil M, Silva Bueno JC (2007) Within-plant signaling by volatiles leads to induction and priming of an indirect plant defense in nature. Proc Natl Acad Sci 104(13):5467–5472

Heiling S, Schuman MC et al (2010) Jasmonate and pphsystemin regulate key malonylation steps in the biosynthesis of 17-hydroxygeranyl linalool diterpene glycosides, an abundant and effective direct defense against herbivores in *Nicotiana attenuata*. Plant Cell 22:273–292

Hendel-Rahmanim K, Masci T et al (2007) Diurnal regulation of scent emission in rose flowers. Planta 226(6):1491–1499

Holm YI, Laakso I (1997) Variation in the essential oil composition of *Artemisia annua* L. of different origin cultivated in Finland. Flavour Fragrance J 12(4):241–246

Hopkins W (2003) Plant physiology. Ed. Deboek Larcier, 513 pp

Horiuchi JI, Arimura GI, Ozawa R, Shimoda T, Dicke M, Takabayashi J et al (2003) Lima bean leaves exposed to herbivore-induced conspecific plant volatiles attract herbivores in addition to carnivores. Appl Entomol Zool (Jpn) 38:365–368

Iijima Y, Davidovich-Rikanati R et al (2004a) The biochemical and molecular basis for the divergent patterns in the biosynthesis of terpenes and phenylpropenes in the peltate glands of three cultivars of basil. Plant Physiol 136(3):3724–3736

Iijima Y, Gang DR et al (2004b) Characterization of geraniol synthase from the peltate glands of sweet basil. Plant Physiol 134(1):370–379

Inigo RPA, De Viana MEL, Catalan CAN, De Iglesias DIA (1981) Ess Deriv Argumari 51:349–351; in Ferhat M Kabouche A Kabouche Z (2014) Comparative compositions of essential oils of three Ruta species growing in different soils. J Mater Environ Sci 5:735–738

Jeaun JM, Annie F, Chrystian JL (2005) Phenolic compounds of plants. pp 203–204

Johnson CB, Kazantzis A (2004) Seasonal, populational and ontogenic variation in the volatile oil content and composition of individuals of *Origanum vulgare* subsp. *Hirtum*, assessed by GC headspace analysis and by SPME sampling of individual oil glands. Phytochem Anal 15 (5):286–292

Keeling CI, Bohlmann J (2006) Genes, enzymes and chemicals of terpenoid diversity in the constitutive and induced defence of conifers against insects and pathogens. New Phytol 170:657–675

Kelemu S, Niassy S, Torto B, Fiaboe K, Affognon H, Tonnang H, Maniania NK, Ekesi S (2015) African edible insects for food and feed: inventory, diversity, commonalities and contribution to food security. J Insects Food Feed 1(2):103–119

Kessler A, Baldwin IT (2002) Plant responses to insect herbivory: the emerging molecular analysis. Annu Rev Plant Biol 53:299–328

Khadhri A, Bouali I, Belkhir S, El Mokni R, Smiti S, Almeida C, Nogueira JMF, Eduarda M, Araújo M (2014) Chemical variability of two essential oils of tunisian rue: *Ruta montana* and *Ruta chalepensis*. J Essent Oil Bear Pl 17:445–451

Kholi RK (1994) Allelopathic implications of eucalyptus in agroecosystems In: Narwal SS Tauro P (eds) Allelopathy in agriculture and forestry. Scientific Publishers, Jodhpur

Kienast H (1885) About the development of containers in the leaves of Hypericum and Ruta

Knudsen J, Eriksson R et al (2006) Diversity and distribution of floral scent. Bot Rev 72(1):1–120

Kordali S, Kesdek M et al (2007) Toxicity of monoterpenes against larvae and adults of Colorado potato beetle, *Leptinotarsa decemlineata* Say (*Coleoptera: Chrysomelidae*). Ind Crops Prod 26 (3):278–297

Lafuente EO (2006) Stratégies d'émission de Composés organiques volatils (COV) par quatre espèces végétales méditerranéennes, 286 p

Lamnaour D (2006) Morphological description, geographical distribution and ecology of *Ruta chalepensis* L. Rutaceae. Guide Med Plants North Afr 12(1):241–244

Levin RA, McDade LA et al (2003) The systematic utility of floral and vegetative fragrance in two genera of nyctaginaceae. Syst Biol 52(3):334–351

Lin FD, Knabe DA, Tanksley TD (1987) Apparent digestibility of amino acids, gross energy and starch in corn, sorghum, wheat, barley, oat groats and wheat middlings for growing pigs. J Anim Sci 64:1655–1665

Loreto F, Schnitzler JP (2010) Abiotic stresses and induced BVOCs. Trends Plant Sci 15:154–166

Maffei ME (2010) Site of synthesis, biochemistry and functional role of plant volatiles. S Afr J Bot 76:612–631

Majdoub O, Dhen N, Souguir S, Haouas D, Baouandi M, Laarif A, Chaieb I (2014) Chemical composition of *Ruta chalepensis* essential oils and their insecticidal activity against *Tribolium castaneum*. Tunisian J Plant Prot 9:83–90

Malik AA, Showkat RM, Javed A (2013) *Ruta graveolens* L. essential oil composition under different nutritional treatments. Middle East J Sci Res 17(7):885–890

Mancuso G, Borgonovo G, Scaglioni L, Bassoli A (2015) Phytochemicals from *Ruta graveolens* activate TAS2R bitter taste receptors and TRP channels involved in gustation and nociception. Molecules 20:18907–18922

Mann J (1987) Secondary metabolism, 2nd ed. Clarendon press Oxford, 374 p

Martin DM, Gershenzon J et al (2003) Induction of volatile terpene biosynthesis and diurnal emission by methyl jasmonate in foliage of Norway spruce. Plant Physiol 132:1586–1599

Mattiacci L, Hütter E et al (2000) Plantodour mediates parasitoid host handling and oviposition in an endophytic tritrophic system. Chemoecology 10:185–192

Mazzei M, Nieddu E, Miele M, Balbi A, Ferrone M, Fermeglia M, Mazzei MT, Pricl S, La Colla P, Marongiu F, Ibba C, Loddo R (2007) Activity of Mannich bases of 7-hydroxycoumarin against Flaviviridae. Bioorg Med Chem 16:2591–2605

Mejri J, Abderrabba M, Mejri M (2010) Chemical composition of the essential oil of *Ruta chalepensis* L: influence of drying, hydro-distillation duration and plant parts. Ind Crop Prod 32:671–673

Mejri J, Bouajila J, Aydi A, Barth D, Abderrabba M, Mejri M (2012) Supercritical CO_2 extract and essential oil of *Ruta chalepensis* L. growing in Tunisia: a natural source of Undecan-2-one. Anal Chem Lett 2:290–300

Merghache S, Hamza M, Tabti B (2009) Etude physicochimique de l'huile essentielle de *Ruta chalepensis* L. de Tlemcen. Algérie Afriq Sci 5:67–81

Michael Smith C (1985) Expression, mechanisms and chemistry of resistance in soybean, *Glycine max* L. (Merr.) to the soybean looper. Pseudoplusia Includens (Walker) 6(3):243–248

Milesi S, Massot B, Gontier E, Bourgaut F, Guckert A (2001) *Ruta graveolens* L.: a promising species for the production of furanucoumarins. Plant Sci 161:189–199

Millar J, Midland S et al (2005) (2,3,4,4-Tetramethylcyclopentyl) methyl acetate, a sex pheromone from the obscure mealybug: first example of a new structural class of monoterpenes. J Chem Ecol 31:2999–3005

Monterde P (1986) Nouvelle flore du Liban et de la Syrie. Ed Beyrouth, Liban, 563p

Moore BD, Andrew R, LKülheim C, Foley WJ (2013) Explaining intraspecific diversity in plant secondary metabolites in an ecological context. New Phytol 201(3):733–750

Muller CH (1966) The role of chemical inhibition (allelopathy) in vegetational composition. Bull Torrey Bot Club 93(5):332–351

Muller WH, Lorber P, Haley B, Johnson K (1969) Volatile growth inhibitors produced by *Salvia leucophylla*: effect on oxygen uptake by mitochondrial suspensions. Bull Torrey Bot Club 96 (1):89–96

Muñoz-Bertomeu J, Ros R et al (2008) Expression of spearmint limonene synthase in transgenic spike lavender results in an altered monoterpene composition in developing leaves. Metab Eng 10(3–4):166–177

Oussalah M, Caillet S, Saucier L, Lacroix M (2007) Inhibition effects of selected plant essential oils on the growth of four pathogenic bacteria: E. coliO157: H 7, *Salmonella Typhimurium*, *Staphylococcus aureus* and *Listeria manocytogense*. Food Control 18:414–420

Paris R, Delaveau P (1965) Possibilités et limites de la Chimiotaxinomie. Bull de la Société Botanique de France 112:143–149

Paschold A, Halitschke R et al (2006) Using 'mute' plants to translate volatile signals. Plant J 45:275–291

Pellmyr O (1986) Three pollination morphs in Cimicifuga simplex; incipient speciation due to inferiority in competition. Oecologia 68:304–307

Penuelas J, Llusia J (2002) Linking photorespiration, monoterpenes and thermotolerance in Quercus. New Phytol 155:227–237

Peñuelas J, Munné-Bosch S (2005) Isoprenoids: an evolutionary pool for photoprotection. Trends Plant Sci 10(4):166–169

Peñuelas J, Staudt M (2010) BVOCs and global change. Trends Plant Sci 15:133–144

Peterson RL, Scott MG, Ellis BE (1978) Structure of a stem-derived callus of *Ruta graveolens*: meristems, leaves, and secretory structures. Can J Bot 56(21):2717–2729

Pichersky E, Gershenzon J (2002) The formation and function of plant volatiles: perfumes for pollinator attraction and defense. Curr Opin Plant Biol 5(3):237–243

Rasmann S, Turlings TCJ (2008) First insights into specificity of belowground tritrophic interactions. Oikos 117:362–369

Rasmann S, Kollner TG, Degenhardt J, Hiltpold I, Toepfer S, Kuhlmann U (2005) Recruitment of entomopathogenic nematodes by insect-damaged maize roots. Nature 434:732–737

Rasool R, Ganai BA, Akbar S, Kamili AN, Masood A (2010) Phytochemical screening of *Prunella vulgaris* L. an important medicinal plant of Kashmir. Pak J Pharm Sci 23:399–402

Raymond M (2005) L'aromathérapie chez le nourrisson et le petit enfant, 101 p

Reyes RE, Gonzales AG (1970) Structure of pinnarin and furopinnarin, two new coumarins from the roots of *Ruta pinnata*. Phytochemistry 9:833–840

Rice EL (1984) Allelopathhy. Ed Academic Press, Orlando

Richter G (1993) Métabolisme des végétaux: Physiologie et biochimie. Ed. Presses polytechniques et universitaires. Romandes, 292 p

Rose USR, Manukian A et al (1996) Volatile semiochemicals released from undamaged cotton leaves. Plant Physiol 111:487–495

Rustaiyan A, Khossravi M, Sultani-Lotfabadi F, Yari M, Masoudi S, Monfared A (2002) Constituents of the essential oil of *Ruta chalepensis* L. from Iran. J Essent Oil Res 14:378–379

Salvo G, Bacchetta G, Ghahremaninejad F, Conti E (2008) Phylogenetic relationships of Rutaceae: new evidence from the chloroplast genome and comparisons with non-molecular data. J Mol Phylogenet Evol 49:736–748

Salvo G, Simon YW, Rosenbau G, Ree R, Conti E (2010) Tracing the temporal and spatial origins of island endemics in the mediterranean region: a case study from the citrus family (Ruta L., Rutaceae). Syst Biol 59(6):705–722

Samate DA (2002) Compositions chimiques d'huiles essentielles extraites de plantes aromatiques de la zone soudanienne du Burkina Faso, 250p

Sangwan NS, Farooqi AHA et al (2001) Regulation of essential oil production in plants. Plant Growth Regul 34(1):3–21

Sharkey TD, Wiberley AE et al (2008) Isoprene emission from plants: Why and how. Ann Bot 101:5–18

Singh N, Luthra R et al (1989) Effect of leaf position and age on the essential oil quantity and quality in lemongrass (*Cymbopogon flexuosus*). Planta Med 55(3):254–256

Singh R (2016) Chemotaxonomy: a tool for plant classification. J. Med Plants Stud 4(2):90–93

Singh HP, Batish DR, Kaur S, Arora K, Kohli RK (2006) α-Pinene inhibits growth and induces oxidative stress in roots. Ann Bot 98(6):1261–1269

Singsaas EL, Lerdau M, Winter K, Sharkey TD (1997) Isoprene increases thermotolerance of isoprene-emitting species. Plant Physiol 115:1413–1420

Smith PM (1976) The chemotaxonomy of plants London. Ed Edward Arnold

Soleimani M, Aberoomand-Azar P, Saber-Tehrani M, Rustaiyan A (2009) Volatile composition of *Ruta graveolens* L. of North of Iran. World Appl Sci 7:124–126

Spichiger RE, Savolainen VV, Figeat M, Jean-Monod DB (2004) Botanique systématique des plantes à fleurs: Une approche phylogénétique nouvelle des angiospermes des régions tempérés et tropicales, 3rd edn. Presse polytechnique et universitaire romande, Lossane, p 413p

Stashenko EE, Acosta R, Martínez JR (2000) High-resolution gas-chromatographic analysis of the secondary metabolites obtained by subcritical-fluid extraction from Colombian rue (*Ruta graveolens* L.). J Biochem Biophys Methods 5(43):379–390

Stevens PF (1984) Homology and phylogeny: morphology and systematics. Syst Bot 9(4):395–409

Tang Z, Yang Y, Yang Y, Xu Y (2011) Chemical composition and biological activity of the essential oil of *Ruta graveolens*. Zhongguo Xiandai Yingyong Yaoxue 28(9):834

Tarayre M, Thompson JD et al (1995) Intraspecific variation in the inhibitory effects of *Thymus vulgaris* (Labiatae) monoterpenes on seed-germination. Oecologia 101:110–118

Tirse M, Benhassaini H, Bennaoum Z, Nitsche S, Chaudanson D, Aufray B, Sail K, Bassou G (2017) Foliar microphytodermal characterization of *Pistacia lentiscus* under different bioclimates from Northwest Algeria. Arab J Med Aromat Plants 3(1):28–42

Ton JD, Alessandro M et al (2007) Priming by airborne signals boosts direct and indirect resistance in maize. Plant J 49(1):16–26

Touche J (1997) Représentativité et reproductibilité des extraits de végétaux aromatiques au niveau du végétal. Rivista Italiana EPPOS 288–294

Tschirch A, Stock E (1933) The resins. Ed Berlin, Borntraeger

Turlings TCJ, Tumlinson JH et al (1990) Exploitation of herbivore induced plant odors by host-seeking parasitic wasps. Science 250:1251–1253

Ulubelen A, Öztürk M (2006) Alkaloids and coumarins from Ruta species. Nat Prod Commun 1 (10):851–857

Unsicker SB, Kunert G (2009) Protective perfumes: the role of vegetative volatiles in plant defense against herbivores. Curr Opin Plant Biol 12(4):479–485

Velikova V, Várkonyi Z, Szabó M, Maslenkova L, Nogues I, Kovács L, Peeva V, Busheva M, Garab G, Sharkey TD, Loreto F (2011) Increased thermostability of thylakoid membranes in isoprene-emitting leaves probed with three biophysical techniques. Plant Physiol 157:905–916

Verzera A, Mondello L, Ragusa S, Dugo G (2000) Essential oil of the leaves of a typical mediterranean plant: Note II. *Ruta chalepensis* L. (Rutaceae). Essenze e derivati agrumari 70:207–210

Voirin B, Bayet C (1996) Developmental changes in the monoterpene composition of *Mentha x piperita* leaves from individual peltate trichomes. Phytochemistry 43(3):573–580

War AR, Paulraj MG, War MY, Ignacimuthu S (2011) Role of salicylic acid in induction of plant defense system in chickpea (*Cicer arietinum* L.). Plant Signal Behav 6:1787–1792

Waterman PG (1975) Alkaloids of the Rutaceae: their distribution and systematic significance. Biochem Syst Ecol 3:149–180

Waterman PG (1983) Phylogenetic implications of the distribution of secondary metabolites within the Rutales. In: Salvo G, Bacchetta G, Ghahremaninejad F, Conti E (2008) Phylogenetic relationships of Rutaceae: new evidence from the chloroplast genome and comparisons with non-molecular data. J. Mol Phylogenet Evol 49:736–748

Wink M (1988) Plant breeding: importance of plant secondary metabolites for protection against pathogens and herbivores. Theor Appl Genet 75(2):225–233

Wink M (2003) Evolution of secondary metabolites from an ecological and molecular phylogenetic perspective. Phytochemistry 64:3–19

Yaacob KB, Abdullah CM, Joulain D (1989) Essential oil of *Ruta graveolens* L. J. Essent Oil Res 1:203–207

Yamaura T, Tanaka S, Tabata M (1989) Light dependent formation of glandular trichomes and monoterpenes in thyme seedlings. Phytochemistry 28:741–744

Zellagui A, Belkassam A, Belaidi A, Gherraf N (2012) Environmental impact on the chemical composition and yield of essential oils of Algerian *Ruta Montana* (Clus.) L and their antioxidant and antibacterial activities. Adv Environ Biol 6:2684–2688

Zeybek N, Zeybek U (1994) Pharmaceutical botany. Ed University Printing House, Izmir

MIX
Papier aus verantwortungsvollen Quellen
Paper from responsible sources
FSC® C105338

If you have any concerns about our products,
you can contact us on
ProductSafety@springernature.com

In case Publisher is established outside the EU,
the EU authorized representative is:
Springer Nature Customer Service Center GmbH
Europaplatz 3, 69115 Heidelberg, Germany

Printed by Libri Plureos GmbH
in Hamburg, Germany